The Universe Next Door

The Universe Next Door

A Complete Guide to Exploring the Skies and Understanding What You See

TERRY HOLT

CHARLES SCRIBNER'S SONS
New York

To my parents

Library of Congress Cataloging in Publication Data

Holt, Terry.
The universe next door.

Bibliography: p.
Includes index.
1. Astronomy—Amateurs' manuals. I. Title.
QB63.H66 1985 520 85-10751
ISBN 0-684-18358-7

Published simultaneously in Canada by Collier Macmillan Canada, Inc.

1 3 5 7 9 11 13 15 17 19 F/C 20 18 16 14 12 10 8 6 4 2

Printed in the United States of America.

Contents

Acknowledgments

There are many people whose assistance has been a great help in the writing of this book. Among the organizations I would like to thank are: the AAVSO, NASA, the Jet Propulsion Laboratory at Caltech, the Lick Observatory, the Palomar Observatory, the Kitt Peak National Observatory, the United States Naval Observatory, and the Royal Astronomical Society of Canada, and the Harvard College Observatory. Several manufacturers of equipment and publishers of texts for amateurs were also important; Meade Instruments, Roger Tuthill, Tele-Vue, Coulter Optical, Sky Publishing, and *Astronomy* magazine lent information or equipment that was helpful in the equipment sections. Among the individuals I would like to thank are Al Nagler, Janet Maffei, and Jerry Stasavage of the Astronomy Department of Cornell University, and Guy Ottewell, who provided the star charts in the appendix. Several other members of that department lent invaluable time and care in reading the descriptive sections of this book. Jonathan Gradie and Damon Simonelli read several chapters each, and patiently explained where I had gone wrong. Mike Skrutskie read over a dozen chapters; without his help, this book would contain more errors of fact or emphasis than it does. Those errors that do remain are, of course, my own.

Finally, I would like to thank Laurie Langbauer, who provided most of the pen-and-ink illustrations.

Introduction

The moon belongs to everyone
—"The Best Things in Life Are Free"

This book offers the essentials you'll need to get started as an amateur astronomer. Once you have mastered the material covered here, you will find yourself prepared to go on to more complex techniques and able to understand more advanced books about astronomy. Whether you want to go on to collect data for scientific research, or simply to observe and enjoy the infinite variety that the universe has to offer, this guide will help you to get started on your way.

The Universe Next Door is based on several principles. The first is that an amateur astronomer wants clear, simple instructions on how to see celestial objects. This book offers detailed explanations to help you to learn your way around the constellations, to find the planets, to build and use a powerful telescope. The second principle is that seeing isn't enough. If your interest in astronomy is to last, you will want to understand more about the sky than the names and positions of the heavenly bodies. This book also explains how the universe works—not all of it, of course, and not in great depth, but in sufficient detail for you to understand the meaning of what your telescope can show you. This explanation is in ordinary, nontechnical language; you don't need a background in the sciences, or a head for figures, to understand and enjoy astronomy.

Another principle behind this book is that, even though the best things in life are free, not everyone interested in astronomy is wealthy—or a skilled machinist. Or an optician. You may have considered taking up amateur astronomy before now, but been put off by statements such as:

> *An inexpensive telescope can be purchased for $300.*

or:

> *If you don't have your own lathe, a precision machine shop can make this mounting for a very reasonable fee.*

or:

> *After you have finished the sixth stage in the grinding process, test your mirror with a Foucault apparatus; your work should be accurate to within one millionth of an inch.*

You don't need to take out a third mortgage, or enroll in a vocational skills course, or resign from the human race to outfit yourself with a good telescope. Several chapters and appendices explain how to build a powerful telescope with off-the-shelf parts. The instructions are detailed, they work, and they require no expensive equipment or materials and no unusual skills. Moreover, you can complete these projects in a weekend or two. I know these things because I've made them all myself—and as a handyman, I specialize in first aid for mashed thumbs. The plans and instructions in this book won't make astronomy free, but they will take the sting out of the price of admission.

Another principle behind this book is that an amateur astronomer does not have to "do science." Many do, but many more do not, and both groups can take their own pleasure in what they do. If your enjoyment in looking at the stars does not make you want to measure things, take records, or rush to publish your results, you are still no less an amateur astronomer. This book does not assume that you want to be one kind of amateur or the other. The basic skills that both kinds need are the same.

Amateur astronomers come in all kinds. Some actually *are* scientists: their academic credentials may not show it, but their diligent collection of data, their rigorous research programs, and their contributions to international astronomical organizations all go to prove it. In fact, amateurs frequently discover new comets, exploding stars, and other important and unexpected developments in the heavens. The universe is too big for all of the professional astronomers to keep track of; they need the help of amateurs. But there is another kind of amateur astronomer, who is more like the average, backyard birdwatcher. Such people watch the skies because the things they see there are beautiful. They watch for the pleasure of understanding a wider world than the one that stops at the horizon. They watch because the universe is not something Out There: it's the world we live in. If you don't know what astronomy has to offer, you're missing something—almost everything.

That knowledge—that the world is far bigger than the average person thinks—is the most important principle behind this book. It's also what unites the serious amateur, the professional, and the casual astronomer. The universe is much bigger than any of us can dream. It is ultimately mysterious (though we learn more about it every day), fantastically beautiful (though primarily a vast wasteland of nothing), and appallingly violent (though it somehow managed to nurture several billion defenseless pink babies into human adulthood). It is our home. As astronomers, we share a curiosity about the world we live in, and the satisfaction that comes from discovering our place in it. Whichever kind of astronomer you want to be, I hope this book helps you with your first steps out, into our universe.

The door is open. Shall we?

The Measure of the Sky 1

In this chapter, we will learn some of the techniques astronomers use to find their way around the sky. Your first close look at a clear, dark sky can be confusing: all those stars, and no helpful labels. Luckily, with the help of some ancient traditions and a few simple yardsticks, you can soon learn your way well enough to find a dim planet, star, or nebula.

Keeping track of the sky involves two basic facts about the way the sky appears to us. First, it looks flat. To an earthbound observer, all of the stars, the planets, the sun, and the moon seem to lie at the same distance—"up." To locate a heavenly object well enough to find it in a telescope, its distance is irrelevant. All we need to know is its position "on" the sky. For this reason, astronomers find it useful to imagine the sky as a transparent sphere, with all of the heavenly objects on it. The earth lies at the center of this *celestial sphere* (Fig. 1.1).

That the earth spins once each day around its axis is at once the most obvious of the facts of life and the hardest to remember. It ought to be obvious, because the sun, moon, and stars all rise in the east and set in the west. But the very words we use reflect the difficulty of remembering these essentials: we say "rise" and "set," when the sun, moon, and stars do nothing of the kind. The sun does not rise: the eastern horizon falls away from it. But as long as the scenery around us appears to be larger than the sun and we feel no motion in the ground, our minds automatically interpret these events in earthly (or *geocentric*) terms. An astronomer mapping the sky maintains this geocentric point of view. We imagine the sky to be just what it seems—a dome clapped over our heads. This dome, rimmed by the *horizon,* is one-half

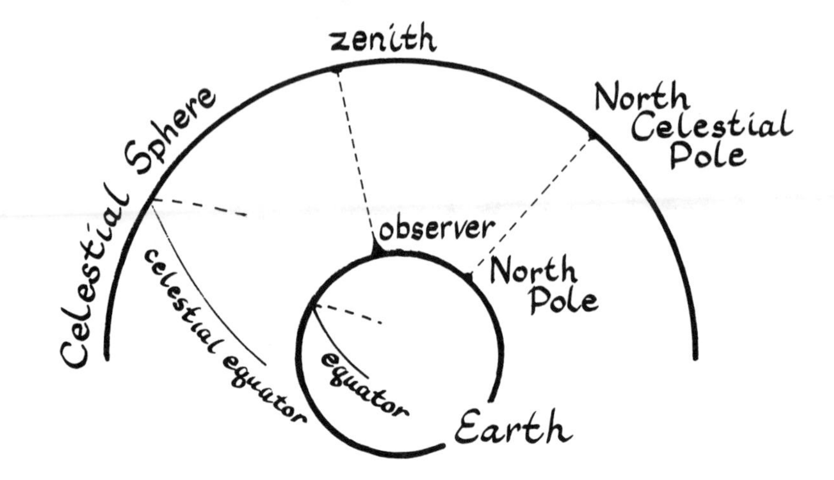

1.1 The celestial sphere is an imaginary sphere around the earth. The celestial equator is a circle directly above the earth's equator; the celestial poles are two points directly above the earth's poles. The zenith is the point directly over the observer.

of the celestial sphere. The other half is, of course, the part seen by someone beyond the horizon, on the opposite side of the earth. We imagine this sphere to be centered on the earth, and map it simply by extending the latitude and longitude of a terrestrial globe to the points above them on the celestial sphere.

The second basic fact of life for astronomers is that, as the earth rotates, the celestial sphere appears to rotate in the opposite direction, just as the scenery seems to rotate around you when you ride a merry-go-round. Because many of the effects of these movements are subtle, we'll leave them for a later chapter. Our immediate purpose is to learn our way around the celestial sphere itself.

ANGULAR MEASURE

When measuring positions on the celestial sphere, the units of length we use are not absolute units, like inches or kilometers. If you want to know the *apparent* distance between two objects on the celestial sphere, you can't use an ordinary unit of distance, like an inch or a mile, because the celestial sphere isn't real. Unlike the solid surface of the earth, it is an illusion. Two objects that seem to lie an inch apart on the sphere could actually be thousands of light years apart in space. And on the celestial sphere an inch is an ambiguous measure: if you hold a ruler close to your eye, an inch on its scale will cover a wide region of the sky; move that ruler to arm's length, and the area spanned by that same inch shrinks. We need a system that gives the apparent size of an object in the sky in units that remain constant at any distance. We use units of *angular measure: degrees, minutes,* and *seconds* of *arc.* Because the lines defining an angle can extend infinitely without changing the angle, these are appropriate units for measuring the celestial sphere. Units of angular measure divide the observer's entire field of view—the entire circle visible around you from side to side, or up and down, or diagonally—into 360 degrees (360°). You can do this with any circle: if you divide a pie into 10 equal pieces, the pieces will each be 36 degrees wide; divide the horizon, for instance, into 24 pieces, and each will be 15 degrees wide.

The degree is the main unit of angular measure; with it, you can accurately describe the distance between two stars, and another observer will know exactly what you mean. The degree is subdivided into 60 minutes (60′), and each minute into 60 seconds (60″). These units are sometimes called *arc-minutes* and *arc-seconds* to distinguish them from units of time.

It's helpful to get some idea of the size of these units. The full moon, for instance, covers an angle of half a degree, or 30 arc-minutes. The smallest angular measure that the unaided eye can detect is about 1 arc-minute; the largest craters on the moon appear about twice that wide. An arc-second is so small that you need a good amateur telescope operating at high power to distinguish it. A golf ball 3 miles away appears about 1 arc-second across.

Knowing that the full moon is about half a degree wide can be a useful gauge. Figure 1.2 shows how you can use your own hand to gauge larger angles. Hold your hand at arm's length against the sky and position the fingers as shown: the distance from the end of your outstretched thumb to the tip of your forefinger is about 20 degrees; the distance across your knuckles when you make a fist is about 10 degrees, and the width of your index fingernail is about 1 degree. You can also use familiar constellations for celestial yardsticks: the Big Dipper (chapter 2 tells how to find it) is about 25 degrees long, and the two stars in the Dipper's bowl that serve as pointers to the pole star are 5 degrees apart. On a larger scale, it is helpful to remember that the distance from the horizon to the point directly over your head is 90 degrees.

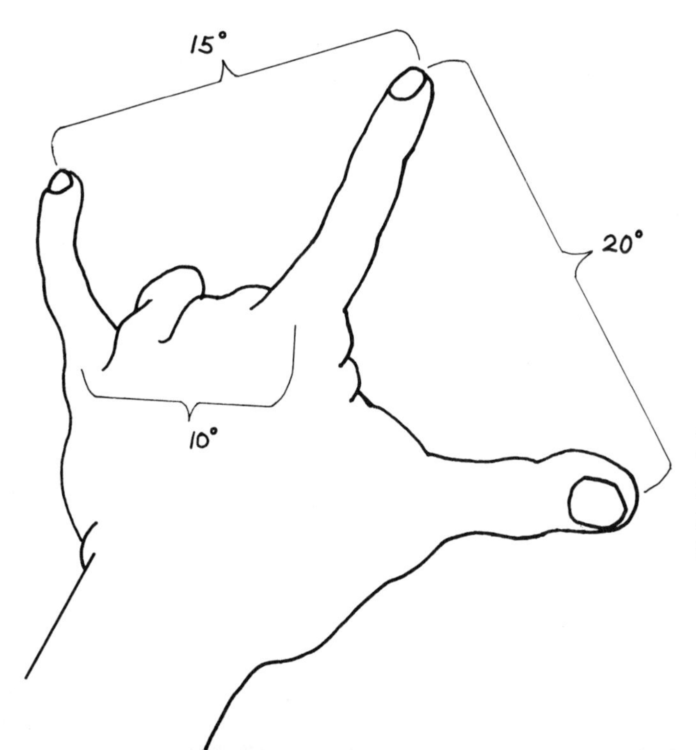

1.2 Angular measurement with your hand. By holding your hand out at arm's length, you can use it as a general gauge of angular measure.

How will you use this ability to gauge angular measures? At first, you will find these units most useful in learning the constellations: if told to look for a star 20 degrees above the horizon, you will have the answer at your fingertips. When locating the planets, knowing that Venus is 40 degrees east of the sun (40 degrees above the western horizon at sunset) will tell you not only where, but when, to look for it. Frequently, the size of a celestial object is given in terms of its *angular diameter,* which is the angular measure of a line across the center of the face of that object. The moon, for instance, has an angular diameter of half a degree. (An equivalent expression is that an object *subtends* so many degrees, which simply means that it fills up a section of sky so many degrees wide.) Knowing the angular diameter of a dim object, such as a galaxy, is a big help when you're trying to locate it in a telescope.

THE CELESTIAL COORDINATE SYSTEM

Knowing how to measure the celestial sphere, you will find it easy to locate objects in relation to recognizable land- (or sky-) marks. Another aid to learning the constellations is the *celestial coordinate system.* This is a locater grid, like those you find on the margins of maps, that will tell you the position of a constellation, star, or planet. The coordinate grid most people are familiar with is the set of latitude and longitude lines that map the surface of the earth. This coordinate system is based on the earth's North and South poles, and two *great circles.* (A great circle is any circle that divides a sphere into two equal halves. The equator is one, dividing the earth into northern and southern hemispheres; the Greenwich meridian is another, dividing the earth into eastern and western hemispheres. A *meridian* is any great circle that passes through the poles; all lines of longitude are also meridians. This is why the Greenwich meridian is sometimes termed the "prime meridian.") In the earthly coordinate system, latitude locates a place in relation to the equator. Longitude gives position relative to the Greenwich meridian. In each case, the units of measurement are degrees, minutes, and seconds of arc.

A similar grid maps the sky. The *north celestial pole,* for instance, is the point that would be overhead if you were standing at earth's north pole; the entire sky seems to rotate around it, just as the earth rotates around its poles. The other benchmarks of the celestial grid are the *south celestial pole,* the *celestial equator,* and the *vernal equinox,* which is the equivalent of the Greenwich meridian. On the celestial sphere, the poles are those points that lie directly above the earth's poles; the equator is the great circle above the earth's equator. Like the Greenwich meridian, the vernal equinox is an arbitrary benchmark; it is the meridian that the sun crosses each year on the first day of spring (see chapter 4 for more on this point).

Lines of *declination* are the celestial equivalent of latitude; they give distance north or south of the *celestial equator,* measured in degrees of arc. Declination is exactly equivalent to latitude: a star with a declination of 15 degrees north will appear directly overhead a place on earth with a latitude of 15 degrees north. The declination of an object at the equator is 0 degrees; at the poles, 90 degrees; in the latitudes of the United States, the stars over your head have declinations in the 30s and 40s.

The celestial equivalent of longitude is called *right ascension.* Like longitude, it measures positions east or west of an arbitrarily chosen meridian. However, it uses different units of measurement. Right ascension (r.a.) is measured in hours, minutes, and seconds. There are 24 hours of right ascension, each of which is divided into 60 minutes, and each of those into 60 seconds. The right ascension of the vernal equinox is zero; right ascension increases to the east of this point, with every 15 degrees along the equator adding another hour. Because the celestial sphere behaves very much like a clock, making one complete rotation every 24 hours, these units are more convenient than the 360 degrees of standard angular measure. They can cause confusion, however, because the size of these units varies depending on the area of the sky; Figure 1.3 shows why.

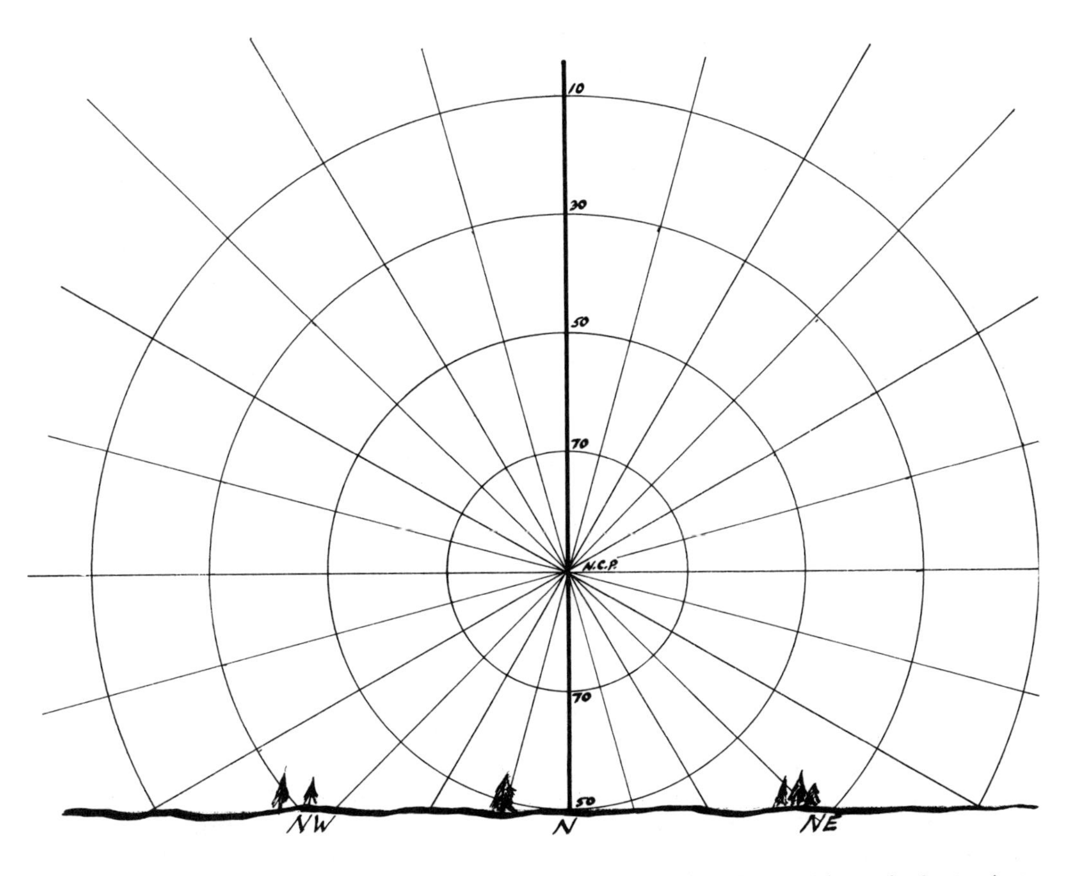

1.3 The celestial coordinate system around the North Pole. Note how the hour circles converge toward the north celestial pole. The altitude of the north celestial pole at this location is 40°.

The celestial coordinate system gives the absolute coordinates of an object. Using this system, any two observers, anywhere on earth, can point their telescopes to the same celestial coordinates and see the same star. There is another coordinate system, however, which is often useful to describe the position of an object relative to your particular place on earth.

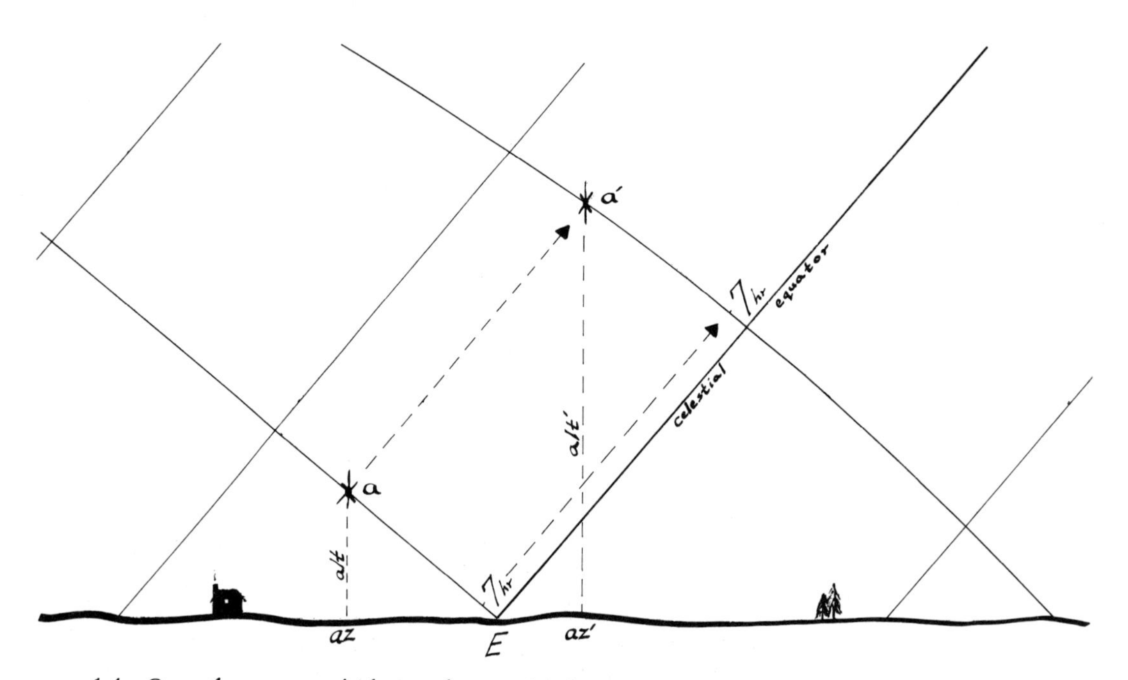

1.4 Over the course of 1 hour, the star (a) rises, moving to (a′). The celestial coordinate grid moves with the sky, but the altazimuth coordinates do not. The star's altitude increases, from alt to alt′, and its azimuth changes from az to az′.

These are the *altazimuth coordinates* (Fig. 1.4). Benchmarks in this system are the *zenith* and the north point on the *horizon*. The horizon is a great circle that divides the half of the celestial sphere which you can see from the half you cannot. Its north point is the point directly to your north. The zenith is the point directly over your head. None of these points is fixed on the celestial sphere: as the earth rotates, different parts of the sky rise above or sink below the horizon; different stars pass through the zenith. For observers at the same place and time (or observing certain celestial events, such as the rising or setting of the planets), however, altazimuth coordinates are much easier to use than celestial coordinates. An object's *altitude* is its angular distance above the horizon. The zenith, for instance, has an altitude of 90 degrees; the horizon's altitude is zero. The *azimuth* of an object is its compass bearing: an object due north of you has an azimuth of 0 degrees; due east, 90 degrees; south, 180 degrees, and west, 270 degrees. Typically, these figures are used something like this: "Look, about 35 degrees up in the east!"

Some persons will find angular measure and coordinate systems helpful in learning their way around the sky. For many others, however, these subjects seem like they're never going to be more than slippery abstractions. If you're one of the latter group, take heart; experience is a much better teacher than any book, and in a few months the degrees and hours, right ascensions and altitudes will all fall into place. Luckily, there's a much more powerful tool for learning the stars, one developed in antiquity, and still in use around the world. To use it, you need only a clear sky, your imagination—and the next chapter.

The Constellations

2

If you were to go outside and try to find the following constellation,

1101+62
1059+56
1151+54
1213+57
1252+56
1322+55
1346+50

you would probably fail. But if you were to go out looking for *this* constellation, your chances of success would be much better:

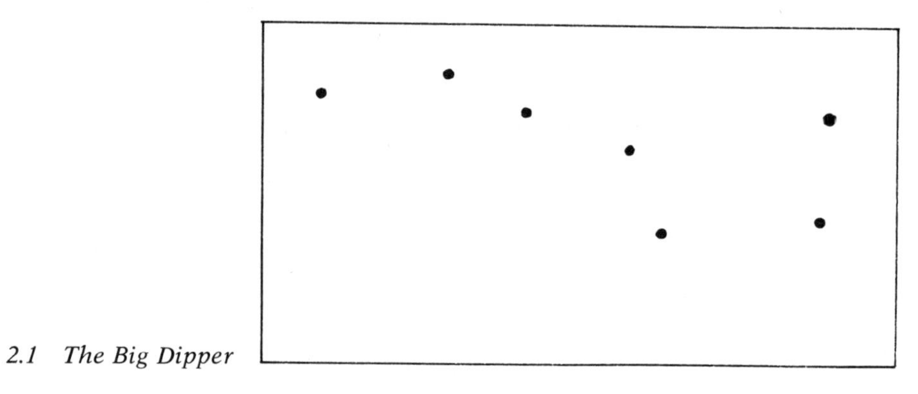

2.1 The Big Dipper

The two constellations are the same. The column of numbers gives the positions, in right ascension and declination, of the stars shown in the illustration: both represent the familiar figure of the Big Dipper. That the second is by far the easier to recognize and remember accounts for the continued use of the constellations. Although these star pictures were probably invented for nonscientific purposes, astronomers today continue to refer to many celestial objects by the constellation in which they appear. When you turn a telescope to the sky, often you will not find your way by the coordinate grid, but by familiar forms and faces: "in the sword of Orion," "in the sting of the Scorpion," or "above the head of Hydra."

Many of the constellations in the sky were first named in ancient times. By the second century, the Egyptian astronomer Ptolemy could list a total of 48 constellations in his *Almagest,* a collection of the astronomical knowledge acquired by previous generations of astronomers. When European astronomy revived as a science in the seventeenth and eighteenth centuries, more constellations were gathered out of the stars neglected by the ancients, especially from the southern skies, which were largely unknown to the ancient Mediterranean cultures.

The official names of the constellations are Latin; they became "official" in the late 1920s, when the International Astronomical Union (I.A.U.) standardized the celestial zoo. There are now 88 constellations, of which 28 are in the northern sky, and 48 in the southern. The remaining 12, called the *zodiac,* lie along the ecliptic. (Briefly, the ecliptic is a circle that divides the sky into two equal hemispheres, just as the equator divides the earth. See chapter 3 for more on this region of the sky.) The I.A.U. also standardized the boundaries between the constellations, so that the celestial sphere is now divided up into 88 distinct territories of varying size. The term *constellation* now applies not only to the imaginary star figure, but to the territory in which it lies.

Some of the most familiar star groupings, such as the Big and Little Dippers, are technically not constellations but only parts of them; the Dippers, for instance, belong to the big and little bears, Ursa Major and Ursa Minor. Such familiar groupings, when not constellations in and of themselves, are called *asterisms.* Other asterisms, and the constellations they belong to, include:

the Northern Cross	— Cygnus
the Sickle	— Leo
the water jar	— Aquarius
the teapot	— Sagittarius
the milk dipper	— Sagittarius
the keystone	— Hercules
the circlet	— Pisces
the kids	— Auriga
the W or M	— Cassiopeia
the Great Square	— Pegasus

Asterisms are useful to know because frequently they are the only part of a constellation that looks like its name. It takes a great deal of imagination to see Sagittarius as a centaur shooting an arrow, but "the teapot" is instantly recognizable. In addition to asterisms, there are other star groupings that stand out in the sky for various reasons, and can serve as guides to recognition. The three brights stars of Orion's belt are one example; the pair of bright stars in Gemini called Castor and Pollux are another. This chapter follows the same principle: rather than describing each constellation—its location, member stars, dimensions, and the myths associated with the figure—the description that accompanies each chart concentrates on steering your attention to the bright stars that will help you to orient yourself. Once you know these, you can use the charts to fill in the spaces between.

LEARNING THE CONSTELLATIONS

In this chapter, we will divide the sky into five regions; these are the sky around the north celestial pole, and the evening skies of fall, winter, spring, and summer. We will ignore the 18 constellations lying entirely south of declination −45 degrees (a negative declination signifies a location south of the celestial equator). They are never visible from the continental United States (except parts of some of them, from south Florida and Texas).

To familiarize yourself with the constellations described in this chapter, choose an evening when the weather forecast is for fair skies. (Chapter 12 tells more about ideal conditions for observing—you might want to browse through it while waiting for nightfall.) When such a night arrives, read over the relevant section of this chapter and study the accompanying chart. You will find that you have to wait awhile for the sky to become fully dark. This is especially so in the summer, when the sun sets late, and never sinks far below the northern horizon (chapter 4 explains more about this). You can use the time during twilight to seek out a good observation post. For the purposes of this chapter, the ideal site is one with few trees or buildings blocking the sky; avoid bright lights as well. Bring along a flashlight to read the charts, but be sure to bring a thick, *red* sock too. With the flashlight inside the sock, the light that filters through should be just bright enough to show the chart—and red light is the best for keeping your eyes adjusted to the dark. The only other equipment you might want to take along is a blanket to lie on.

ALL-SKY CHARTS

To use the charts in this chapter, follow the instructions in the accompanying text, identifying the bright guide stars. Memorize the patterns of the brightest three or four stars in each chart, using their distance from the horizon and each other as your guides. Then start to fill in the patterns of the dimmer stars between them. The more detailed charts in the appendix will help you to learn the names of these stars, as will the monthly charts in the astronomy magazines, and the much more detailed sky atlases (see chapter 11 and the Bibliography for more on these). The charts in the appendix also show the names of asterisms and stars mentioned in this chapter; refer to these if you need help following the instructions here, but you will probably find the simpler charts in this chapter a closer match with the sky on your first night out.

Start by choosing the chart marked for the time of night and season of the year when you will be observing. Each chart shows the appearance of the entire sky at a given time on a given date. However, you can use these charts at other times, on other dates, by remembering a few simple rules about the motions of the celestial sphere. On any one night, the sky moves into the west by one hour of right ascension per clock-hour. For any given time of night, it moves into the west by 2 hours of right ascension per month. If one chart, for instance, shows the sky as it appears at 9 P.M. on April 15, it will also show the sky as it appears at 11 P.M. on March 15, 1 A.M. on February 15, and so on. If you are out at 1 A.M. on April 15, you will find that the summer chart fits the skies visible at that hour. The same is true of any of the all-sky charts you will find printed in the monthly astronomy magazines.

Because each chart shows the entire sky, you will find it easier to use if you concentrate on one quarter of the chart at a time. Turn the chart so that the edge marked *N* (for north) is down: the lower half of the chart now matches the stars that will appear before you as you face north. The center of the chart marks the zenith. To look at the western sky, turn the *W* side down and face west, and so on, around the horizon.

THE NORTH CIRCUMPOLAR STARS

We choose the northern sky first for a particular reason. In the rest of the sky, stars come and go from season to season, but the stars of the northern sky are always visible to anyone within the continental United States. The reason for this has to do with the earth's rotation, which we'll go into in a later chapter—for now, we'll take advantage of the fact to give you a surefire entry into the celestial sphere. Each set of charts shows the sky around the north celestial pole, so choose whichever chart most closely matches your observing time.

The first star group to find is the Big Dipper. An asterism within Ursa Major, the great bear, its seven stars are the most prominent group in the northern sky. Only in the fall, when evening finds them low on the horizon, are they at all difficult to locate. Use your outstretched hand to get an idea of the correct scale: the Dipper spans a region 25 degrees across. Once you have found it, you can use the Big Dipper to start orienting yourself within the celestial coordinate system. It lies between 11 and 14 hours right ascension, and between 50 and 60 degrees declination. The most useful feature of the Big Dipper is the pair of stars that mark the outside edge of the Dipper's bowl. A line between them points north to Polaris, the North Star, which lies about a full dipper-length north of them. (Remember that north is toward the north celestial pole—in early summer, when the dipper lies above the north celestial pole, north will be *below* the pointers.) Polaris belongs to Ursa Minor, the small bear. It also marks the end of the handle of the Little Dipper. This Dipper is a good deal dimmer—you will need very good skies to make out all of the stars in its handle.

After the Big and Little Dippers, the easiest northern constellation to recognize is Cassiopeia, the queen in her chair. Find her on the opposite side of Polaris from the Big Dipper. The *W* or *M* asterism marked on the chart makes a conspicuous grouping of medium-bright stars less than 10 degrees across.

The other stars of the north circumpolar skies are harder to pick out, but once you know the Dippers and Cassiopeia, you can fill in the gaps between them, using the charts. Look for the houselike shape of Cepheus and the long curves of Draco threading between the Dippers, and you can fairly say you know the northern sky. Later, you can learn the two remaining circumpolar constellations, Lynx and Camelopardalis. These are both modern groupings, and dim.

Circumpolar Sky Data:

Bright stars: Polaris, Kochab
Asterisms: Big Dipper, Little Dipper, *W* of Cassiopeia

THE SPRING STARS: 10:30 P.M. Standard Time, mid-April

The Big Dipper is our key to the stars of spring (Fig. 2.2), as it was for the circumpolar constellations. In the more southerly regions of the sky, more bright stars appear, but in the spring, the evening skies contain fewer bright stars than at any other time of year. There are only three bright stars of spring, which simplifies the task of identifying them. The Big Dipper also helps, by providing pointers to them, as it did for the North Star.

About 45 degrees south of the bowl of the Big Dipper, Leo the Lion strides westward across the sky, covering about 30 degrees from nose to tail. The asterism called "the sickle," which looks like a backward question mark, marks the Lion's mane; Regulus is the bright star at its base.

The next bright star of spring is Arcturus, 3 hours of right ascension east of Regulus, and 7 degrees north. Find it by extending the *curve* (not a straight line) of the Big Dipper's handle 35 degrees to the south. Arcturus is the brightest star of the constellation Boötes, the Charioteer. It is the brightest star north of the celestial equator: if you look carefully, you can see that its light has a yellow-orange color. Arcturus is remarkable for being one of the few bright stars that actually changes its position on the celestial sphere

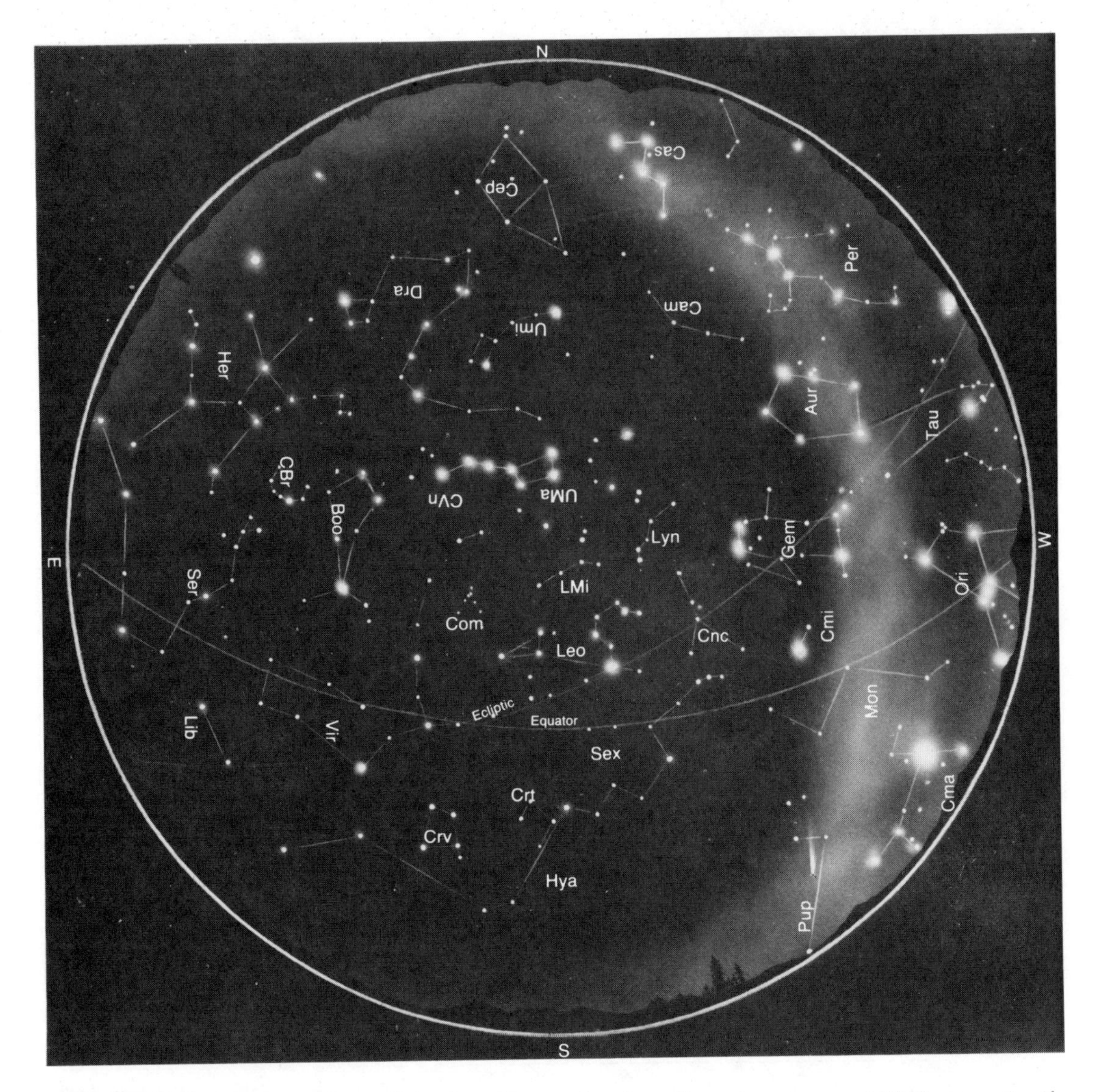

2.2 The night sky for spring. This chart shows the sky as it appears around 10:30 on an April evening. To use it, hold the page so that the compass direction in which you are facing is down: the chart will then match the orientation of the sky. The middle of the chart will be directly over your head; the top of the chart will be behind you. Starting from the Big Dipper in Ursa Major (UMa), follow the directions in the text to the other bright stars and constellations of spring. Appendix 8 gives the full spellings of the constellation names abbreviated here.

(relative to other stars) at a pace fast enough to have become apparent in historical times. We know from this shift that Arcturus is moving at a high speed through the rest of the stars we see when we look in that direction—like a hawk diving through a flock of sparrows. The color of Arcturus's light is especially vivid when you compare it to the third bright star in the spring sky, Spica. You can find Spica in Virgo, the Virgin, by extending the curving line from the Big Dipper to Arcturus farther south. Its light is white, tinged with an electric blue, like the arc of an electric welder seen far away.

Outside the region bounded by Leo, Virgo, and Boötes, the spring sky has only one conspicuously bright star. This is Alphard, "the lonely one," which marks the heart of Hydra. It shines 20 degrees south and slightly west of Regulus. Hydra is the largest constellation on the celestial sphere; the easiest part of it to locate is the small ringlet of dimmer stars marking its head, about 20 degrees west and 7 degrees south of Regulus. The rest of the southern spring sky contains a few, dim constellations you will learn more easily after you know the brighter ones.

The huge area bordered by Virgo in the south, Leo in the west, the Big Dipper in the north, and Boötes on the east contains several constellations. Directly north of Virgo is the constellation Coma Berenices, dimly visible as a faint sparkling, like a spider's web. On clear nights, especially if you're using binoculars, this haze appears as a large cluster of very dim stars. Its appearance gives rise to the constellation's name, which means "Berenice's Hair."

There is one dim constellation that you ought to try to locate now. About 20 degrees west and slightly north of Regulus lies Cancer, the Crab. Although its stars are so dim that they are very difficult to see through urban or suburban skies, this is a useful constellation to know. Cancer belongs to the zodiac, the region of the sky where the planets always appear. Like Virgo and Leo, Cancer is sometimes brightened by a temporary visitor; if there is a fourth bright "star" in the spring sky in any of these three constellations, it is no star at all but a planet. To find Cancer, look 15 degrees above the head of Hydra for a pair of dim stars about 3½ degrees apart on a north-south line. These (believe it or not) mark the body of the crab.

Spring sky data:

Bright stars: Arcturus, Spica, Regulus.

Key constellation: Big Dipper. Follow the arc of its handle 30 degrees to Arcturus ("arc to Arcturus"), then extend the same curve another 30 degrees to Spica ("Spike to Spica"). Regulus lies 60 degrees west of Arcturus.

Zodiacal constellations: Cancer (Cnc), Leo (Leo), Virgo (Vir).

Other constellations: Canes Venatici (CVn), Coma Berenices (Com), Corvus (Crv), Crater (Crt), Hydra (Hya), Leo Minor (LMi), Sextans (Sex).

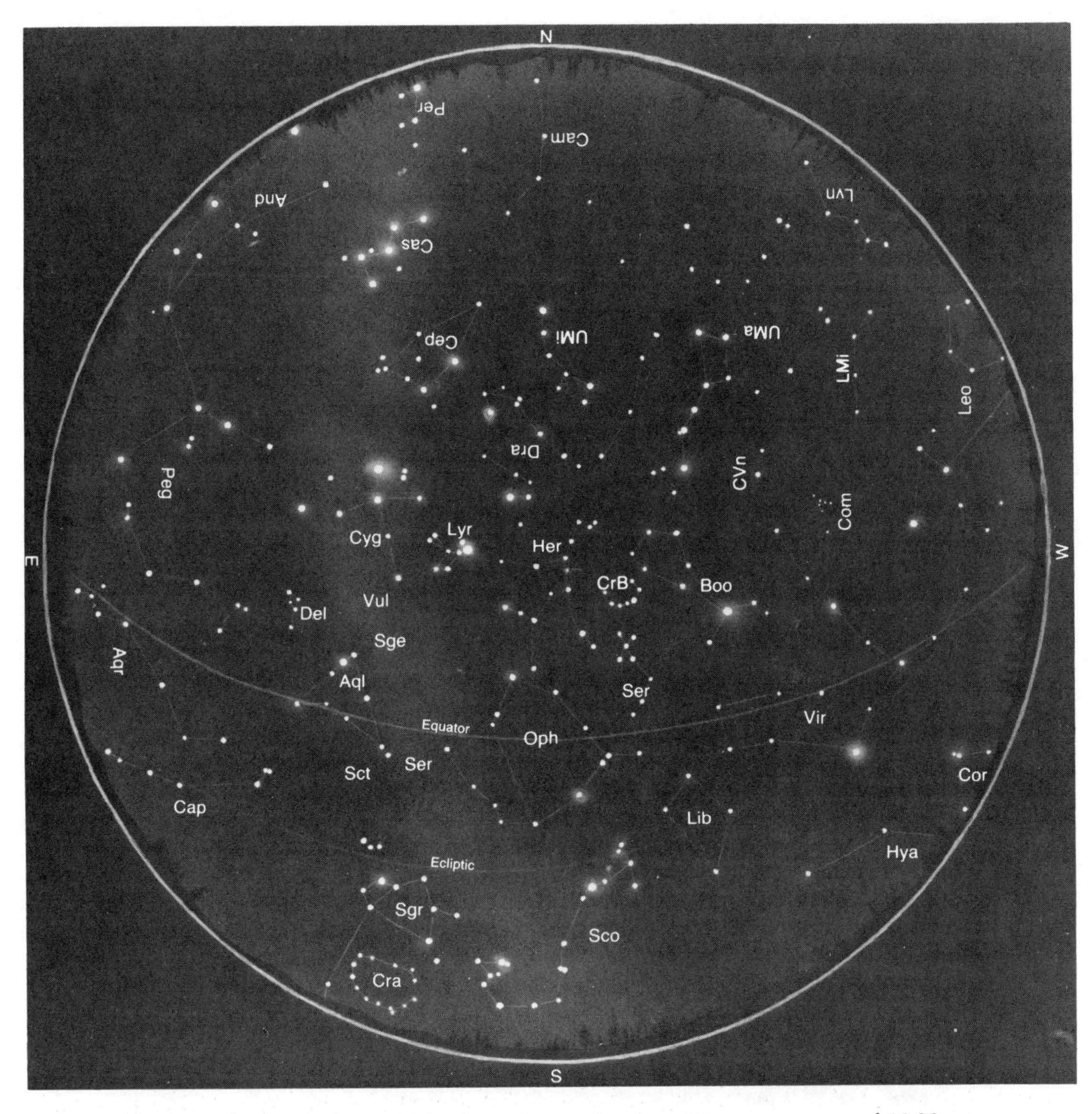

2.3 The night sky for summer. This chart shows the sky as it appears around 10:30 on a July evening. To use it, hold the page so that the compass direction in which you are facing is down; the chart will then match the orientation of the sky. The middle of the chart will be directly over your head; the top of the chart will be behind you. Starting from the summer triangle (Lyr, Aqu, and Cyg), follow the directions in the text to the other bright stars and constellations of summer.

THE SUMMER SKY: 10:30 P.M. Standard Time (11:30 P.M. Daylight Time), Mid-July

The summer sky (Fig. 2.3) has four bright stars, three now rising in the northeast, and one low in the south. The three in the northeast are the stars of the summer triangle. They form an unmistakable, almost-right triangle, some 35 degrees long. Vega, the star at the triangle's right angle, is highest in the sky. Let's start our exploring from there.

Vega, the fifth brightest star in the sky, dominates the small constellation Lyra, the Lyre. Lyra's compact parallelogram figure, about 7 degrees long, is easy to make out. More than thirty degrees south and 1 hour east of Vega lies Altair. The second brightest star in the summer triangle, Altair is in Aquila, the Eagle. The last star of the summer triangle to rise, far in the northeast, is Deneb. Deneb is in Cygnus, the Swan, and also marks a prominent asterism, the Northern Cross. Although the dimmest of the bright stars of summer, Deneb also lies farthest away in space—more than 30 times farther than Vega, almost 100 times farther than Altair. It is, in fact, one of the most distant stars visible to the naked eye. The rest of the Northern Cross is easy to see, composed of medium-bright stars.

The Northern Cross is your best pointer to the most important feature of the summer sky—even though, on many nights, you won't be able to make out what it's pointing at. This feature is the Milky Way galaxy, which on very dark, clear, and dry nights appears as a ghostly haze that stretches from the northeast to the south. This haze is actually the light of perhaps 100 billion stars, so distant, so crowded, that their fierce burning is only this faint shimmer. This galaxy is also the home of the sun and every other star you can see with the naked eye (with one significant exception—see the next section of this chapter). On good nights, you can explore the bright and dark clouds of the Milky Way with or without binoculars. Later sections of this book explain more about this vast, mysterious island universe and tell how you can see it—and others like it—in more detail.

The two really distinctive constellations of the summer sky are Sagittarius the Archer and Scorpius the Scorpion. They are both centered around declination −30 degrees, low in the southern sky. The center of the Milky Way Galaxy lies between them, enriching the region with a golden haze of stars. Find Scorpius by marking Antares, the bright red star whose name means "rival of Mars." It is by far the brightest star in this region of the sky; that, and its ominous red color, earned it its name. Scorpius is also a zodiacal constellation, the middle of the three zodiac groups in the summer sky. The tail of the Scorpion has a distinctive, fishhook shape—some of which may be hidden beneath your horizon.

Sagittarius, an hour east of Scorpius, is also a zodiacal constellation. The chart shows the Teapot, an asterism of Sagittarius that appears to pour out upon the tail of Scorpius. The western stars in this asterism are also called the Milk Dipper, because the figure they form seems about to scoop into the thickest star cloud of the Milky Way, the Great Sagittarius star cloud that marks the hub of our galaxy, just northwest of the Teapot's spout.

The rest of the summer sky is a combination of very large and very small constellations, containing several rather bright stars, but few distinguishing asterisms. One unmistakable, though dim, asterism, is the keystone of Hercules. Look for its four, medium-bright stars about a third of the way from Vega to Arcturus.

South of Hercules sprawls the enormous, dim constellation Ophiuchus, the Serpent Bearer. A very irregular pentagon of medium-bright stars, it is so large you will find it difficult to make out, at first. Your search will be complicated by the constellation Serpens, the Snake, which twines through Ophiuchus. Serpens is an undistinguished grouping of dim stars; its most peculiar feature is that it is split, like the state of Michigan, into two separate areas, with Ophiuchus between them. The two parts of Serpens, called Caput (head) and Cauda (tail), lie to the west and east of Ophiuchus, respectively. If the pattern marked on the charts does not appear obvious to you in the sky, don't worry; Serpens is much easier to see after you know the surrounding constellations well.

South of Serpens Caput lies another undistinguished grouping, Libra, the Scales. Libra is a zodiacal constellation, but its stars are quite dim and are only easily visible if you're far away from city lights.

The rest of summer's constellations, though faint, are all compact, so their figures are easier to recognize. The most distinctive is Delphinus, the Dolphin. Look for it 15 degrees east and slightly north of Altair: the tiny 3 degrees by 1 degree diamond-shaped grouping of medium-bright stars is unmistakable, and along with a dim, tail star, 3 degrees south, looks like a small fish jumping from the Milky Way.

Between Cygnus and Sagittarius are several small, dim constellations, but the region is so packed with the stars of the Milky Way and the total population of the summer sky is so large, you may want to leave these groupings for next summer. By then, the rest of the groupings will have fallen into place, and you will be more familiar with the tools necessary to ferret the smaller ones out: pattern recognition, a good eye for angular measure, and the ability to read the more detailed charts available in sky atlases and astronomy magazines.

Summer sky data:

Bright Stars: Vega, Altair, Deneb, in the northeast; Antares, in the south.

Asterisms: summer triangle, Teapot and Milk Dipper, Keystone of Hercules (Her).

Milky Way: passes from northeast to south.

Zodiacal constellations: Libra (Lib), Scorpius (Sco), Sagittarius (Sgr).

THE FALL CONSTELLATIONS: 10:30 P.M. Standard Time, Mid-October

The skies of fall (Fig. 2.4) are dominated by the Great Square of Pegasus, the Flying Horse (Peg), almost directly overhead. Although it contains no very bright stars, this asterism is so centrally located in an otherwise dim region of the sky that it stands out clearly. The Great Square marks the spring equinox node more clearly than any other asterism—the 0 hour circle passes through its eastern side. This is the point on the celestial sphere where the hours of right ascension start—the Greenwich meridian of the sky. We can also use the widely spaced stars of the square as pointers to steer us to the other important constellations of fall. The square appears in the sky as a slightly irregular rectangle, 18 degrees by 14 degrees. The brightest star in Pegasus is not part of the square. Enif lies about 20 degrees west and 5 degrees south of Markab, the square's southwest corner. Its name means "the horse's mouth"; you can trace the strong, equine curve of Pegasus's neck arcing south through fairly bright stars from Markab to Enif. The eastern side of the square marks the meridian of the spring equinox, the zero point of right ascension. Alpheratz, the star in the northeast corner of the square, actually belongs to Andromeda. This constellation curves away east and north of the square.

To find Capricornus, the Sea Goat, extend a line twice the distance from Alpheratz through Markab; about 40 degrees southwest of Markab, the sea goat will appear as a roughly *v*-shaped collection of medium-bright stars, about 20 degrees from end to end. Capricornus is a zodiacal constellation. You can also locate it by finding its brightest star, Deneb Algiedi, in the eastern tip of the constellation.

About 20 degrees southeast of Deneb Algiedi is the loneliest star in the sky—Fomalhaut, in Pisces Austrinus, the Southern Fish. Fomalhaut is the eighteenth brightest star, but the blankness of the sky around it makes it seem even brighter than it actually is. Pisces Austrinus is a difficult constellation to make out, both low in the sky and dim.

Starting from Alpheratz, in the Great Square, look for the two strands of medium-bright stars that curve northeast almost 30 degrees toward the eastern edge of Cassiopeia. These are the main stars of the Andromeda, the Maiden in Chains. Of these two strands, the more southerly has the brighter stars. On a very dark, clear night, use binoculars to scan the sky above the dimmer chain of stars, about 12 degrees northeast of Alpheratz. If you see a faint haze there, about the size of the moon, you have found the Great Galaxy in Andromeda, which is another galaxy, as large as the entire Milky Way. It is 2.2 million light years from earth; if you can detect it without binoculars, it is the most distant object you will ever see with your unaided eyes.

About 20 degrees east of the Great Square, find Aries, the Ram. It is essentially a pair of brightish stars, less than 5 degrees apart. The very compact grouping makes Aries easy to learn; these is no pair of stars so bright so close together between Sagittarius and Orion.

The last conspicuous constellation of fall is also the last to rise; Perseus,

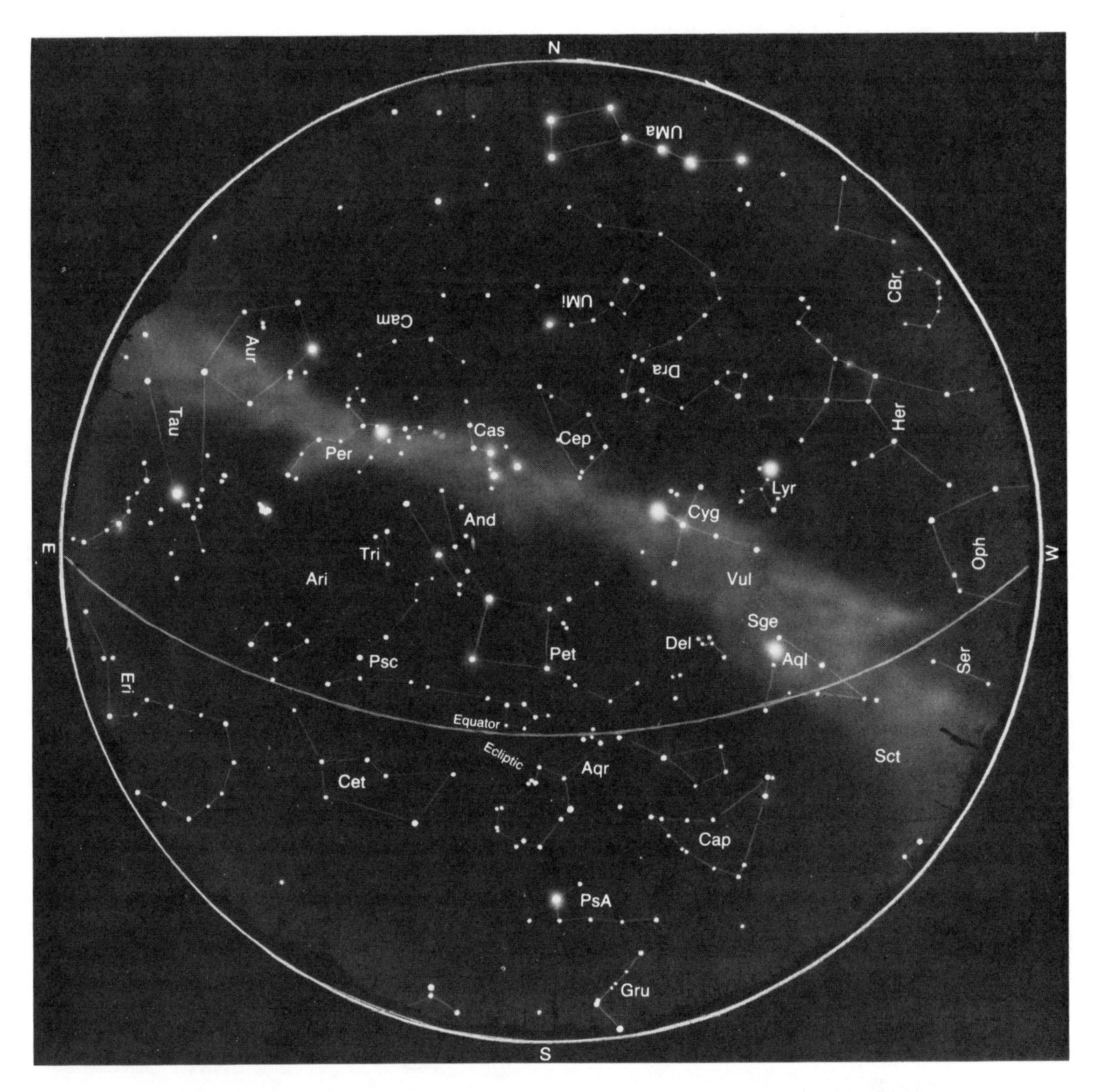

2.4 The night sky for fall. This chart shows the sky as it appears around 10:30 on an October evening. To use it, hold the page so that the compass direction in which you are facing is down; the chart will then match the orientation of the sky. The middle of the chart will be directly over your head; the top of the chart will be behind you. The text explains how to use the Great Square of Pegasus (Peg) as your guide to the other constellations of fall.

slayer of monsters, is about 45 degrees up in the northeast. Perseus is a straggling, two-branched group of medium-bright stars. The first thing your eye will notice is its brightest star, Mirfak. Find it by extending the curving line of Andromeda's bright stars about 20 degrees farther into the northeast. Perseus lies in one of the dimmer regions of the Milky Way.

The rest of the constellations are less distinct than those above, but several of them are worth learning for their status as zodiacal constellations. These are Pisces, the Fishes; and Aquarius, the Water Carrier. Luckily, each contains a distinct if dim asterism. Find Pisces by locating the Circlet, about 10 degrees south of the Great Square. You'll see it as a ring of dim stars, about 5 degrees across. North of Fomalhaut, extending west as far as Aquila, sprawls Aquarius. Its most prominent part is the dim, *y*-shaped asterism called the Water Jar, about 15 degrees west of the Circlet of Pisces.

To find Cetus, the Whale, look for the brightest star 20 degrees south of Aries; this is Menkar. The rest of the constellation extends south and west. The western end of Cetus contains its brightest star, Deneb Kaitos, the only bright star east of Fomalhaut in the fall sky.

If you can learn to navigate the watery skies of fall, you can find your way in any region of the sky. Mastery of this area comes slowly, usually after you have learned the brighter stars of the northern fall constellations; clear dark skies are an enormous help as well. You may find that one fall season, with its frequently cloudy weather, is not enough to become completely familiar with the southern part of the sky. If that is your case, take heart; the ice-clear nights of winter are coming.

Fall sky data:

Bright stars: Fomalhaut, Hamal, Mirach, Enif, Mirfak.
Key asterism: The Great Square of Pegusus.
Other asterisms: Circlet, Water Jar.
Zodiacal constellations: Capricornus, Aquarius, Pisces, Aries.
Milky Way: in the north, through Cassiopeia and Perseus.

THE WINTER CONSTELLATIONS. 10:30 P.M. Standard Time, Mid-January

The evening sky (Fig. 2.5) has more bright stars in winter than at any other time of year. They form a pattern some people call "the Heavenly *G*," which starts at Aldebaran, in Taurus, curves around the top of the *G* via Capella, Castor, and Pollux, and around the bottom through Procyon, Sirius, and Rigel, then forms the crossbar at Betelgeuse. It's possible to miss this pattern in the sky because it's so large: at 70 degrees from top to bottom, some of the stars can easily be obstructed by trees or buildings. The Milky Way

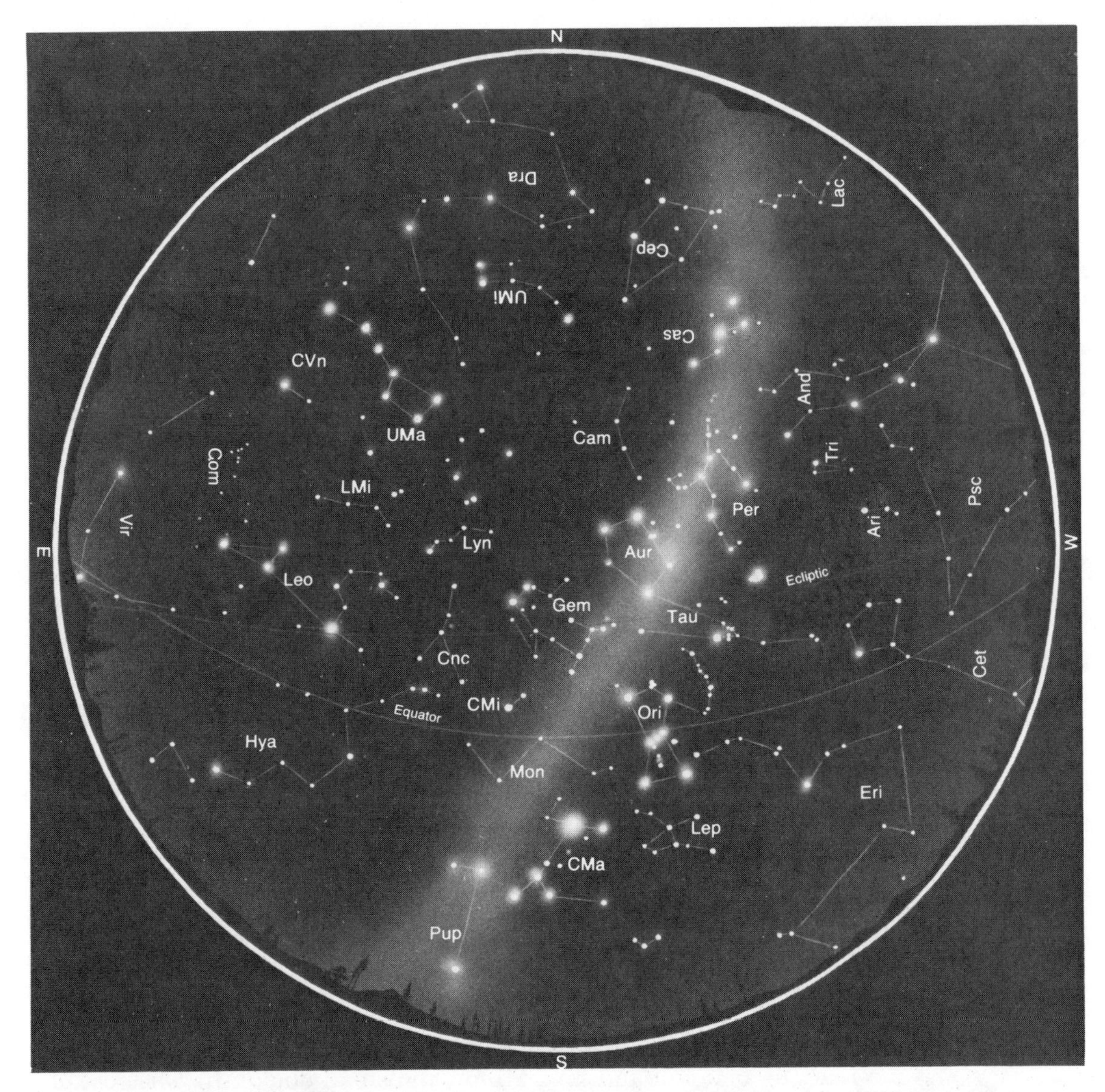

2.5 The night sky for winter. This chart shows the sky as it appears around 10:30 on a January evening. To use it, hold the page so that the compass direction in which you are facing is down; the chart will then match the orientation of the sky. The middle of the chart will be directly over your head; the top of the chart will be behind you. The text explains how the constellation Orion (Ori) will guide you to the other bright stars of winter.

also returns to the zenith in winter, and we can see a nearby, densely populated region of it called the Orion Arm of our galaxy.

The winter sky is also blessed in having its brightest, most-recognizable constellation—Orion, the Hunter—centrally placed. This large, bright gathering stands high in the south on winter evenings. He spans the celestial equator, his midsection about 50 degrees above the southern horizon. That midsection is marked by the second best-known asterism in the sky (after the Big Dipper), the three stars of Orion's belt. These three stars lie on a line 3 degrees long, slanted slightly down on the eastern side.

The brightest star in Orion (Ori) is *usually* Rigel, 10 degrees below and west of the belt. It is the seventh brightest star in the sky, and marks the giant hunter's foot. Contrast Rigel with Betelgeuse, about the same distance above and east of the belt. Not only is Betelgeuse a much redder star, but its light is variable: at times, it can shine more brightly than Rigel. (The odd name is arabic for "armpit of the giant." Pronounce it, without apology, "beetle juice.") The chart shows two more bright stars completing the figure of Orion. These are Bellatrix and Saiph. Bellatrix marks the shoulder of the arm that holds Orion's shield; this is an arc of dim stars, plainly visible only on clear nights. The arm upraised from Betelgeuse holds a club, marked by the two medium-bright stars. Saiph marks the eastern leg. Orion has suffered from a good deal of ridicule on account of his tiny head, comprising the little triangle north of a line connecting Betelgeuse and Bellatrix. One of the joys of pointing out Orion to your friends is that he comes so well equipped. In addition to the head, belt, club, and shield, he also bears a short sword, which hangs at his belt. The middle of this chain of dim stars may appear slightly fuzzy, especially if you look at it through binoculars. It is actually a nebula—to be discussed in more detail in chapter 29.

When you know Orion, the rest of the winter sky is just a short hop away. Sirius, in Canis Major (CMa), the great dog, is the brightest star on the celestial sphere. Find it about 15 degrees southeast of Saiph, at the bottom of the Heavenly *G.* Sirius marks the collar of the dog, which is leaping after Orion, its master. Although Venus, Mars, and Jupiter (and of course the sun and moon) can all shine brighter, no other star even approaches the brightness of Sirius.

Continuing counterclockwise around the Heavenly *G,* the next bright star we reach is Procyon, in Canis Minor, the small dog (CMi). The rest of Canis Minor isn't much more than the dimmish star 5 degrees northwest of Procyon. The figure made by these two stars doesn't resemble a small dog so much as (in the words of H. A. Rey) "a tail without a dog."

North of Procyon on the Heavenly *G,* two bright stars lie close together. These are Castor and Pollux, of Gemini (Gem), the Twins. Like most twins, these two are difficult to tell apart. By position, "Castor is faster"—the more westerly, it leads the duo across the sky each night. Castor may be faster, but Pollux is the brighter, and its light a polleny yellow compared to Castor's pure white. Castor and Pollux mark the heads of the twin-figure of Gemini. The bodies of the twins appear as parallel strands of medium-bright stars that form stick figures, with their feet resting on the Milky Way. Like Cancer to its east, Gemini is a member of the Zodiac.

The northernmost star in the Heavenly *G* is bright Capella, in Auriga (Aur), the Goatherd. Capella is the sixth brightest star in the sky, and shines the color of a goat's eye—a yellow gold light. The star's name means, in fact, "the little she goat." Auriga's herd also includes the dim asterism, "the kids," a narrow, 3 degrees-long triangle of dim stars, 3–5 degrees southwest of Capella. The rest of the constellation is a bright, roughly pentagonal figure, about 15 degrees long, of medium-bright stars. It passes directly overhead on midwinter evenings. The southernmost star, at the point of the pentagon, is called El Nath and actually belongs to Taurus.

The bright orange eye of Taurus (Tau), the Bull, is Aldebaran, the thirteenth brightest star. It completes the top of the Heavenly *G*. Being the westernmost member of the *G*, Aldebaran is the first of the bright stars of winter to rise; you will find it in the northeast around 9:00 P.M. in mid-October. A zodiacal constellation, Taurus also threads the Milky Way through its long horns, which are marked by the moderately bright El Nath, and the somewhat dimmer star 10 degrees to its south.

Taurus is the bull that charges eternally at Orion's upraised shield, forever oblivious to the giant's club about to descend upon its skull. The star figure actually shows only the bull's head and shoulders, its forelegs, and its long, curving horns, which follow the lines laid out by the *V*-shaped asterism of dim stars that mark the face of the bull, around Aldebaran. This asterism is called the Hyades, and it is actually a nearby star cluster, like Coma Berenices. Looming northwest of the Hyades are the bull's massive shoulders, where lies the beautiful cluster of stars known as the Seven Sisters, or the Pleiades.

Taurus is the last of winter's many conspicuous constellations, but some of the dimmer ones are worth learning, especially those that lie along the track of the Milky Way. If your southern horizon is clear, Lepus the Hare (Lep) forms a distinct, butterfly shaped pattern of yellow and orange stars below the legs of Orion, west of Sirius. A more interesting region of the sky lies due west of Orion, between Betelgeuse, Procyon and Sirius, where lies the constellation Monoceros, the Unicorn. Unfortunately, the unicorn's brightest star, 15 degrees south of Procyon, is so dim that it's hard to see if your skies aren't fully dark, and the rest of the stars in this large region are even dimmer. The Milky Way runs directly through this area.

Due east of Canis Major lies the northern section of Puppis (Pup), the poopdeck of Argo Navis, the celestial ship. Argo, an enormous constellation of ancient times, was broken into more manageable parts when the modern constellations were established. Puppis is one of these, the stern of a ship that sails south along the Milky Way, largely hidden in the south circumpolar region from northern latitudes. The most prominent stars of Puppis should clear your south horizon near midnight in mid-January; they're of medium brightness, and form a pattern distinctly like an italic *V*, 5–10 degrees east of the hindquarters of Canis Major. The last of the winter constellations accessible to mid-northern observers is Eridanus (Eri) the River, which is a dim chain of stars curling from the moderately bright star about 5 degrees north of Rigel, which long ago marked Orion's knee, and still looks as if it ought to. Winding west and east until it drops below the south horizon, Eridanus ends far south at declination -57 degrees, in bright Achernar, the ninth brightest star on the celestial sphere.

Winter sky data:

Bright stars: Sirius, Capella, Rigel, Procyon, Aldebaran, Betelgeuse, Pollux, Castor.
Key Constellation: Orion.
Asterisms: Heavenly *G,* Orion's belt, the Kids, Hyades, Pleiades.
Zodiacal constellations: Taurus, Gemini.
Milky Way: north to south, through Auriga, Gemini, Taurus, Orion, Monoceros, Canis Major.

Because this chapter has covered so many thousands of stars and to grasp at once. Fortunately, the sky moves at a slower pace than your reading eye, giving you a full year to find your own way among the stars. Take it a season at a time, starting indoors with the charts, and then moving outdoors to compare the diagrams to the sky. Start with the bright stars, and then the more conspicuous asterisms and brighter constellations, leaving the dim wide spaces of the sky until the last. At that stage, you will find binoculars helpful—but not necessary. By the time you reach the hardest groupings of each season, the sky above will no longer be an unknown wilderness: much of it will be filled with familiar stars and figures. As in finishing a jigsaw puzzle, the last few spaces fall into place almost by themselves.

Stellar Magnitude and Color

3

In the previous chapter, you probably noticed that many of the brighter stars have names: Rigel, Deneb, Antares. Most of these names were given by the Arabic astronomers who kept the science of astronomy alive during Europe's Dark Ages; they are perhaps the most useful way to refer to the bright stars. But many of the dimmer ones are yet unnamed, and we had to refer to them as "the medium-bright stars near . . ." Such a system of identification is clumsy, and astronomers long ago developed two alternative systems of classifying the stars. The oldest system goes back thousands of years, and categorizes stars by brightness. Several hundred years ago, more elaborate systems, assigning individual labels to the stars, were developed by astronomers in Europe.

STELLAR MAGNITUDE

The most obvious characteristic of a star is its brightness. The Greek astronomer Hipparchus first used this feature to classify the stars. You have probably noticed that there are about two dozen stars in the sky that dominate all of the others. They are the first to come out at night, and the last to fade at dawn. Hipparchus called these the stars of first *magnitude.* Stars somewhat dimmer he ranked as second magnitude. Still dimmer stars are third magnitude, fourth, fifth, and finally sixth, the dimmest stars the average eye can see in a good sky. The most important fact to remember about this system is that the higher the number, the dimmer the star: *high* magnitude means *low* brightness. If you substitute the terms "first class," "second class," and so on, the relationship becomes easier to remember. Also, there are far more dim, high-magnitude stars than there are bright, low-magnitude stars.

With the introduction of electronic devices for measuring brightness, astronomers found that a typical first-magnitude star is 100 times brighter than a sixth-magnitude star. Astronomers have since developed a precise scale of magnitudes. In this system, a first-magnitude star is 2.51 times brighter than a second, a second is 2.51 times brighter than a third, and so on. The difference in brightness between two stars two classes apart—a second and a fourth, for example—is 2.51 squared, or 6.3 times. Three classes apart is 2.51 cubed, or 15.81 times, and so on. We can extend this system beyond sixth magnitude, to describe stars too dim to be seen with the naked eye. A typical amateur telescope can show stars of magnitude 12.7. This is 1/476 the intensity of a sixth-magnitude star. The 200-inch telescope at Mount Palomar can detect stars of the twenty-fourth magnitude—1/15,636,047 the brightness of a sixth-magnitude star.

Numbers lower than 1 are used to describe celestial objects brighter than a first-magnitude star. The planet Venus, for instance, is at times more than 100 times brighter than first magnitude, giving it a *negative* magnitude, lower than −4. The sun, on this scale, is magnitude −27, and the full moon is −12. Several stars are actually brighter than first magnitude—a total of fifteen in all. Of these, four are so bright as to have negative magnitudes: Sirius, the southern stars Canopus (magnitude −.72) and Alpha Centauri (−.27), and Arcturus (−.06). Vega, at magnitude +.04, just misses belonging to this group. These negative values have no special meaning: the zero on the scale of magnitudes is a purely arbitrary figure, 2.51 times brighter than the average brightness of all the stars called first magnitude by Hipparchus centuries ago.

The magnitude system, when used in the old fashioned way that most amateurs do, can occasionally be misleading. Any stars between first and second magnitude tend to be "binned" together as "first-magnitude stars," even though they may be much closer to second magnitude; and so on. Calling Polaris a star of first magnitude is something of a polite exaggeration: at magnitude 1.99, it looks much more like a second-magnitude star than a first. It is useful to know some stars that actually are of a particular magnitude: Spica and Antares are close to magnitude 1.0; Polaris's light is slightly variable, but it is a good standard for second magnitude. As your familiarity with dimmer stars increases, through your use of star catalogs and atlases, you can develop a list of stars of standard brightness down to the limit of your vision (Fig. 3.1).

BAYER LETTERS

The stars are also labeled according to a system developed around A.D. 1600 by the Bavarian astronomer Johann Bayer. In this system, each star in each

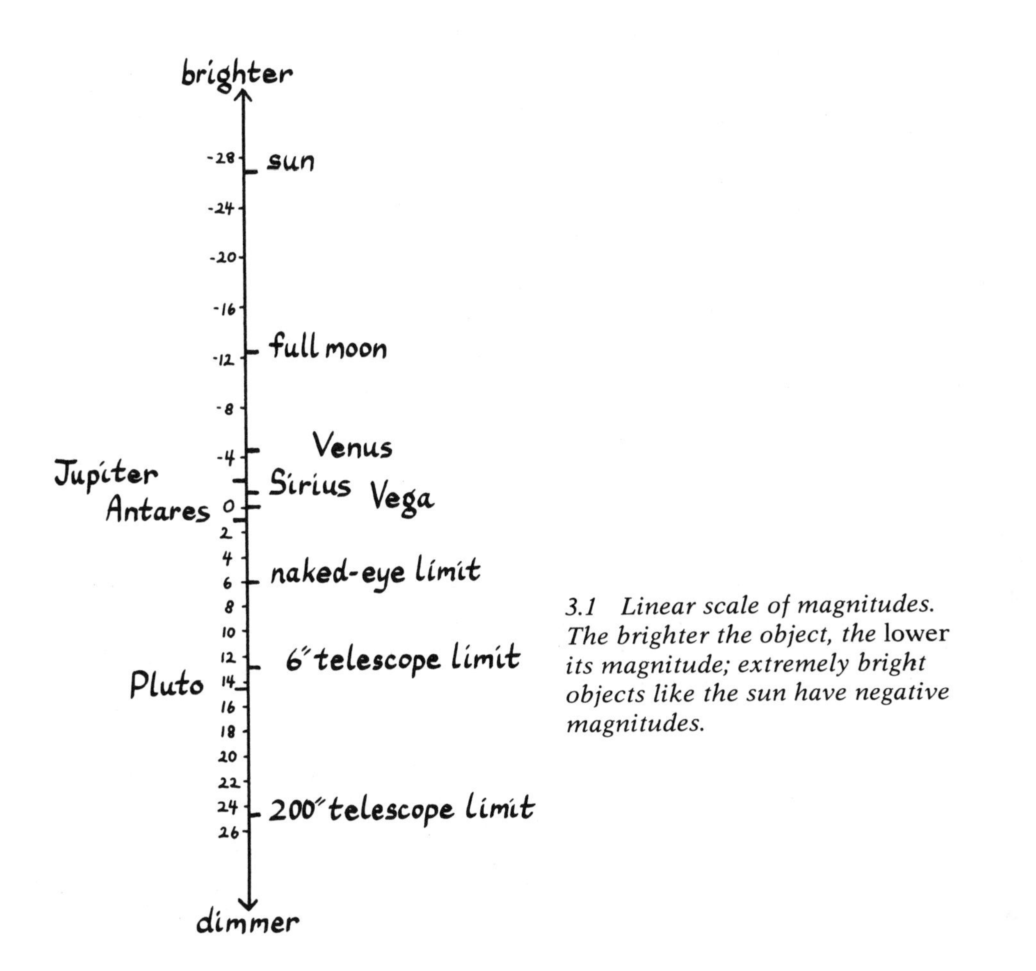

3.1 Linear scale of magnitudes. The brighter the object, the lower *its magnitude; extremely bright objects like the sun have negative magnitudes.*

constellation is identified by a letter in the Greek alphabet, followed by the constellation's name. Polaris, for instance, is called alpha (α) Ursae Majoris. (The name of the constellation is usually slightly changed to the possessive, or "genitive," a grammatical form that is the Latin equivalent of an *'s*: alpha Lyr*ae*, for instance, simply means "Lyra's alpha star.") The Bayer letters, as they are still called, were not assigned at random. Alpha is usually the constellation's brightest star, beta its second brightest, and so on. This system is convenient because it tells you immediately not only the star's approximate location but also its brightness: you know without looking that a star labeled *alpha* or *beta* is a bright star, while a *chi* or *omega* is dim.

The entire Greek alphabet looks like this:

	Upper Case	**Lower Case**		**Upper Case**	**Lower Case**
alpha	Α	α	nu	Ν	ν
beta	Β	β	xi	Ξ	ξ
gamma	Γ	γ	omicron	Ο	ο
delta	Δ	δ	pi	Π	π
epsilon	Ε	ϵ	rho	Ρ	ρ
zeta	Ζ	ζ	sigma	Σ	σ
eta	Η	η	tau	Τ	τ
theta	Θ	θ	upsilon	Υ	υ
iota	Ι	ι	phi	Φ	φ
kappa	Κ	κ	chi	Χ	χ
lambda	Λ	λ	psi	Ψ	ψ
mu	Μ	μ	omega	Ω	ω

Don't worry about learning the entire Greek alphabet at once—I still have trouble remembering zeta from xi. You will grow more familiar with the alphabet as you spend more time with charts and stars. Until you start dealing with the dimmer stars, the first five lower-case letters will be adequate.

Some constellations are so large that there aren't enough Greek letters to name those stars visible to the naked eye. After omega, any visible stars remaining are numbered sequentially, starting with 1 at the western edge of the constellation, and increasing toward the east. These numbers, called *Flamsteed numbers* after the system's inventor, have no connection with the brightness of the stars; they serve merely to name them, and give some idea of their position. As with Bayer letters, they usually occur with the Latin genitive of the constellation: "67 Aquilae," for instance. The multitudes of stars too dim to be seen without a telescope are cataloged in numerous ways, usually by a series of letters or numbers, referring to the particular catalog in which the listing occurs. Most amateurs find the Bayer letters and Flamsteed numbers adequate for most of their purposes.

STELLAR COLOR

Another difference you may have noticed among the stars is their color. Although all stars shine with a light that is more white than anything else, many stars show a subtle tint. The color you are most likely to notice first is a reddish orange, making some stars look like a glimpse of a distant fire. Antares is one such star, its redness accentuated by its riding low in the south on summer nights. (The light of any star close to the horizon is further reddened by the same atmospheric filtering that reddens the setting sun.) Betelgeuse is another bright, red star. The tint of Rigel and Spica may be harder to see at first, since it is closer to pure white. If you look carefully, you may notice traces of blue in their light. Other stars, like Capella (alpha Aurigae), and Procyon (alpha Canis Minoris), have a tint of yellow. This shade warms to orange in Arcturus (alpha Bootis) and Aldebaran (alpha Tauri).

The colors of the stars are of major importance in some fields of astronomy, and later sections of this book will go into the subject in more detail. While learning your way around the sky, you will find the delicate tints visible to the naked eye another useful aid to memory. Remember, however, that these color effects are subtle. Once you know they are there, they will become unmistakable—so much so that you will be surprised when you point them out to friends, and they obstinately refuse to see them. Your friends will be wrong, but let their apparent color blindness be a reminder to you that the "vivid crimson," "mauve," and "teal green" tints attributed to stars by some observers are tributes more to the observers' imaginations than to the glory of the heavens. If your eyes see only reddish, yellowish, and bluish tints where the guidebooks insist on technicolor, don't assume that you are at fault. Even the subtle tints that most people see are wonderful enough.

· THE 23 BRIGHTEST STARS ·

Name	*Constellation*	*Season*	*Color*	*Magnitude*
1. Sirius	Canis Major	winter	white	−1.6
2. Canopus	Carina	winter	yellow	−0.7
3. Arcturus*	Bootes	spring	orange	−0.1
4. Capella	Auriga	winter	yellow	−0.0
5. Vega	Lyra	summer	white	0.0
6. Rigel Kent.*	Centaurus	spring	yellow	0.1
7. Rigel	Orion	winter	blue	0.1
8. Procyon	Canis Minor	winter	yellow	0.4
9. Betelgeuse	Orion	winter	red	0.4
10. Achernar*	Eridanus	winter	blue	0.5
11. Hadar*	Centaurus	spring	blue	0.6
12. Altair	Aquila	summer	white	0.8
13. Aldebaran	Taurus	fall	orange	0.9
14. Spica	Virgo	spring	blue	0.9
15. Antares	Scorpius	summer	red	0.9
16. Acrux*	Crux	spring	blue	1.1
17. Fomalhaut	Pisces Austrinus	fall	white	1.2
18. Pollux	Gemini	winter	orange	1.2
19. Deneb	Cygnus	summer	white	1.3
20. Beta Crucis*	Crux	spring	blue	1.3
21. Regulus	Leo	spring	white	1.4
22. Castor	Gemini	winter	white	1.4
23. Adhara	Canis Major	winter	blue	1.5

*Not visible from mid-northern latitudes.

Keeping Time

4

For astronomers, time is of the essence. Because we observe a universe constantly in motion, our understanding of events often depends on an understanding of time. The uses of time in astronomy can be as simple as knowing when the moon will rise tonight, or as complex as measuring the size of a quasar at the ends of the universe. But the cosmos provides us with no fundamental units of time. Over the past few millennia, we have had to invent them, using the motions of our planet as our clock. Our understanding of time starts, then, with the motions of the earth. Along the way, you will find the answers to some of the questions that may have puzzled you as you learned the constellations: why different constellations appear at different times of year; why the length of the day changes; why the sun rises and sets at different points on the horizon; why some constellations hang in the sky year-round, and others come and go.

THE MOTIONS OF THE EARTH

Rotation

Astronomers define rotation as the circular motion of an object around its axis: a wheel rotates, so does the earth. The earth rotates into the east, counterclockwise around its north pole. One complete turn on the earth's axis equals one day, and during this period, we see the celestial sphere rotate once, into the west, around the north celestial pole. Of course, the rotation of the earth is real; the rotation of the celestial sphere is only an illusion. But for the purposes of knowing where to look and when to see a particular object, it is often more convenient to speak of the motion of the sky as if it were real; in any case, these two motions are often mirror images.

The speed of earth's rotation depends on your latitude: someone standing at the North Pole, for instance, will simply turn in place, like a spindle on a turntable, at one turn per day. Points on the equator, however, must travel around the entire circumference of the earth every 24 hours. The speed of rotation at the equator is thus a little more than 1,600 kilometers (1,000 miles) per hour. At 40 degrees latitude, the earth carries you into the east at about 1,300 kilometers (800 miles) per hour. The same variation in speed occurs in the sky. Objects at the celestial equator travel a full 360 degrees of arc per day, at 15 degrees per hour. Polaris, the pole star less than 1 arc-degree from the north celestial pole, hardly moves at all; its hourly movement is a mere 13 arc-minutes.

Because the earth itself always blocks your view of half the celestial sphere, your location determines what stars you can see. It also affects the way the stars seem to move. To an observer at the poles, the celestial equator is on the horizon. The pole star hangs overhead, while at the horizon, stars move westward without rising or setting. An observer at the North Pole could never see stars like Antares or Rigel that are south of the celestial equator. But if you were to observe from the earth's equator, you would have the celestial equator overhead: the entire celestial sphere would wheel by every 24 hours (Figs. 4.1a and 4.1b).

If you move north from the equator, the sky changes: the south celestial pole sinks below the horizon. In the northern sky, Polaris rises as we travel; every night it hangs higher above the north horizon. If its altitude at the pole is 90 degrees, and 0 at the equator, it stands to reason that for every degree of latitude we move north, Polaris will climb a degree of arc above the horizon. One of the more useful rules to remember in astronomy is that the altitude of Polaris above your northern horizon is always equal (within a degree) to your latitude.

What happens to the visibility of the other stars as we head north? Stars around the south celestial pole become invisible. Twenty degrees north of the earth's equator, no star within 20 degrees of the south celestial pole can rise above the horizon. The farther north you go, the more stars disappear in the south, until at the pole the entire southern celestial hemisphere is gone. Whatever stars cannot rise at your northern latitude are said to be in the *south circumpolar region.*

The opposite occurs in the northern region of your sky. From 20 degrees north latitude, stars within 20 degrees of the north celestial pole *never set.* They circle counterclockwise around the pole, moving from the northeast to above the pole star, then sloping into the northwest. There they do not set, but swoop below the pole, skimming the horizon until they climb again into the northeastern sky. Such stars are called *north circumpolar stars,* or (for northern hemisphere observers) simply *circumpolar.* Your own set of circumpolar stars depends on your latitude (Fig. 4.2). At 40 degrees north lati-

4.1a Star trails. The rotation of the earth during these long exposure photographs caused the images of the stars to trail, showing the apparent motion of the celestial sphere. These are trails around the celestial pole (Official U.S. Navy photograph).

4.1b Star trails at the celestial equator (Lick Observatory photograph)

tude, for instance, stars within 40 degrees of the north celestial pole are circumpolar; any star of declination greater than 50 degrees will never set. These include all the stars in the constellations Draco and Cassiopeia, and all but one in the Big Dipper. A circumpolar star crosses the celestial meridian twice each 24 hours, once when it is above the pole, and once when it passes below. These points are called a star's *upper* and *lower culmination* (Fig. 4.2).

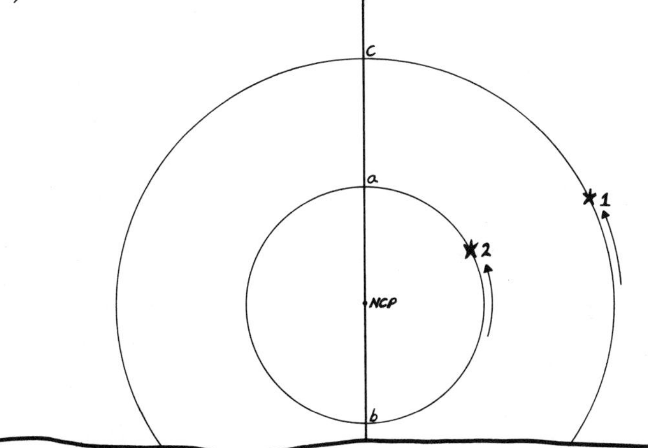

4.2 Circumpolar motion of stars. Star #1 is not circumpolar: too far from the North Celestial Pole (NCP), it rises in the northeast, and sets in the northwest, transiting at (c). *Star #2 is circumpolar, circling the NCP close enough to skim the horizon at lower culmination* (b); *it reaches upper culmination at* (a).

The paths of noncircumpolar stars take varying courses. A star like Vega, the bright star in the summer constellation Lyra, has a declination of about +39 degrees; from 40 degrees north latitude, Vega misses being circumpolar by 11 degrees. This means that it sinks a maximum of 11 degrees below the northern horizon, and rises well to the north of northeast. When it reaches the celestial meridian, it is almost directly overhead. When a noncircumpolar star crosses the meridian, it is said to *transit.* From the meridian, Vega curves around to the northwest, setting finally far north of west. Because it never goes very far below the horizon, Vega never sets for long; it is above the horizon (though its light may be lost in the glare of daylight), for about 22 hours each day.

To find the altitude at which a star transits, subtract your latitude from 90 degrees, and add the star's declination (if its declination is north), or subtract (if its declination is south). This altitude is called the *co-latitude.*

A star on the equator rises due east of you—even if you are not on earth's equator. It sets due west of you again, 12 hours after it rose. Stars north of the equator all describe curving paths, centered on the North Pole and curving away from the equator; the farther from the equator, the tighter the curve, and the longer the star stays above the horizon.

For stars south of the celestial equator, the opposite is true. Their paths curve away from the equator, but are centered on the south celestial pole. From the northern hemisphere, they appear to rise low in the southeast, arc up to transit at a low altitude, then sink a few hours later below the southwest horizon. A star on the borders of the south circumpolar region will appear only for an hour or so, when it is about to transit. If you observe from 40 degrees north latitude, stars with south declinations greater than 50 degrees never rise at all; a star such as Shaula, the bright star in the tail of Scorpio (declination −37 degrees 4 minutes) transits less than 13 degrees above the southern horizon, and sets less than 5 hours after rising.

A written description of these motions is bound to be confusing at first. An easy way to become quickly familiar with them is with a *planisphere.* This is a disc printed with a map of the celestial sphere. The disc rotates within an envelope cut out to match your horizon. By spinning the disc, you can watch the stars rise and set, or circle the pole, and soon the turning of the celestial sphere will become second nature to you. Scales printed around the edge of the disc also allow you to "set" the sky to show its appearance at any time on any given date—useful for planning an evening's observing. There are several planispheres available; the best is sold by Sky Publishing Company; see the Bibliography under Charts for details.

Revolution

In astronomy, *revolution* has a particular meaning: it refers to the orbital motion, roughly circular, of one body around another. A communications satellite, for instance, revolves around the earth. And of course the earth, and all the other planets, revolve around the sun. But just as the rotation of the earth makes the sky seem to revolve around us once a day, the revolution of the earth makes the sky seem to revolve around us—once each year. We see this slow revolution as the constellations move west from season to season, and by the sun's apparent motion eastward around the sky. You can see the slow process by which the sun and constellations slide around the sky. Go outside the next clear night, and find a bright star (refer to the star charts in chapter 2, to make sure it's not a planet—the brightest "star" in the sky is often Venus, Jupiter, or Mars). Rest your head against a wall, tree, or other stable support, and time the moment when your star passes some other permanent marker—a treetop, telephone wire, or building. If you go out the next night, and place your head on the same spot, you will find that the star passes your marker 4 minutes earlier. In 2 weeks, you will need to go out 1 hour earlier to catch this passage.

Over the course of a year, the sun passes completely around the celestial sphere, moving east by two hours of right ascension each month. As it moves, it passes between us and the constellations of the zodiac. This motion makes the constellations appear to vanish, month by month, into the west. One month, when the sun appears "in" Gemini, Cancer appears low above the western horizon just after sunset. A month later, the sun has moved into Cancer, so that in the evening sky, Cancer has vanished, and Leo is about to dive into the west; in the morning sky, just before dawn, Gemini is rising. When the sun is positioned between the earth and a constellation, that constellation is said to be *in conjunction* (from the Latin for "join together") with the sun; it is invisible then. When a constellation and the sun are on opposite sides of the celestial sphere, they are *in opposition*; the constellation rises at sunset, transits at midnight, and sets at dawn. Figure 4.3 shows how the apparent drift of the constellations comes about.

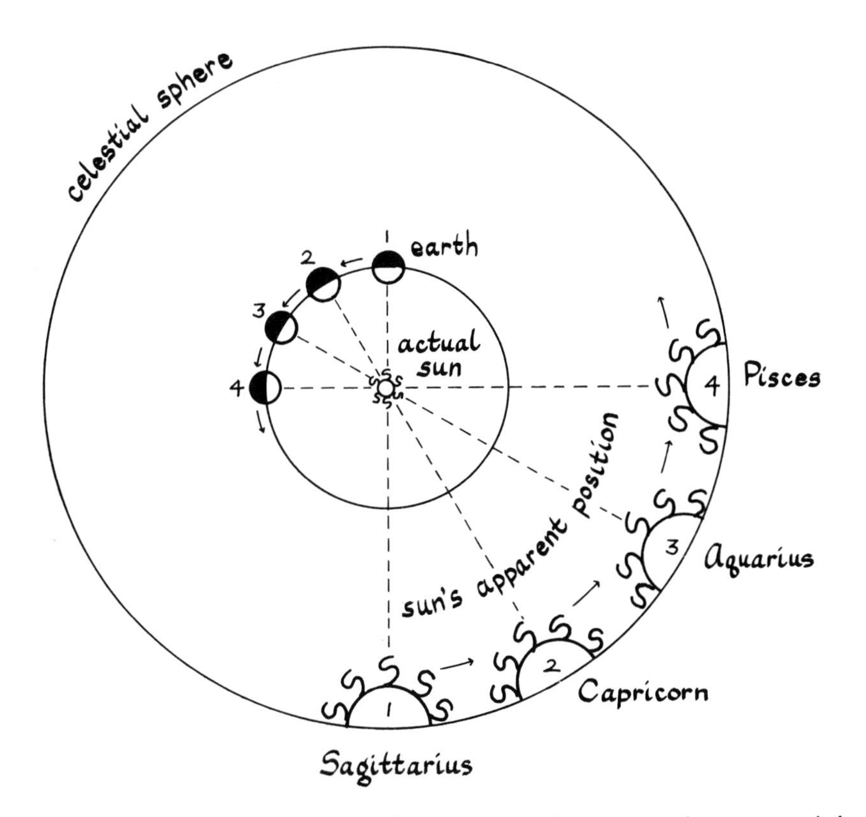

4.3 The sun's apparent motion around the celestial sphere, over the course of three months. The apparent motion is caused by the earth's motion in its orbit, causing the sun to shift relative to the distant stars.

We have been describing this process as if it occurs monthly, but of course it is a steady process. Two hours of right ascension each month, divided by 30 days, comes out to a 4-minute-per-day change (Fig. 4.4). If the sun moves east 4 minutes a day, you might expect that noon would happen 4 minutes later each day. But because we set our clocks by the sun, we consider that the sun remains stationary and that the stars move *west* and transit 4 minutes earlier each day. Those 4 minutes, adding up night after night, are enough to account for the month-by-month march of the constellations into the west.

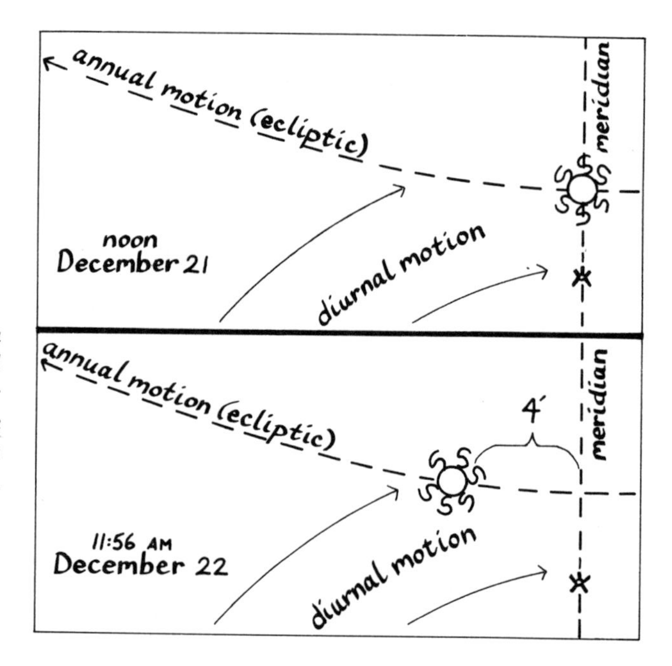

4.4 A star and sun transiting on successive days. The earth's orbital motion makes the sun appear farther east each day. Relative to the sun, the star transits 4 minutes earlier each day.

The Ecliptic

The other conspicuous annual motion of the sun is its change in declination north and south of the equator. The actual cause of this apparent motion is the 23.45-degree tilt of the earth's axis. You may have heard this tilt mentioned before, and wondered, tilted with respect to what? Since we use the earth's axis as our reference point in our mapping of the celestial sphere, it is hard to see what larger benchmark the poles could tilt from. That benchmark is the plane of the earth's orbit around the sun—the imaginary disc around which the earth traces its annual course. This plane is called the *ecliptic.* It is the single most important guidepost in the solar system, because all of the planets orbit the sun in roughly the same plane. It appears on the celestial sphere as a curving line, swooping north of the equator for half its length, south of the equator for the other half. The ecliptic passes through the constellations of the zodiac (and clips the corners of a few others). Because it slants from the equator, once each year the sun's path takes it to a declination of +23.45 degrees. Six months later it stands low in the south, at declination −23.45 degrees. When the sun's declination is high, it stays above the horizon more than twelve hours, giving us the long days of summer. Low in the south, it rises late and sets early, giving us the long nights of winter. In between these extremes, the sun crosses the equator. On these dates, days and nights are both 12 hours long. The north and south points of the ecliptic are called the *solstices* (summer and winter, respectively). The points where the eliptic crosses the equator are the *equinoxes,* fall and spring. The spring equinox is the zero point of right ascension. These milestones in the sun's annual passage occur at about the same dates each year: spring equinox, 20 March; summer solstice, 21 June; fall equinox, 20 September; winter solstice, 21 December (Fig. 4.5).

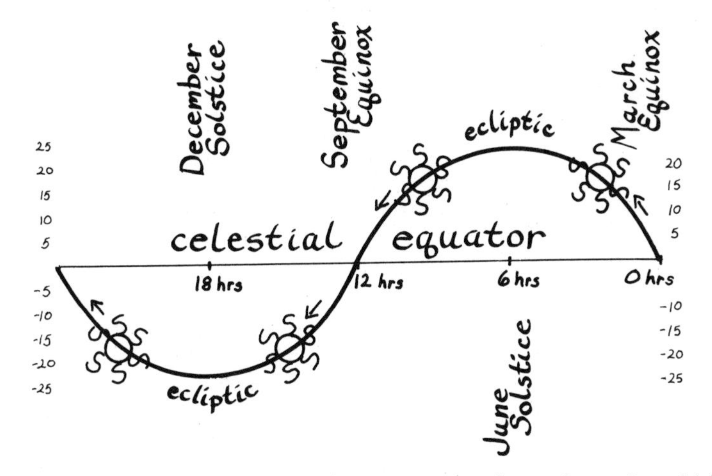

4.5 Map of the equatorial regions of the celestial sphere, from 0 to 24 hours right ascension, between declination 25° north and 25° south, showing the sun's annual motion around the ecliptic.

Solstice is Latin for "sun stands still." Figure 4.5 shows the reason for this. For several weeks around each solstice, the slope of the ecliptic is virtually flat. The sun continues to move into the east at approximately the same rate each day, but its motion north or south appears to slow until, at the solstice, it stops, and reverses itself. Since the distance of the sun north or south of the equator determines the length of the day, the day's length stays constant at these times. Around the equinoxes, when the ecliptic's path is steepest, the sun moves sharply north or south each day, and the days lengthen or shorten dramatically.

We call the ecliptic "the sun's path," but of course it is actually the path of the earth around the sun. If you take Figure 4.5 and roll the page up so its ends meet, you can see that this path is the shape of one circle (the ecliptic) intersecting another (the celestial equator). The circle of the ecliptic marks the orientation of earth's orbit in space. Because the orbits of the other planets share this orientation, their passage around the sky appears within a few degrees north or south of the ecliptic.

Precession and Other Motions

The spring equinox on March 20 is sometimes also called the First Point of Aries. This is because, 2,000 years ago, the equinox found the sun in that constellation. Today, the sun stands in Pisces on March 20. What happened? As the earth spins and revolves, it also wobbles exactly as a spinning top wobbles before it falls. This causes the celestial poles to sweep out a broad spiral across the sky. The entire celestial coordinate system moves in relation to the plane of the earth's orbit, changing the point at which the ecliptic crosses the celestial equator. This motion, called *precession,* is a result of the gravitational pull of the sun and moon, and shows itself only slowly. Each year, the First Point of Aries moves west by 50 arc-seconds, or 3⅓ arc-seconds of right ascension. The right ascension and declination of every object change proportionately, requiring large-scale star charts to be updated at 50-year intervals, or more often for more precise tasks. This is the only effect of precession you are likely to notice in your lifetime.

Over the long run, however, precession causes extreme changes in the sky. The north celestial pole wanders among the stars along a circle 23½ degrees in radius, taking almost 26,000 years to complete one circle. The center of this circle is the north pole of the ecliptic (which is the solar system's equivalent to the north celestial pole, having the ecliptic for its equator). Different eras have different pole stars, or none, as the celestial pole wanders from point to point on the northern sky. At the time of the First Dynasty of Ancient Egypt, Thuban, in the constellation Draco, stood just 10 minutes from the pole; around A.D. 12,000, the pole will pass 4½ degrees away from Vega. We are fortunate now to have the brightest star on the precessional circle, Polaris, so close to the pole. For the next 120 years, the pole will precess even closer to the North Star, until, in A.D. 2102, they will lie only 24 arc-minutes 31 arc-seconds apart.

In addition to precession, the earth performs a number of other, relatively minor motions. Called *nutation, variation in latitude,* and *planetary precession,* they are caused, respectively, by the pull of the moon, shifting weather patterns on the earth, and the pull of the other planets. The largest annual effect of any of these is only about 0.5 arc-second, and you are not likely to notice them. Finally, every now and then a major earthquake knocks us about half-a-second off kilter, and we take a year or so to stagger back upright.

MEASURING TIME

Solar and Sidereal Time

A day, of course, is the time it takes for the earth to rotate once. But how do we know when one rotation ends and another begins? We can use the sun, marking its transits on one day and the next. This is the *solar day*, and is the one we use to set our clocks. We can also measure a day by marking two transits of a star. This is called the *sidereal day*, and is often used by astronomers. Because a given star transits 4 minutes earlier each day, the sidereal day is 4 minutes (actually, 3 minutes 57 seconds) shorter than the solar day; the 24 hours, 60 minutes, and 60 seconds into which it is divided are all shorter by a proportionate amount. Figure 4.6 shows how this works. (Sidereal time is not used to measure anything but the passage of time; when an astronomer speaks of something traveling so many kilometers per second, for instance, the units of time are solar, not sidereal.)

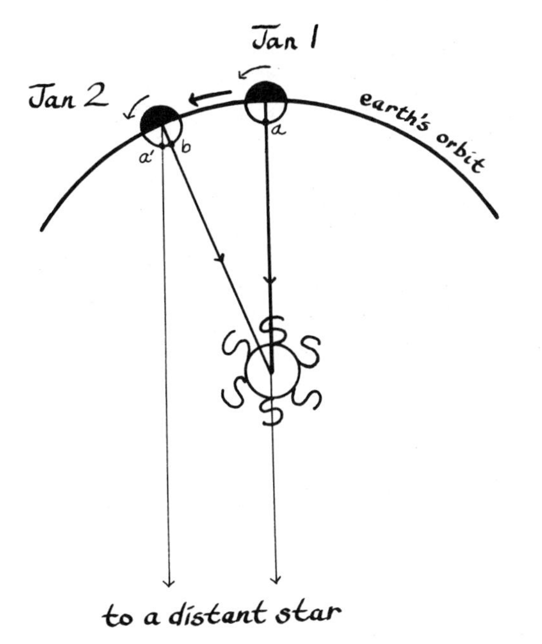

4.6 Solar and sidereal days. Between two days, the earth moves along its orbit, changing the angle between the sun, earth, and the distant stars (the angle is exaggerated here by about 25 times). Sidereal time measures the earth's rotation relative to the stars; point a *is sidereal noon on the first day,* a' *is sidereal noon on the next. The changing angle between the sun, earth, and the stars makes the point of solar noon change from* a *to* b*—adding about 4 minutes to the solar day.*

Just as solar time is determined by the position of the sun, sidereal time is determined by the position of the First Point of Aries. When the spring equinox is 1 hour west of the meridian, the sidereal time is 0100 (i.e., 1 A.M.); 22 hours west of the meridian (that is, 2 hours east of it), and the sidereal time is 2200 hours (10 P.M.). In other words, the position of an object on your meridian equals your sidereal time. If you want to look at a constellation that has a right ascension of 2200 hours, if you go out to observe at 2200 hours sidereal, you will find it on the meridian; if you go out at 1000 hours, it won't be in the sky at all.

One complete revolution around the sun equals one year, or 365.2422 days. Once again, we need to decide how we mark the end of one year, and the beginning of another. Here, the solution seems straightforward: we mark the passage of the sun through the First Point of Aries. Called the *tropical year,* this seems a logical system, because the sun's position on the ecliptic determines the seasons, the seasons govern the raising of food, and that's what calendars were originally all about. But several complications arise.

That 0.2422 day, for instance, makes life hard for calendar makers. If we just ignored it, using a 365-day or 366-day calendar, the seasons would quickly shift around the year; rules about when the first frosts come in the fall would need constant updating. In our present system, we let that odd fraction accumulate for four years, then account for it each leap year by adding a twenty-ninth day to February. This *Gregorian Calendar* maintains an average *calendar year* equal to the tropical year. The system of adding a leap day isn't entirely enough, however—it lacks about 12 minutes per year of the required 365.2422 figure. So, every year divisible by 100, but not by 400 (for example, 1900 but not 2000), we skip a leap year.

Precession also causes some confusion in our measurement of the year. The First Point of Aries is constantly moving west. The tropical year, then, is about 20 minutes shorter than the *sidereal year*—the year measured, like the sidereal day, with respect to the sun's passage across the stars. The tropical year is 365 days, 5 hours, 48 minutes, 46 seconds, to be exact. The sidereal year is 365 days, 6 hours, 9 minutes and 9.5 seconds long. Neither of these years is used in our calendar. The sidereal year ignores the position of the sun along the ecliptic, and so, like the tropical year, ignores the rhythms of human life.

Universal, Local, and Mean Time

In addition to sidereal time, astronomers also use a system called *universal time* (UT). This is Greenwich mean time (GMT)—the solar time as measured at Greenwich, England. When you measure solar time, you measure it for your location only: noon in the United States is evening in Europe, and dawn over the Pacific. The position of the sun gives *local time.* Universal time is a useful standard for dealing with global events such as eclipses.

Universal time is written in 24-hour notation, so that 1 A.M. becomes 0100 hours, and 10 P.M., 2200 hours. To convert to universal time, *add* the following corrections to the local time in your zone:

Zone	*Meridian*	*Standard*	*Daylight*
Eastern	(75°)	5 hours	4 hours
Central	(90°)	6	5
Mountain	(105°)	7	6
Pacific	(120°)	8	7

If the result from this addition is greater than 2400, subtract 24 hours, and add 1 day. For example, 2200 eastern standard time (EST) on 9 June + 0500 hours = 2700. Subtract 2400, and the universal time is 0300 hours *on 10 June.* To convert from universal to local time, reverse the process.

The most basic way to measure solar time is to find the moment when the sun transits the meridian. A vertical stake in the ground will throw its shadow due north at that moment, telling you that noon has arrived—this is the principle behind sun dials. The time of noon measured in this way, however, may not agree with what your clock tells you—even though your clock keeps solar time. There are two reasons for this. The first is that our clocks are set to standard time zones. Your true local time may vary from the time in your zone by up to fifteen minutes. To correct for this variation within zones, you add four minutes for every degree of longitude you lie *east* of the standard meridian for your zone, and *subtract* four minutes for every degree you lie *west* of the standard meridian. The meridians appear in the above table.

The second reason that your watch may not be synchronized with the sun is that, because of the tilt of the ecliptic and variations in the earth's orbital speed, the length of the solar day changes over the course of the year. The sun transits later than mean solar noon from January to April, slightly before noon in May and June, late again in July and August, and fast from September through December. The *equation of time* is the difference between mean (clock) time and actual (solar) time; some almanacs list it as the "correction to sun dial"; astronomical ephemerises (see chapter 9) usually give the figure for each day of the year.

The Julian Day Calendar

In some fields of astronomy, it is useful to be able to tell quickly how many days have elapsed between two events. Astronomers simplify this calculation by using the *Julian Day Calendar.* This calendar has neither years nor months: each day is simply one number higher than the one before, all the way back to Julian Day 1, which was 1 January, 4713 B.C. The Julian Day begins at 1200 UT; at noon UT on 1 January 1985, the Julian date was 2,446,067. You can use that date as a benchmark to arrive at today's Julian date; many astronomical calendars also list it alongside the regular date.

The Facts of Light

5

The telescope performs three basic functions. It gathers light. It forms an image. It magnifies the image. Each of these is important, and all are interrelated. In each case, however, the telescope is a tool for working with light. To understand a telescope and to get the most out of it, you need to know the facts of light.

THE ELECTROMAGNETIC SPECTRUM

Light is energy, one of a variety of energies called *electromagnetic radiation,* which include X rays, ultraviolet, infrared, and radio. Despite their different names these are all forms of light. The relationship among them is like that among notes on a piano. The lowest octaves of light are radio, next microwave, then infrared. The middle octave is visible light. In the higher ranges come ultraviolet, X rays, and finally, at the highest keys, gamma rays. The entire sequence is known as *the electromagnetic spectrum.*

This name is derived from the *visible spectrum:* the array of colors we see when light shines through a prism. The prism spreads light out so that its different wavelengths appear. Blue light is high energy, short wavelength, high frequency; red light, at the opposite end of the spectrum, is low energy, long wavelength, and low frequency (frequency equals waves per second). The order of colors in a spectrum is always the same: red, orange, yellow, green, blue, violet. The boundaries between these colors are never distinct; one fades into the next.

If we could see frequencies higher than violet, and lower than red, and had a device to extend the spectrum beyond the visible range, we would have the electromagnetic spectrum: red fades into infrared, which grades into microwave, and so on down to radio waves a mile long. At the other end of the spectrum, beyond violet, ultraviolet merges into X rays, and finally into gamma rays of such intense frequencies that they can kill. Astronomers have developed instruments that can detect all forms of electromagnetic energy. This book, however, is concerned primarily with what you can see with your own eyes—the visible spectrum.

THE PHYSICS OF LIGHT

The speed of light in a vacuum is 299,793 kilometers per second (186,000 miles per hour). This speed is constant, making it one of the few absolutes of nature. Because its speed through a vacuum is such an important constant, astronomers use it in a measure of distance, the *light year,* which is the distance light travels over the course of a year, more than nine trillion kilometers. According to current theories, nothing in the universe can go faster than light. This speed is so high that we are accustomed to thinking of it as infinite. When we look across astronomical distances, however, 300,000 kilometers per second can seem slow. The light of the full moon leaves its surface more than a second before it reaches your eye; light from the sun takes 8 minutes to reach the earth; to look out in space is to look back in time.

Like other forms of radiation, light is subject to an *inverse square law* of intensity. As light spreads out from its source, it weakens. The amount of weakening is determined by the *square* of the distance from the source. If a candle at a distance of one foot has an intensity of 100 units, the same candle 2 feet away is one quarter as bright—25 units. One hundred feet away, that candle is one ten-thousandth as bright—0.01 units. Brightness, in other words, is equal to one divided by r squared ($1/r^2$) (r is the standard symbol for distance in astronomy).

Light is both a particle and a wave. Actually, light "is" simply light. It sometimes *behaves* like a particle, and sometimes like a wave. Most of the time, light follows the laws of wave behavior. Occasionally, light acts as if it also existed in small packages (or *quanta*) of energy. Imagine the waves on a shore suddenly appearing to you as a mass of Ping-Pong balls bouncing off each other and the beach, and you'll get some sense of the difference.

Quanta of light are called *photons;* they are the fundamental unit of electromagnetic energy. A photon has no mass nor is it electrically charged; it is pure energy. When a photon of infrared light lands on your skin, it warms you. A more energetic photon can power a small chemical reaction in your eye, which you see as light. The amount of energy in most photons is extremely small and is related to its frequency. High-frequency photons pack more energy than low-frequency photons, which is why gamma rays can kill you but infrared just makes you warm.

Waves

The frequency of a photon depends on the wave nature of light. Picture a series of uniform waves. The tops of the waves are crests; the valleys in between are troughs. One *cycle* of a wave is the span from crest to crest—the basic form that the wave repeats (Fig. 5.1). Within any given cycle, all waves have velocity, wavelength, frequency, and amplitude. The *frequency* of a wave is the number of cycles that pass a stationary point each second. It is measured in *hertz* (*hz*); one hertz is one cycle per second. A *wavelength* is the distance between two successive crests. *Amplitude* is one-half the height from a trough to a crest, and is related to the amount of energy the wave contains. The velocity (V), wavelength (Greek letter *lambda:* λ), and frequency (f) of any wave are usually related by the formula:

$$V = \lambda \times f$$

Because the velocity of light waves is usually constant, this formula means simply that long waves have low frequencies and short waves have high frequencies. Light waves are very short—around 5,000 angstroms long (one angstrom equals one ten-billionth of a meter). The speed of light is about 300,000 kilometers per second, so visible light has a frequency around 600 trillion hertz. (Compare this with the 440 hertz of the standard A note used in tuning musical instruments.)

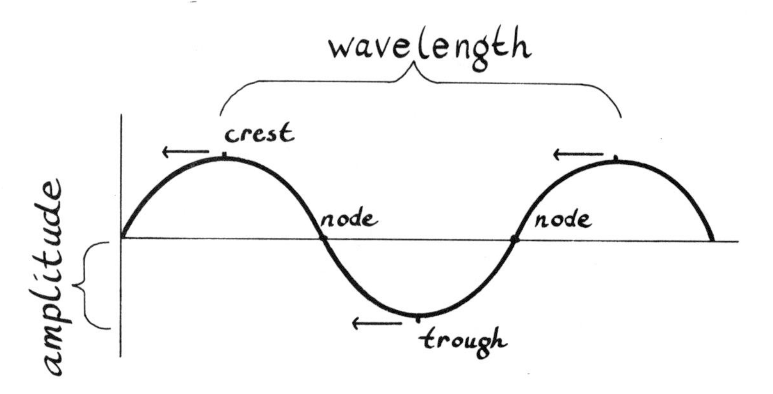

5.1 A wave, showing amplitude and wavelength. One cycle of a wave is the portion between two crests. Frequency is a measure of the number of cycles passing a given point each second.

Waves have several typical properties, some of which are important in visual astronomy. These properties include: reflection, refraction, interference, diffraction, and polarization. We are most interested in the first two of these. As mirrors reflect, and lenses refract, they can also gather light, form and magnify images. Both of these phenomena follow simple rules. When a wave strikes a reflecting surface at an angle, like a billiard ball bouncing off the rail, it rebounds at the same angle. "Angle of incidence equals angle of reflection" is the *law of reflection.* In Figure 5.2, angle *a,* the angle of incidence, equals angle *b,* the angle of reflection.

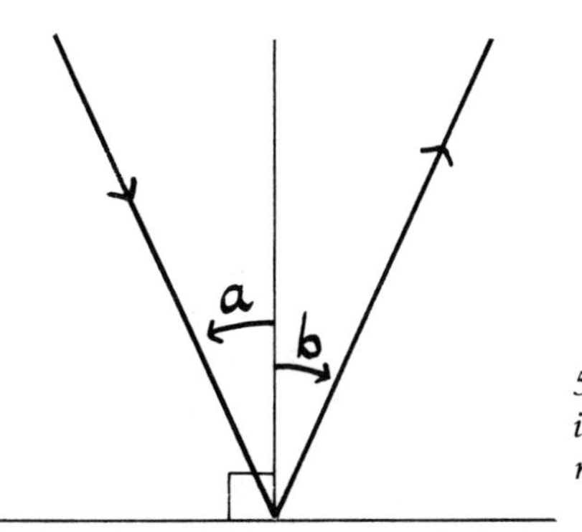

5.2 In reflection, the angle of incidence a *equals the angle of reflection* b.

When a wave moves from one medium (such as air) into a denser one (such as glass) at an angle, it dives more steeply into the denser material. When it moves from a dense to a thin medium, it bends away from the thinner medium, tending to "hug" the thicker one. The angle at which it bends (the *angle of refraction*) depends on both the optical densities of the two materials (their *refractive indexes*), and on the wavelength of the wave. When passing from air into most kinds of glass, shorter waves bend less than longer waves. This is how a prism produces its familiar spectrum. As different wavelengths bend differently on entering and leaving the prism, they depart from the prism spread out along a broad swath. In Figure 5.3, the light wave approaching a denser medium in angle *a* is refracted to the smaller angle *b.*

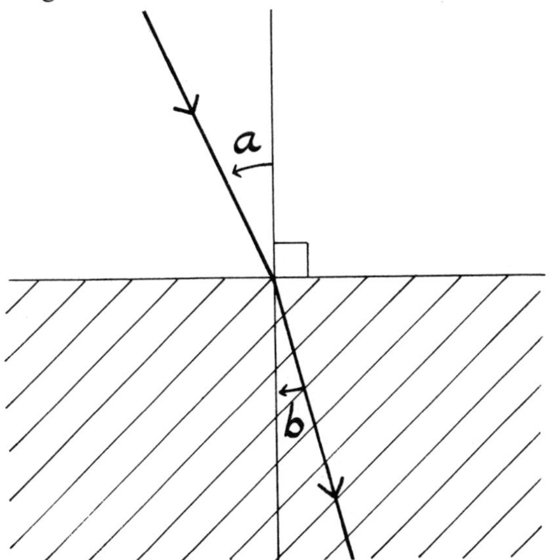

5.3 In refraction, a light ray entering a material at an angle a *is bent to a new angle* b, *by an amount determined by the material's index of refraction.*

Pitch stones in a calm pool, and you'll see *interference.* When the crests of two waves collide, they merge briefly to form a single wave higher than either of the original two. This is called *reinforcement,* or *constructive interference.* Two waves interfering in this way are said to be *in phase* with each other. If the crest of one wave should meet the trough of another, they cancel each other out, forming a momentary calm by *destructive interference.* Two such waves are said to be 180 degrees out of phase. When this happens with sound waves, the result is silence; in light, darkness. Various kinds of interference can happen among waves of the same or of different frequencies.

Diffraction is important to astronomers because it can subtly bend a ray of light, distorting the image in a telescope. Diffraction is explained by *Huygens's principle.* The causes of diffraction depend on a principle of wave-propagation, which says that any point along a wave acts as if it were an individual wave generator. If there were only one such wave-generating point, the wave would spread out from it in an expanding ring, like the ripple from a stone tossed in a calm pool of water. But if you were somehow to drop several hundred stones in that pool, all at once and all on a line, the expanding rings would combine to form straight wave fronts. The ripple from each stone would nudge the one beside it into line. If you imagine a drunken marching band, in which each marcher is kept on course only by the shoulders of the marchers on either side, you get the picture: an individual marcher will tend to stagger around in circles. Marchers at the edge of the band will tend to stray off to either side. This straying is called diffraction.

In principle, any wave acts as if it were constantly being re-formed by the action of millions of infinitely small point sources—or stones, or marchers. Diffraction makes that effect visible in light waves whenever an obstruction—a wall, or a part of the telescope—cuts off part of a wave. The point on the cut-off edge of the wave is free to radiate off to one side, because there are no more point sources on that side to keep it in line. A beam of light blocked off in this fashion will show a slight scattering, off toward the side on which it was obstructed (Figure 5.4).

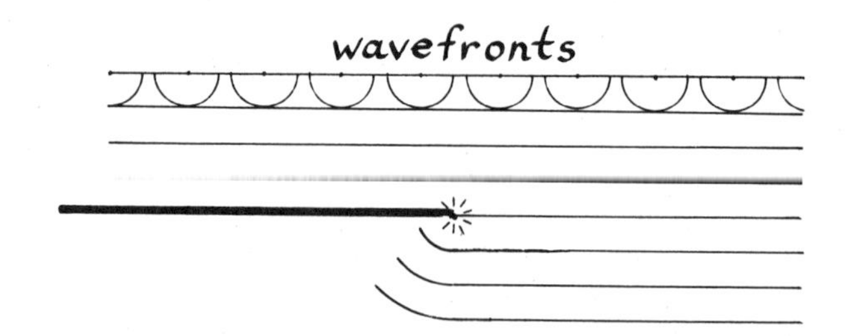

5.4 Huygen's principle of diffraction. Light waves behave as if each point on a wave is a tiny emitter. The combined emissions of each point form the next wave. When a wave passes an obstacle, the point at the edge of the light ray emits in a semicircle ahead of itself, causing the wave to spread out around the obstacle.

The amount of diffraction depends on wavelength, so light diffracts very little, but when magnified in a telescope, the effect is often visible. If two diffracting points are spaced so their waves interfere (as often happens), diffraction will cause patterns of increased or decreased brightness in an image.

Light waves usually vibrate in different planes. Figure 5.1 simplified this, showing a wave vibrating in a single plane (the sheet of paper). It is possible to filter out all waves except those parallel to a given plane; such light is said to be *polarized.* Light can be polarized naturally, and by special filters. As with polarized sunglasses, most amateurs use these filters to cut the glare from very bright objects.

Most objects absorb some wavelengths and not others. We see this as color in everything around us. An opaque object that absorbs all wavelengths except the red ones, for instance, looks red to us. A glass that absorbs all wavelengths except blue is said to *filter out* long wavelengths, or to *pass* short wavelengths. Amateur astronomers use colored filters to add contrast to dull images—blue spots on a planet, for instance, stand out brightly in a blue filter.

The Telescope: Basic Functions

6

FORMING AN IMAGE

All optical telescopes share one basic function: they form an *image*. There are two kinds of image, virtual and real. In a *real image,* waves of light from the imaged object converge, or *focus,* to a point. Curved lenses or mirrors can form a real image. A flat mirror, because it bends all rays equally, does not focus; its image is virtual. A real image can be projected on a screen; you can touch the point where the image appears. A virtual image cannot; we see it, but its apparent location behind the mirror is an illusion, existing only in Wonderland. The reflection we see in an ordinary wall mirror is a virtual image.

When a mirror with a curved, concave surface reflects, it forms a real image. The image appears halfway between the mirror and its center of curvature, the center of the sphere of which the mirror's surface forms a part. The image is upside down, and its size depends on the distance between it and the mirror; in most telescopes it is around half an inch across. A lens can have both its front and back surfaces curved, and can consist of more than one piece of glass. A lens with the proper combination of curves will also form a real image. The distance from lens to image depends on the steepness of its curves.

The three major kinds of optical telescope form images by reflection, refraction, or both. *Refracting telescopes* (*refractors*) use lenses made of flawless glass. *Reflecting telescopes* (also called *reflectors* or *Newtonian* telescopes) use a concave mirror. *Catadioptric telescopes* refract light through a thin, curved glass "correcting plate" and then reflect it from a convex mirror. The main light-collecting and light-focusing element in a telescope is called its *objective.* In a catadioptric, which is basically a modified reflector, the mirror, not the correcting plate, is called the objective.

OPTICAL ELEMENTS

The most important characteristic of any telescope is its *aperture,* the diameter of the objective. This figure is so important that we use it to describe a telescope, calling it simply "a six-inch," "a twelve-inch," and so on. Standard sizes have developed for the apertures of amateur telescopes. The common sizes expressed in inches and millimeters (mm), are:

	2″	(50 mm)	
	2.4″	(60 mm)	
refractors	3.1″	(80 mm)	
	4″	(100 mm)	
	6″	(150 mm)	
	8″	(200 mm)	
	10″	(250 mm)	reflectors
	12.5″	(315 mm)	
	13.1″	(332 mm)	
	14″	(355 mm)	
	17.5″	(445 mm)	

Refractors are generally smaller, appearing commonly in the 6 inch range and under. You rarely see reflectors smaller than 3 inches, and 4 inches is usually the minimum. Reflectors are the largest of amateur scopes—one manufacturer offers scopes up to 29 inches. Catadioptrics generally come in the same sizes as reflectors, ranging from 3 inches to 14 inches. In general terms, any refractor under 3 inches is considered small; 3 inches to 4 inches are medium, and a 6-inch refractor is relatively large. A reflector under 6 inches is small, one 6 to 8 inches is medium, and one 10 inches and up is large.

After light has been refracted or reflected by the objective, it forms an image at the *focal plane;* this location is often called the *prime focus,* or *first focus.* The distance from objective to focal plane is called the *focal length* (Figures 6.1 and 6.2). It determines the size of the image that the telescope forms. The *focal ratio* is the focal length divided by the aperture, and is written *f*/*x*. A 6-inch scope with a 30-inch focal length, for instance, is an "*f*/5" system. The *f*-ratio determines the telescope's photographic *speed,* which is a photographic term for the efficiency of the telescope as a light-gatherer. Telescopes *f*/5 or shorter are "fast," and, if equipped with a camera, can take a photograph with a shorter exposure than average; *f*/10 and longer systems are said to be slow. Speed matters only in cameras, but the terms have come to be used to describe telescopes in visual use as well.

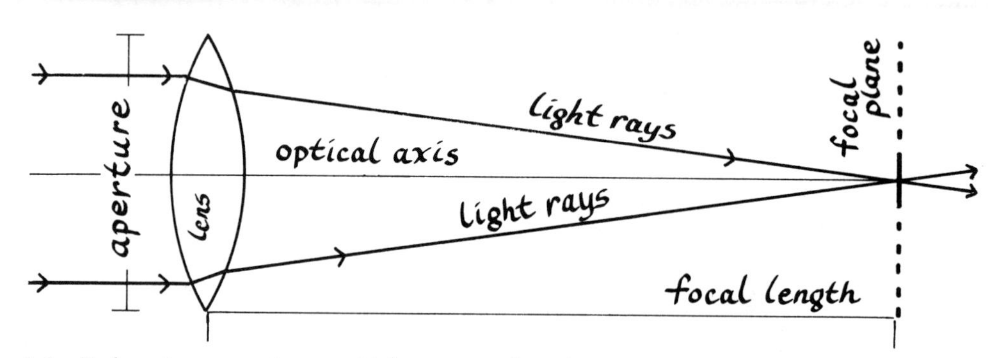

6.1 Refraction to an image. Light enters the (objective) lens, and is bent so that the rays converge at the focal plane. The aperture is the diameter of the objective. The focal length is the distance from the objective to the focal plane. The optical axis marks the center of the cone of refracted light; it is a line perpendicular to the center of the objective.

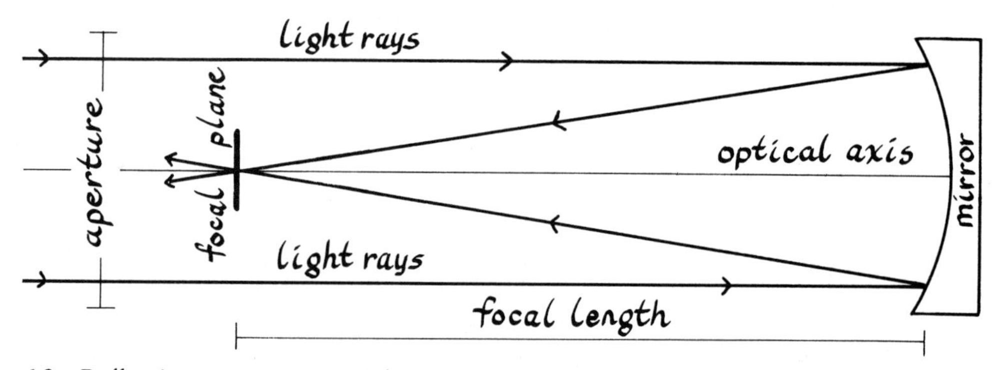

6.2 Reflection to an image. Light strikes the surface of the (objective) mirror, which is curved, so that the reflected rays converge at the focal plane. The aperture is the diameter of the objective. The focal length is the distance from the objective to the focal plane. The optical axis marks the center of the cone of reflected light; it is a line perpendicular to the center of the objective.

If you know the aperture and focal length of a telescope, you can approximate the size of the image it will produce at its prime focus. If F is the focal length in inches, and θ (theta) is the angular diameter (in degrees) of the object you're observing, then its image* will be

$$\theta \times F/57.3$$

For example, in a 6-inch *f*/5 telescope, the 0.5-degree-wide moon appears about ¼ inch across. To see such an image well, we need to magnify it. The magnifying glass is the telescope's *eyepiece,* also called the *ocular.*

OPTICAL PERFORMANCE

The phrase "optical performance" describes the qualities of any telescope that influence the kind of images it forms, the kind of objects it shows best, and how easy it is to use.

Light Grasp

To get enough light to see the image, you need to collect it, and the more the better. If you wanted to use a bucket to collect a lot of rain at once, you would use a wide-mouthed one, not necessarily a deep one. The same is true when you're catching photons. How much light you collect depends on the area of the surface you set out to catch it. Since the area of a circle is π (Greek *pi*) times *r* squared, where *r* equals the radius of the aperture, doubling the aperture of a scope quadruples its light-grasp. An 8-inch scope has *four* times the grasp of a 4-inch. Not all of this light gets to your eye. Lenses absorb some light, and lose more by reflection at their surfaces. A reflector loses light because its optics are not 100% reflective, and also because some of its parts shade the primary. Small refractors are slightly more efficient than reflectors; over 8 inches, the reverse is true. In either kind of scope, special optical coatings can cut light loss.

If the pupil of your eye is too narrow to admit all the light from the eyepiece, you also waste light. The maximum aperture of your eye is about 7 millimeters, or 0.3 inches. (This figure decreases slightly with age.) The right combination of eyepiece and focal length can produce a beam of light just the size of your fully dilated pupil.

The most common measure of a scope's light-grasp is the magnitude of the faintest star it will show under ideal conditions, called the telescope's *limiting magnitude.* The average unaided eye has a limiting magnitude of about 6. The table shows the limits of magnitude for the common sizes of objective (in inches), under good, but not ideal conditions:

Size of Objective (inches)	***Limiting Magnitude***	***Size of Objective (inches)***	***Limiting Magnitude***	***Size of Objective (inches)***	***Limiting Magnitude***
2 ″	10.6	6 ″	12.9	13.1″	14.6
2.4 ″	10.9	8 ″	13.5	14 ″	14.7
3.1 ″	11.5	10 ″	14.0	17.5″	15.0
4.25″	12.1	12.5″	14.5		

*This formula applies only to celestial objects; anything within a few miles of the observer requires a more complex equation.

As you can see, there's a diminishing return on increased aperture—the leap from naked eye to 2 inches is greater than the jump from 2 inches to 17 inches!

Resolution and Definition

These qualities determine the sharpness of a telescopic image. *Resolution* is the ability to separate (or "resolve") two adjacent *point images.* It is expressible as the smallest gap detectable between two points: a typical amateur telescope can resolve stars 0.7 arc-seconds apart. *Definition* refers to the clarity of *extended images.* Extended images include the surface of the moon, planets, or a large galaxy; all stars are *point images.*

Resolution depends on aperture. Some of the light in a telescope is always subject to diffraction. Diffraction causes a star's image, which should be an infinitesimally small point, to appear as a tiny blurred dot, called the *Airy disc.* The Airy disc makes the star appear as if it had an angular diameter of a second or two. Severely diffracted light forms concentric rings (*Airy rings*) of light and dark around the star. The innermost ring is the brightest, and the outer rings fade rapidly with increasing distance. In any one telescope, the size of the Airy disc and rings is irreducible, even if you remove all diffracting obstacles from the light path, because diffraction will still occur at the edges of the objective.

The only way to improve resolution is to increase aperture. If you double the aperture, you improve resolution twofold. Quadruple it, and the resolution improves fourfold. But no matter how much you improve the quality of the optics, nor how large you make them, the aperture of a telescope sets a firm upper limit on the quality of its image: you can improve any telescope so far, but no farther.

The formula for finding the resolving power of a telescope is called *Dawes's limit*: 4.56 divided by the aperture (diameter of objective lens). This formula assumes that the objects being resolved are of similar, moderate brightness; very bright stars have wider Airy discs, and at dim magnitudes other factors intervene to make detection difficult.

The practical value of Dawes's limit is that it tells you how close two objects, such as stars, can be before they merge into a single blur. With a 6-inch telescope, for instance, you can see dark sky between stars as close as 0.76 arc-seconds apart. This makes all the difference in the appearance of clusters of densely packed stars, and especially in your ability to detect double stars.

Dawes's limit does not apply so stringently to extended images. It is possible to see craters on the moon below limiting resolution; contrasting lines, such as dark valleys on the bright lunar plains, or the gaps in Saturn's rings, are detectable even when their angular widths are one-third Dawes's limit.

· DAWES'S LIMITS OF RESOLUTION FOR STANDARD APERTURES ·

Aperture Size (inches)	*Limit of Resolution (arc-seconds)*
2	2.3
2.4	1.9
3.1	1.5
4.25	1.1
6	.76*
8	.57
10	.46
12.5	.36
13.1	.34
14	.33
17.5	.26
29	.16
200	.0228

Definition depends to some extent on resolution, but also on the overall quality of the optics. Flaws in the optics can include bumps or pits in the surface, or a large-scale defect in the curve, or *figure,* of the surface. Most optical surfaces have a precise, geometric form—usually a sphere or paraboloid. Any deviation from the ideal figure will distort the image. *Rayleigh's limit* defines the smallest defect that will have a noticeable effect on the quality of the image. Its value is different for refractors and reflectors. For refractors, the limit is one-half of a wavelength of green light—about 2500 angstroms. For reflectors, the limit is one-eighth of a wavelength of green light—about 625 angstroms. Beyond these limits, the distortion of the image will be smaller than the Airy disc, and so will not be noticeable.

Such an optical surface is *diffraction-limited*—its performance is restricted only by this unavoidable diffraction. Manufacturers usually specify the quality of their optics by guaranteeing them to be accurate to within a given fraction of a wave: "one-eighth wave optics" is a common way of saying that a mirror is diffraction-limited; "null-tested" means essentially the same. Watch out for the sign "±" before the fraction of a wave, as in "accurate to ±⅛ wave." This actually means that surface irregularities as much as one-eighth of a wave *below or above* the desired surface may be present, giving a mirror accurate to only one-fourth wave.

Definition is hard to define. It can't be predicted purely by aperture. Yet it can be the most important quality of a telescope; even light grasp is

*This value represents the maximum resolution attainable, on most nights, through the unstable atmosphere.

wasted without it. You learn to recognize the difference between good and bad definition with experience—usually by comparing different telescopes.

Magnification

Magnification is the enlargement of an image. If a telescope doubles the apparent size of an object, its magnifying *power* is said to be 2×; ten times, the power is 10×, and so on. Magnification makes distant objects look closer; seen in a telescope at 10×, a bird 100 feet away seems 10 feet away.

The magnification (M) of any telescope depends on the focal length of the objective (Fo) and of the eyepiece (Fe):

$$M = Fo/Fe$$

For a given telescope, magnification increases as the focal length of the eyepiece decreases. The power of any given eyepiece depends on the focal length of the telescope.

There are limits to a telescope's capacity for magnification. The upper limit depends mostly on aperture. When you magnify an image, the light in the small disc at the focal plane is spread thinly over a larger area. If you magnify an image too much, it becomes too dim to see. Dawes's limit on resolution determines how much detail your objective can produce. Magnifying an image beyond the amount necessary to see all of the detail in it is like magnifying a newspaper photograph: all you see are larger, blurrier dots. The general rule for estimating a telescope's upper limit of magnification is 60× for every inch of aperture. Very dim objects, however, cannot be enlarged anywhere near that amount. Atmospheric turbulence and other factors make 30× per inch a more reasonable limit on most nights.

The lower limit of magnification is a matter of light conservation. The light beam from the eyepiece gets wider as magnification decreases. If it is so wide that your fully expanded pupil cannot take it all in, you waste some of that light. A rule of thumb for this limit is 3× for every inch of aperture.

· HIGH AND LOW MAGNIFICATION LIMITS ·

Aperture Diameter (inches)	***High***	***Low***
2	120×	7×
2.4	144×	8×
3.1	186×	10×
4.25	255×	14×
6	360×	20×
8	480×	27×
10	600×	33×
12.5	750×	42×
13.1	786×	43×
14	840×	47×
17.5	1,050×	59×

As usual, the preceding table assumes good conditions; average conditions will cut the upper limit in half—more for the larger apertures.

Most beginners push magnification too high. The images they get are dim, blurry, and usually roiled by atmospheric turbulence. You will do better to lower your expectations. Generally, anything over 80× is high power, 30×–80× is medium, and below 30× is low. A good many amateurs spend most of their time observing just at the lower limit. They enjoy the bright, wide-angle images, with more stars visible at a single glance than there are at higher powers. For comet hunting, in particular, low power is a necessity.

Visual Efficiency

Visual efficiency is the visual equivalent of photographic speed. It determines the brightness of an image in a particular telescope working at a particular magnification. The efficiency depends on the aperture divided by the magnification. A 6-inch telescope, at 18×, has an efficiency of 0.33; a 12-inch at 300×, only 0.04—the 6-inch will transmit images almost ten times brighter, under those conditions. This is why high magnification can be so difficult to use, and why people observing large, dim objects, such as galaxies or comets, prefer low powers.

Field

The *field* of an eyepiece is the area of sky visible in it. A wide-field eyepiece shows more than a narrow field. There are *true fields* and *apparent fields.* The true field is the actual size (in angular units) of the area of sky visible in the eyepiece. A true field of 0.5 degrees, for instance, will encompass the full moon almost entirely. The apparent field is the angle your eyeball rotates through while looking from one edge of the field to another. Apparent field is built into the eyepiece; fields in different kinds of oculars range from a cramped 15 degrees to a spacious 82 degrees.

True field (*fdt*) in any eyepiece equals the apparent field (*fda*) divided by magnification:

$$fdt = fda/M$$

At 20×, in an eyepiece with a 40-degrees apparent field, the true field will be 2 degrees—enough to hold four full moons end to end. At 190×, the same design of eyepiece has a true field only 0.21 degrees across.

Field of Full Illumination

This is the area of the focal-plane image that receives 100% of the light the objective. The parts of a telescope can sometimes shade the edges of the image. The size of the shaded area, and how much light it loses, depend on many aspects of telescope design. Even in well-designed telescopes, you can look through a wide-field eyepiece and see beyond the fully illuminated field. The outermost regions of such an eyepiece will be about a third dimmer than the field of full illumination—more in poorly designed telescopes. Oddly, a loss of one-third the available light at the edges is rarely noticeable. In most well-made telescopes, the field of full illumination is about three-fourths the size of the focal-plane image.

Exit Pupil

Also known as the *Ramsden disc,* this is the focal disc produced by the eyepiece, comparable to that produced by the objective at the telescope's focal plane. You can calculate its size (P) by dividing your telescope's aperture (A) by the magnification (M):

$$\text{exit pupil} = \text{aperture/magnification};\ P = A/M$$

This is the beam of light that must fit into your eye's pupil if you are to avoid wasting light.

For each ocular, the exit pupil appears a set distance from the eyepiece. That distance is called the *eye relief.* Eye relief varies with eyepiece designs, and within a given design, by focal length. As a general rule, high-power oculars have uncomfortably short eye relief—they are hard to look through without squinting, and this can be a strain over extended periods. If the eye relief is excessively long, however, it's hard to find the exit pupil. The trick is to bring your eye close to the eyepiece, then back away, keeping the image centered until it reaches its maximum brightness and size.

Field Orientation and Focus

The image in an astronomical telescope can be reversed (right is left), inverted (up is down), both, or neither. The key to keeping your bearings is knowing how your telescope presents its image (Fig. 6.3). Reflectors almost always both invert and reverse. Refractors and catadioptrics, when used with erecting prisms, reverse. Binoculars do neither. To learn your scope's orientation, observe some lettering through it (a billboard is good for this). No matter how well you learn this lesson, though, you will tend to get mixed up from time to time while looking at the stars. To reorient yourself, hold the telescope still. The stars *always* drift toward the west. Many astronomical guides take advantage of this rule by stating that an object is on the "preceding" (western) or "following" (eastern) side of a guide star. To find north, move the telescope toward the south point of your horizon; as the telescope moves south, look in the eyepiece: the direction toward which the stars move across the field is the north.

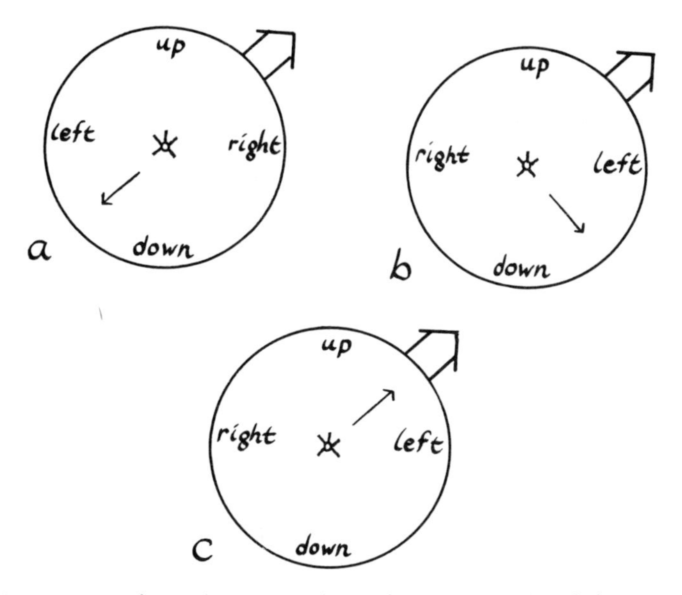

6.3 Image orientation. These diagrams show the orientation of three common telescopic fields. The small arrow shows the apparent motion of the field produced by an actual motion (large arrow) of the tube. (a) *erect, correct image, as in a spotting scope or binoculars;* (b) *erect, reversed image, as in a refractor with an erecting prism;* (c) *inverted, reversed image, as in a reflecting telescope.*

The exact focal length of a telescope depends on the distance of the object. Most telescopes use eyepieces of varying focal lengths. These two facts require that we *focus*: move the eyepiece back and forth to position it the right distance from the focal plane. Nearby objects focus farther from the objective than distant ones; pull the eyepiece close toward you for close objects, push it away for far ones. Back off for low power, move in for high power.

OPTICAL ABERRATIONS

Telescopic images are prone to six kinds of flaws, called *aberrations*. They are: *spherical aberration, coma, astigmatism, curvature of field,* and *distortion.* The sixth, *chromatic aberration,* occurs only in lenses. All aberrations can be avoided by a well-made mirror. Chromatic aberration in a lens can be vastly reduced.

Reflecting telescopes would work perfectly if they formed their images only with light entering close to and parallel to the objective's *optical axis,* the line connecting the center of the objective and its image at the focal plane. But in real telescopes, light reaches the image from far off the optical axis, at a variety of angles. This complicates the geometry of reflection. Most aberrations result from either a failure to achieve sharp focus over the entire field, or a warping of the image. Aberrations become noticeable when they exceed Rayleigh's limit. Of the kinds affecting reflectors, spherical aberration, astigmatism, and coma affect focus. Two other kinds of aberrations, curvature of field and distortion, warp the image.

There are two easy tests that will show the presence of each of these two classes of aberration (although they may not identify the precise nature of it). To find aberrations of focus, aim your telescope at a medium-bright (second to third magnitude) star on a clear, still night. Focus the star sharply: you should be able to see the Airy rings, or the faint spikes caused by diffraction around the supports of the secondary mirror. Then defocus the telescope by moving the eyepiece in (see chapter 8). Notice the appearance of the unfocused image of the star. It should be round, and evenly illuminated. Then defocus in the opposite direction, by the same amount. These *intrafocal* and *extrafocal* images should be identical. If they are different sizes, less than spherical, or of different brightnesses, your optics suffer from some aberration. To make certain the problem is in the telescope itself, repeat this test using different eyepieces.

To detect curvature of field or distortion, set up your telescope during the day, and position a sheet of graph paper so that it just fills the field (Fig. 6.4). If the lines on the graph paper appear straight across the telescopic field, you have neither of these defects.

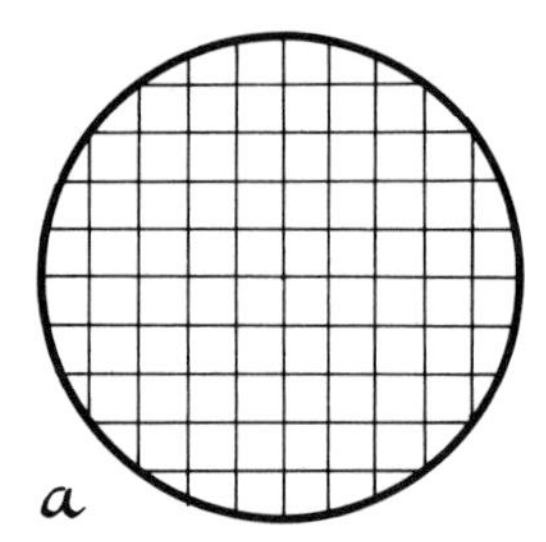

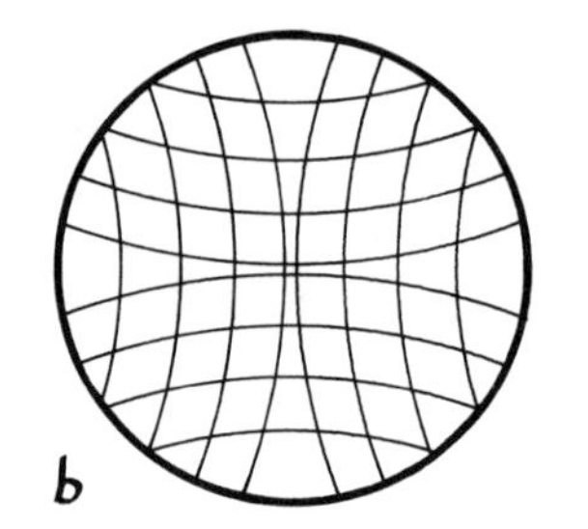

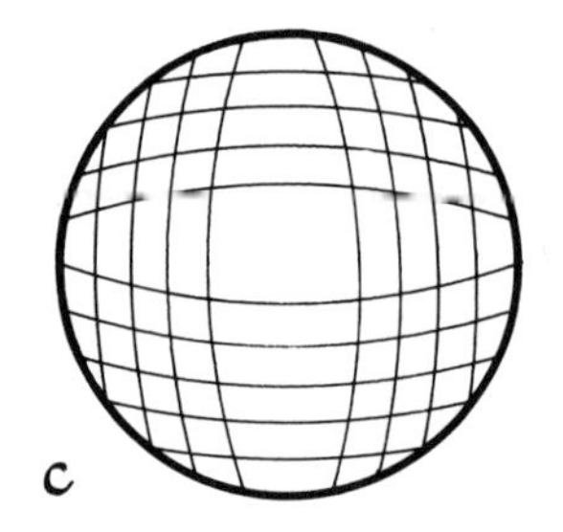

6.4 Optical distortion. A sheet of graph paper, seen through normal and distorted optics. (a) *undistorted.* (b) *positive distortion.* (c) *negative distortion.*

To detect coma, look at the outer regions of the field on a clear night. The stars there should be round. If they are teardrop- or fan-shaped, and all the fans spread outward from the center of the field, your system has coma. Coma is inevitable in systems shorter than *f*/5; the shorter the *f*/ratio, the larger the region of the field that suffers, until, at around *f*/3, the whole field shows fan-shaped stars. Coma can also occur when the telescope's optics are misaligned. When this happens, the images of stars will not fan outward from the center of the image, but will all fan the same direction across the field. Poorly mounted optics can also give a false result on these tests. Always check mountings and alignment before making any test for aberration.

Chromatic aberration affects only lenses. Chromatic aberration arises because a lens is also a prism, and will focus different wavelengths at different focal lengths. A sharply focused red image will have a fuzzy blue halo around it, and vice versa. These halos, and the defect in general, are sometimes called simply *chromatism*. The defect is not only distracting, but it reduces resolution, definition, and to some extent the light efficiency of a telescope. The effect is inevitable, but it can be reduced by making a refractor's objective out of two different lenses, each with a different refractive index. The combination is called an "achromatic doublet," and usually looks like the lens in Figure 7.1 in chapter 7. The two lenses in a doublet cancel out each other's chromatism, but not completely. The image of a white star in an achromatic doublet is usually tinged with yellow, and shows a faint purple halo of unfocused blue and red light, called the *secondary spectrum*. To reduce secondary spectrum, most manufacturers keep the focal lengths of their doublets at *f*/10 to *f*/15, taking advantage of the minimizing effects of high focal ratios.

Kinds of Telescopes 7

Refractors, reflectors, and the hybrid catadioptric telescopes all come in several forms, from no-frills basics to expensive exotics. Each form has its special strengths and weaknesses; no single telescope is truly all-purpose. The later chapters of this book include information on the telescopic appearance of specific celestial objects. Those chapters assume you will be using a simple 6-inch reflector, because such a telescope has a fairly typical light-grasp. The 6-inch reflector is also inexpensive to buy, simple to build, and will show well a wide variety of astronomical objects. I recommend it as a first serious telescope for the amateur. But any such recommendation is a matter of taste. Before you choose a telescope, examine the other options open to the beginner.

REFRACTORS

Refractors take their name from the lens that serves as the objective, which focuses light by refraction. Amateur refractors typically have apertures of 2 to 6 inches, and focal ratios between *f*/10 and *f*/15. The most common is a 2.4-inch of around *f*/11, giving a focal length of slightly over 2 feet.

Most amateur refractors have an *achromatic* objective lens, similar to that shown in Figure 7.1. Also called a *doublet,* it consists of two lenses, each with a different refractive index, and each with a different curvature. Both lenses suffer from chromatic aberration, but of a roughly equal and opposite kind; the aberrations cancel out, greatly reducing chromatism. The lenses in small doublets are usually stuck together with a transparent glue;

these doublets are called *cemented doublets.* Larger doublets have a slight space between them, and are called *air-spaced.*

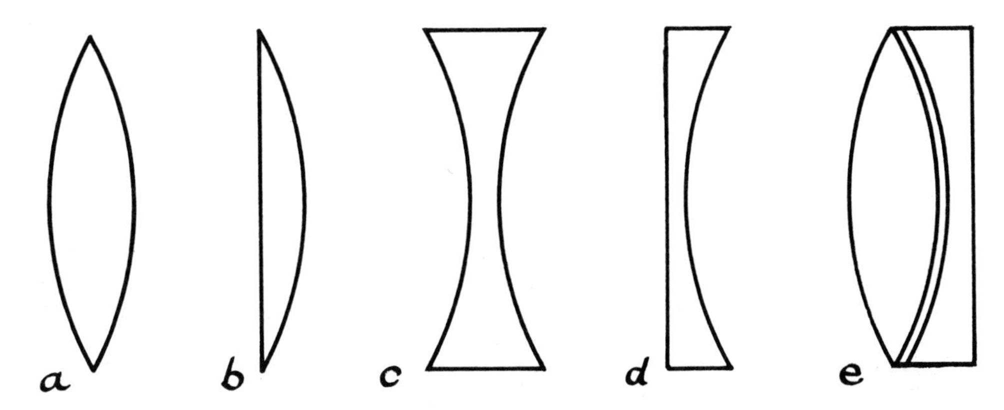

7.1 Kinds of lenses. (a) *double-convex.* (b) *plano-convex.* (c) *double-concave.* (d) *plano-concave.* (e) *an achromatic doublet.*

The optical and mechanical parts of a telescope, excluding its mounting, are called its *tube assembly.* A refractor's tube assembly consists of the cell, the tube itself, the dewcap, and the focusing mount (Fig. 7.2).

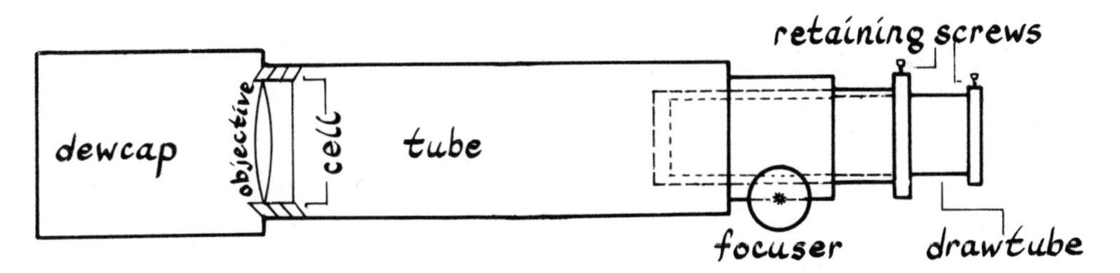

7.2 Schematic of a refractor, showing the principal optical and mechanical parts. The drawtube moves in and out by sliding within a collar, which moves also in and out on a rack and pinion mechanism. The focal ratio of this refractor is much shorter than the typical instrument.

The *cell* holds the objective within the tube. In larger refractors, the cell's position in the tube is adjustable, allowing you to *collimate* the optics—the technical term for centering the objective's optical axis in the tube. In almost all refractors manufactured for amateurs, the collimation is set permanently at the factory.

The tube itself simply holds the parts of the telescope together. It also prevents stray light and air currents from interfering with the image, and protects the optics from dust. Its airtight tube and permanent collimation make a refractor, compared to the open-ended reflector, practically maintenance-free.

The *dewcap* is an extension of the tube. It protects the objective from the dew that forms on cool, exposed surfaces, as well as from stray light. A dewcap should have a matte black finish, and be about twice as long as it is wide. It should also be slightly wider than the objective so as not to "choke" the incoming light.

At the lower end of the tube is the *focusing mount,* which holds the eyepiece at the objective's focal plane. The eyepiece fits snugly in the long, narrow *drawtube.* In most focusing mounts, a rack-and-pinion gear moves the drawtube in and out. These gears should turn as smoothly as possible; any stiffness will make focusing difficult. Gears should not be too loose, however, or the eyepiece will slip off focus.

Refractors require a number of accessories. The *zenith prism* is either a prism or a one-eighth-wave flat mirror. It intercepts the light just before it enters the eyepiece, and bends it at a 45-degree angle. The observer can then look at objects near the zenith without craning his or her neck. For any extended observation (except, perhaps, comet-hunting along the horizon), it is practically a necessity.

A 2-inch refractor has about 100 times the light-grasp of your unaided eye. Such a telescope is nothing to sneer at. As recently as a decade ago, every amateur's first telescope was a small refractor. Japanese imports of decent optical and variable mechanical quality flooded the market, and the price was right. I bought mine (if I recall correctly) for $35. I still have it, and I still use it. I would recommend such a telescope to any amateur astronomer beginning today, but that same telescope now costs around $200; with fancy mountings, the cost rises. The 2.4-inch (60 millimeter) and 3.1-inch (80 millimeter) cost $300–$400 and around $500, respectively. The small refractor is no longer the most economical way to start observing.

A small refractor is best suited for magnifying small, bright objects: the sun, moon, and planets. The vast majority of celestial objects, however, are dim and diffuse. Bright images and wide fields are the greatest single aid to a beginning observer, and the small refractor does not shine in either department. If you can afford the high cost, however, if convenience is of overriding importance, and if you observe through skies where the light of dim objects is drowned out by artificial light, but the moon and planets still shine brightly, your first telescope should probably be a 60 mm. refractor.

Your best policy, if you choose to buy a small refractor, is to buy from a reputable manufacturer, through mail order. They are such simple devices that there is no need to have repair service available locally, so send away to whichever dealer offers the lowest price on a given brand. Under no circumstances should you consider the telescopes offered at discount or department stores: they offer a considerable savings over the cost of a good telescope, but their mountings are so poor, their eyepieces so limited, and their accessories so rudimentary that they are worse than having none at all. Spend the money on a good pair of binoculars instead. A partial list of manufacturers offering good, small refractors is Unitron, Meade, Edmund Scientific Company, and Celestron (see Appendix 8, which lists suppliers); the latter two offer extremely portable small refractors, also useful as spotting scopes. This list includes only the big names in the field; a brief search of the advertisements in the popular astronomy magazines will net you a dozen names of smaller manufacturers.

If you can, sample the different brands available. If you live in or near a large city, college, or university, you should be able to get in touch with the local astronomy club by calling the nearest university astronomy department or planetarium. The members of such an organization will be glad to invite you to their next star party, where you will have a chance to compare firsthand the different instruments on the market. Keep in mind the tests for quality described in the preceding section. But even more useful than any single test for aberrations is your own visual impression of the difference between the images produced by two similar telescopes on the same test object. The best objects for testing, of course, are actual celestial objects: planets, stars, and nebulae. Once you learn how to locate typical members of these groups, as later chapters of this book explain, you will be well prepared to go out and comparison-shop for optics.

One special variety of refractor that overcomes some of the type's traditional handicaps is the rich-field refractor, which is simply a refractor with a very short focal ratio, around *f*/5. Such an instrument suffers some unavoidable aberrations, but can compensate by offering extraordinarily bright, wide images. For a more detailed discussion of rich-field telescopes, see the section of this chapter under that head.

To sum up: the refractor's virtues typically include high magnification*; clear, sharp images, undisturbed by stray currents of air within the tube; and low maintenance. Its drawbacks are commonly high price, restricted field of view, dim images, ungainly long *f*/ratios, and secondary spectrum. The objective's tendency to collect dew, which a dewcap cannot entirely prevent, is also a nuisance, especially if you live in an area prone to cool nights following warm days of high humidity. The high magnification and dim, narrow fields are especially hard on beginners; practiced observers do not find them much trouble. Since the price increases of the 1970s, the refractor's traditional role as a sensible first telescope, which it deserved by virtue not only of its low price but also its simplicity, has changed. Now that a refractor represents a sizable investment, many amateurs are buying or build-

*The comparisons in this chapter of the magnifications possible with different kinds of telescopes refer to the range of magnifications typically available in each kind, as determined by the typical range of focal lengths. All telescopes are subject to the same maximum and minimum limits of useful magnifications; some kinds tend to perform better at the upper or lower ends of that range.

ing them in the 3-inch to 5-inch range as second or third telescopes, at costs of $50 to $500. They appeal to those interested in the solar system or double stars—the small, bright objects that require definition more than light-grasp. For such uses, the refractor remains one of the best telescopes.

REFLECTORS

The reflecting telescope is the most common type of scope for both professionals and amateurs. There are several reasons for this, the most important of which is cost; inch per inch of aperture, a reflecting telescope is anywhere from one-quarter to one-tenth the price of any other. The reflector offers other advantages as well. Because it can be made in short focal ratios without chromatic aberration, it can more easily offer the advantages of a fast system: bright, wide images, portability, and smaller, less expensive mounts. The typical *f*/ratio for a 6-inch is *f*/5 to *f*/8; larger scopes are generally faster. The advantages of short focal ratios are especially important to a beginner, who will find the large, bright fields a help in finding and seeing a larger variety and number of objects.

The refractor's objective must be of flawlessly transparent glass, with four carefully ground and polished surfaces. The reflector uses a mirror for its objective; the glass need not be perfect, and only the reflective side need be ground to perfection. In most reflectors, this mirror is made of aluminum-coated Pyrex, overcoated with a substance such as silicon monoxide for protection. Pyrex is necessary because it will not expand or contract as the temperature changes; the aluminum is applied by vaporizing a tiny quantity in a vacuum chamber; this coating is extremely thin and delicate, so most manufacturers provide a protective coating, to resist scratches. The reflecting surface is usually ground in a paraboloid, and is accurate to better than one-eighth wave. Such an objective will cost $60 to $75 for a 6-inch mirror, less than $300 for a 12.5-inch.

If cost were the only factor in choosing a telescope, the reflector would be everyone's choice. Convenience is important to many people, however, and a long-standing complaint against the reflector is that it has more optical components than the refractor, making it less reliable. Why the extra components? The prime focus of a concave mirror falls inside the telescope's tube: the reflector needs a second mirror to deflect the light path to the outside of the tube. This extra mirror is called the *secondary*; the objective is called the *primary*. The parts that support the secondary mirror make the reflector more complex; they also interfere with the light on its way to the objective, making the reflector's definition slightly inferior to the refractor's. Although the reflector is a bargain, like most bargains, it isn't perfect.

In its basic form (Fig. 7.3), the reflector uses a flat secondary, which lies at a 45-degree angle to the optical axis (hence its other name, the *diagonal*). This puts the prime focus out the side of the tube, near its upper end. In this

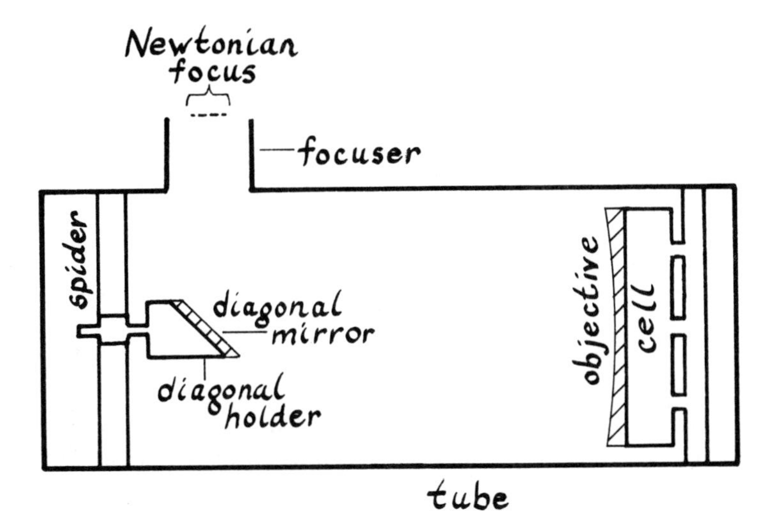

7.3 Schematic of a Newtonian reflector. The diagonal mirror redirects the light from the objective mirror out the side of the tube; the repositioned prime focus is called the Newtonian focus.

configuration, the reflector is called a *Newtonian reflector*; the deflected focus is sometimes called the *Newtonian focus*. The secondary mirror's supports are called the *spider* and *diagonal holder*. The spider is simply four thin metal vanes; they suspend the diagonal holder in the center of the tube. The vanes are thin in order to minimize diffraction of the light that must pass them to reach the objective. They should be stiff, so as not to shake; a shaking diagonal will blur the image of the finest optics beyond recognition.

The diagonal holder is most often a dowel of light wood cut off at a 45-degree angle, the cut faced with cork or felt. The secondary attaches to this face with either tiny clips, a tight collar, or silicone glue. At the skyward end of this dowel, it bolts to the spider; three or four screws set around that bolt adjust the secondary's angle relative to the light path. If the diagonal is not precisely oriented in the tube, the light path will not deflect precisely into the eyepiece. The need to adjust this orientation (also called *collimation*) is the reason for the reflector's reputation as a "touchier" instrument than the refractor. The process is easy to learn, however, and takes no more than a few minutes, every month or so.

A question that often puzzles observers, even some experienced ones, is, Why doesn't the secondary block off part of the image? The reason is that there is no one-to-one correspondence between the parts of the image (say, of the moon) and the parts of the objective: the top of the objective does not receive light only from the top of the moon. If this were the case, you could not see the entire moon without an objective as big as the moon itself. What does happen is that light from the whole moon falls on each point of the ob-

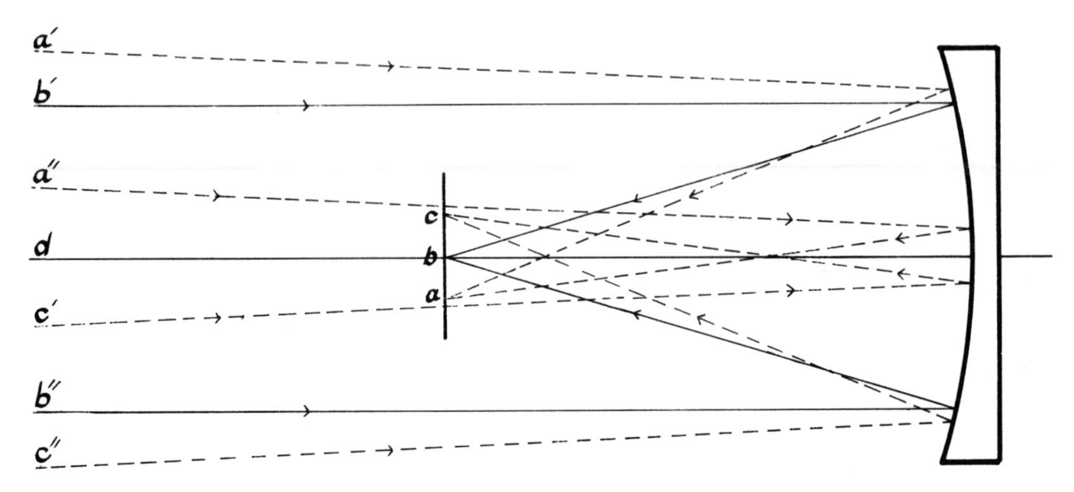

7.4 Why the diagonal mirror of a reflecting telescope does not cause a dark spot in the center of the image. Light from points a′ *and* a″ *enters at an angle to the optical axis, from the upper region of the field; it converges on point* a *on the focal plane. Light from points* b′ *and* b″ *enters parallel to the optical axis* d, *and falls on the center of the field, at point* b. *Light from* c′ *and* c″ *enters from the lower edge of the field, and focuses at point* c. *Blocking off the edge of the aperture (points* a′ *and* b′*), or the center* (a″ *and* c′) *will not blank out* a, b, *or* c *at the focal plane, but merely dim them slightly.*

jective's surface. If you were to cover half of the objective, the whole moon would still be visible from the uncovered half; the image would be half as bright, but still complete. Suspending a secondary mirror in the incoming light path of a reflector does not cause a dark silhouette to appear in the view. It merely dims the image slightly—losing about as much as a refractor loses to absorption of light within the lens (Fig. 7.4).

Like the refractor, the reflector also requires a cell for its objective. The cell supports the mirror from the back, where there are three spring-loaded adjusting screws to collimate the primary. Collimating the primary is somewhat easier than adjusting the secondary. In apertures 10 inches and smaller, the mirror rests on three raised spots within the cell, providing a stable tripod (often called "3-point suspension"). Larger mirrors require 9-point and sometimes 18-point suspension, so that the mirror doesn't flex under its own weight. To prevent flexure, mirrors are usually one-sixth as thick as they are wide.*

A reflector's tube is nothing more than a mechanical support for the optical parts; it can be simply a pipe or a plank. Tubes do offer the advantage of protection from stray light, air currents, and dust. To absorb stray light, the

*See, however, the section on Dobsonian telescopes later in this chapter.

interior should have a coating of a nonreflective, black paint. Because the tube is open-ended, some air currents are unavoidable. They become a problem only when the open tube becomes a chimney for heated air. The most likely source of heat is the tube itself. Metal tubes (except for aluminum) tend to hold heat, and are more likely to cause a problem than fiberglass, cardboard, or wood. The material should also be as stiff as possible, without being too heavy. Cardboard, wood, and fiberglass (in order of increasing cost) are all excellent materials, offering light weight, sturdiness, low thermal conductivity, low cost, and ease of construction.

The one remaining part of the reflector is the focusing mount, which is the same as in the refractor.

For anyone who is already seriously interested in astronomy—or thinks he or she may become more interested in time—a medium-sized Newtonian reflector makes sense as a first telescope. A 6-inch, *f*/5 is a near ideal compromise: optically, it offers respectable light-grasp, wide fields, and a good range of magnifications; physically, it is small enough to mount cheaply, transport conveniently, and handle easily; it is also a bargain. Its disadvantages are the needs to collimate it and to clean the optics from time to time. Its upper limit of magnification is usually lower than that of the other types of reflectors.

The *Cassegrain reflector* overcomes this limitation by compressing several yards of focal length into a tube a few feet long. The Cassegrain's secondary reflects its light back to the primary, where it exits through a hole bored through the center of the objective. The secondary also has a convex curve, which increases the focal length. The amount by which the secondary increases the focal length is called its *amplification factor.* The *effective focal length* of a Cassegrain is therefore the focal length of the primary multiplied by the amplification factor of the secondary. A typical configuration is an *f*/4 primary and a 4× secondary, giving an effective focal length of *f*/16. The precise curves necessary to accomplish this are usually a paraboloidal primary, and a hyperboloidal secondary, which are harder to make than the Newtonian's parabola. This is one reason why Cassegrain optics are more expensive than an ordinary reflector's. The other reason is that the figure of the secondary, because it magnifies, must be even more accurate than that of the primary. The layout has much in common with the Newtonian: the primary rests in an adjustable cell, and the secondary in a spider-mount assembly. The major differences are the position of the drawtube, and the two light baffles. Baffles are tubes placed at critical points around the light path; they prevent stray light from interfering with the image (Fig. 7.5).

The result is a telescope which, in an 8-inch aperture, for example, packs 128 inches of focal length into a tube less than 3 feet long. If you are interested in observing small, dim objects, a Cassegrain offers the high magnifications of the refractor and the large aperture of a reflector at a cost somewhere between the two, and with a portability neither can match. These strengths come with their corresponding weaknesses, including

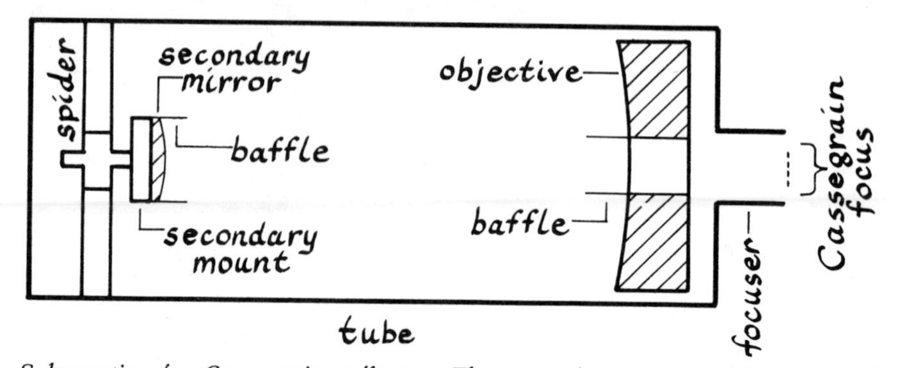

7.5 Schematic of a Cassegrain reflector. The secondary mirror redirects the light back toward the objective, where it reaches the Cassegrain focus via a hole in the objective's center. Baffles are thin tubes that prevent stray light from entering the optical path.

narrow fields, low contrast, poor definition, and tricky collimation. These drawbacks haven't kept the Cassegrain from being popular, especially with advanced amateurs. In fact, a modification of the design is the best-selling telescope in the world: the Schmidt-Cassegrain, which is one of a class of telescopes called catadioptric.

CATADIOPTRICS

This mix of refractor and reflector comes closest to being all things to all people, combining portability, convenience, large aperture, and high magnifications. It is also, unfortunately, expensive to buy and difficult to build; an 8-inch Schmidt-Cassegrain (the most popular aperture) lists at about $1,000; necessary accessories easily add another several hundred. This price almost always includes the electric drive necessary for astrophotography, so the high initial purchase price can be a bargain, if you have the money to spend.

The first catadioptric was the *Schmidt camera,* invented in 1930. Schmidt found a way to take advantage of the wide, coma-free images of a spherical mirror, and yet avoid spherical aberration. A lens placed at the sky end of a reflector's tube corrects the spherical mirror; the lens itself is spherically aberrant by an equal but opposite amount (that is, an over-corrected mirror for an undercorrected lens, or vice versa). This lens, called a *correcting plate,* is very thin, with an extremely shallow curve, and thus introduces no noticeable chromatic aberration. The result is a telescope that can be as fast as *f*/1, with fields as wide as 20 degrees. Unfortunately, the Schmidt's focal plane is not a plane but a curved surface. Specially curved photographic film can compensate for this curvature, but the eye cannot. Hence the name, Schmidt *camera.*

The Schmidt can be modified for visual use by altering the curve of the primary slightly, and using an equally but oppositely aspherical secondary. The secondary is simply glued to the center of the correcting plate, eliminating the need for a spider and improving image quality slightly. The layout is Cassegrain. A typical Schmidt-Cassegrain, an 8-inch, has a focal ratio

7.6 Catadioptric telescopes. These Schmidt-Cassegrains are the most popular style of telescope manufactured for the amateur observer. They are shown here with auxiliary equipment, including equatorial wedges, field tripods, and photoguide telescopes. (Courtesy Meade Optics)

around *f*/10, packing more than 2 meters of focal length into a tube little more than twice as long as it is wide. And the images are *good,* too.

With this design, a large-aperture telescope becomes truly portable (Fig. 7.6). The Schmidt is also more easily mounted, complete with an electrical drive that allows it to track celestial objects automatically. Moreover, an accessory called a *telecompressor* ($50–$150, depending on quality), can cut the focal ratio by one-half to one-third, giving wide fields and fast photographic speed. There are, at present, three major manufacturers offering

Schmidt-Cassegrains (Meade, Celestron, Bushnell), all of whom offer complete lines of accessories.

There is an arguably better catadioptric than the Schmidt-Cassegrain, the *Maksutov*. The Maksutov replaces the Schmidt's correcting plate with a thicker, spherically curved *meniscus plate*, in which both surfaces have the same, parallel curve. The Maksutov's primary is also a simple sphere, as is the secondary. Moreover, the secondary's curve is exactly the same as the meniscus plate's: the secondary is usually made by aluminizing the center spot on the inside surface of the plate. The Maksutov's focal ratio can be as short as *f*/8; commercially produced models are usually twice as long, giving magnifications even greater than the Schmidt-Cassegrains. Image quality is superb. Like the Schmidt-Cassegrains, most commercial Maksutovs are part of a complete system of accessories, of extravagantly high mechanical and optical quality. The leader in this field is the Questar Corporation, with 3.5-inch, 7-inch, and 12-inch models, costing from several thousand dollars to around fifty thousand. Celestron offers a Maksutov as their smallest (3.5-inch) catadioptric.

The attractions of the Schmidt-Cassegrain designs are compelling. It's no wonder they have become the widest-selling amateur telescopes. Their one drawback for the beginner, however, is their price. Unless you have money to burn, it's probably a better idea to start out with a less expensive instrument, saving a major outlay like this one for later.

RICH-FIELD TELESCOPES

A rich-field telescope (RFT for short) is any telescope with a focal ratio less than *f*/5 (more or less). They have become increasingly popular in recent years for a number of reasons, including their portability, low cost (for reflectors), and ease of use. All of these virtues are a result of the rich-field telescope's short focal length. The shortness makes the telescope easy to pack and carry, lighter, and therefore easier to hold steadily on lightweight, easily constructed mountings. Low focal ratios also mean low inherent magnifications. To someone who has not used a telescope, this sounds unpromising, but experienced observers know that the difficulty of observation increases dramatically with increased magnification. Not only is a short-focus telescope easy to handle, it also gives extremely bright, wide-angle fields—up to 3 degrees across. The wide field means that a larger area of the sky appears in the eyepiece; your chances of finding an object (which is often the most difficult part of observing, especially for the beginner) increase proportionately.

I'm sold on the rich-field telescope, especially as a first telescope. There is no other design that is so easy to use, and enables you to see such a variety of celestial bodies. It is also inexpensive, and easy to build; the Appendix includes plans for building an *f*/5 Newtonian. The advantages of the rich-field telescope have their drawbacks, as every design must. High magnification may be hard to use, but it can be important, and this is not one of the RFT's main strengths. An extremely short-focus (4-millimeter) eyepiece on an *f*/5,

6-inch telescope will give only 190×. "Only," however, is misleading here; 190× is more magnification than the atmosphere will allow on most nights. The other major disadvantages of the RFT also stem from its short focal length, which leaves it prone to a number of aberrations, mainly coma in reflectors, and chromatism in refractors. If you buy or build an RFT, check its optics carefully: all aberrations become more severe, or harder to avoid, at shorter focal lengths. All things being equal, in apertures up to 7 inches, the refractor makes a better RFT. The main objection to the rich-field telescope refractor is cost: the lens alone costs over $500; a mirror, around $70.

DOBSONIANS

The Dobsonian telescope is essentially a Newtonian reflector; ingenious design and low-cost materials combine to make a large-aperture telescope that is easy to build and use. The Dobsonian's biggest innovation is its thin mirror, which is only half as thick as usual on other telescopes. This lightweight mirror is less expensive, and makes the telescope easier to mount. To prevent flexure, the cell supports the mirror not only from the back, but also from the side. Such a cell, however, can only support one side of the mirror: one side becomes, of necessity, the "down" side. This rules out Dobsonian support in an equatorially mounted telescope, in which the tube must be free to rotate around its optical axis. It does lend itself to the altazimuth mounting, which not only keeps the same side of the mirror down no matter which way you point it, but is also extremely easy to build.

The Dobson mount is just a simple, effective altazimuth. The telescope rotates in azimuth (horizontally) around a dowel set in a sheet of formica. The telescope's weight rests on three strips of Teflon. The Teflon acts as a set of extremely smooth, frictionless bearings. The altitude axis (around which the telescope rotates vertically) is a pair of large, vinyl pipe flanges, which turn in semicircles cut out of a plywood fork; these semicircles are also lined with Teflon. You can build one of these mounts in an afternoon, with very few tools and skills, for around $25. Appendix 2 gives plans and instructions.

One major manufacturer of optics for amateurs (Coulter Optical) offers complete Dobsonian telescopes in large apertures. A 13.1-inch costs about a quarter of the price of an equatorially mounted 12.5-inch reflector. The same company also offers a 17.5-inch Dobsonian for around the price of an 8-inch Schmidt-Cassegrain, and has just introduced a mind-boggling 29-inch instrument for around $3,000. They also sell the optics for these instruments at about half the price of the complete telescopes. Instruments this large represent more than an investment, however; they also commit you to a certain amount of heavy lifting (in the case of the 29-inch, they commit you to building an observatory). A large Dobsonian, like the Schmidt-Cassegrain, is probably best for your second telescope. But for an amateur who wants to get the most observing for the money, the Dobsonian is the wave of the future (Fig. 7.7).

7.7 A reflecting telescope. This large-aperture Newtonian reflector, on an altazimuth mounting, is also known as a Dobsonian. (Courtesy Coulter Optical)

Deciding what kind of telescope to buy (or even whether to buy one at all) is finally a personal matter. The above discussion should help, but there are other sources of information you should consult before you put your money down. Go to the library and look through the books and magazines available there. Look for recent books, preferably those written for American audiences; practices, prices, and other factors change rapidly, and terminology varies overseas. An amateur astronomy club is also a useful source of opinions; most amateurs have good reasons for owning the telescopes they do, and they'll be glad to try to convince you, too.

The popular astronomy magazines are extraordinarily useful. *Astronomy* frequently prints articles especially for the beginner confronting a telescope purchase. If you can, consult the index for back issues, and you can quickly accumulate a surprising amount of good, solid advice. *Sky and Telescope* has a regular column, called "Gleanings for ATM's," in which Amateur Telescope Makers describe their own telescope-building projects. If you decide to build a telescope, you will find there many ingenious ideas to adapt to your own project. Finally, the best thing you can do as a prospective shopper right now is to read on; the more you know about the purposes to which you will put your telescope, the better you will be able to arrive at an informed opinion.

Eyepieces

8

Eyepieces are an essential part of the optical system, since they magnify the small image at the telescope's prime focus. They are lens systems, and are therefore liable to the same aberrations as refracting telescopes. All aberrations become more severe at the short focal lengths found in eyepieces, so eliminating aberration is the main challenge in eyepiece design. This is generally accomplished by using extra lenses to balance the aberrations of one lens against another. Eyepieces come in a number of styles, of which four have become most popular among amateurs. These are (in order of increasing complexity) the Kellner, Orthoscopic ("Ortho" for short), Plössl, and Erfle. Generally, more lenses improve the image—and raise the price.

Just what constitutes a good eyepiece image? The first claim a manufacturer makes for an ocular is usually "flatness": the field must not be distorted or curved. Another important feature is sharpness: all parts of the field must focus at once. Chromatic aberration is another risk in eyepieces; three lenses or more usually avoid it. Other criteria of a good eyepiece include dark field, bright images, and freedom from internal reflections (caused when light bounces about inside the lens system, causing "ghost" images of bright objects to flash across the field). These three requirements are met by good mechanical construction and choice of materials: good glass, *fully coated* with magnesium fluoride, and well-blackened, baffled barrels cut stray light and reflections. Beware of eyepieces advertised merely as "coated"; unless *all* internal air-to-glass surfaces are coated, you'll be haunted by ghosts. Finally, the eyepiece should have a comfortably long eye relief: 8 millimeters is about the comfortable minimum (16 millimeters for eyeglass wearers). Few eyepieces, however, have an eye relief greater than 80% of their focal length. This is one of the factors contributing to the difficulty of using high-power eyepieces.

Until you have actually seen a good image, it's hard to evaluate an eyepiece. Most of the qualities of an eyepiece are determined by the design; if you choose one of the most popular types from a major manufacturer, even bottom-of-the-line models will be satisfactory, if you don't ask too much of them. When the day comes that you can afford an expensive design, do; although it is possible to be well-satisfied with the view in any eyepiece, it is always possible to improve a telescope's performance by improving the ocular.

In the meantime, inspect your purchases by testing them for the more apparent flaws. The tests for aberration in objectives described previously will also show up problems in eyepieces. If you don't know whether an aberration is in the objective or the ocular, switching oculars or objectives, or rotating the ocular will help to isolate the trouble. Chromatic aberration will be evident when observing medium-bright (second to third magnitude) stars. Extremely bright objects will flare or glitter in almost any ocular, and they are a stringent test for ghosts. If you see faint "echoes" of a bright object, you have a ghosted ocular. Coma is not much of a problem in eyepieces.

As with objectives, the choice of an eyepiece involves some compromise. For instance, ghosts are unavoidable in certain inexpensive designs that are otherwise excellent on dim, large objects. The best test is simply to use a new eyepiece on a selection of objects whose appearance you already know well, and compare the results with the standard set by the eyepiece(s) you own. A good selection of test objects would include the whole range of the celestial bestiary, emphasizing those on which you would most likely use that eyepiece. Test a high-power eyepiece on the moon, planets, and double stars; a low-power on dim, extended nebulae, or large open clusters of stars.

The gross characteristics of eyepieces depend on the number of separate lenses, or *elements.* The simplest are *single-element* eyepieces, and are useful only in toys. *Two-element* eyepieces were the next developed, and in their more sophisticated forms are still useful today. In any multi-element eyepiece, the element nearest the objective is called the *field lens*; the element closest to the observer is called the *eye lens.* The exact arrangement of these elements depends on the kind of eyepiece; each kind of eyepiece comes in a range of focal lengths (Fig. 8.1).

Most eyepieces are identified by a few letters and numbers, usually engraved in a circle around the eye lens, or at the upper circumference of the barrel. The letters will be an abbreviation of the design name: *K, Ke,* or *Kel,* for a Kellner, *Or* for an Orthoscopic, and so on. The numbers specify the ocular's focal length, usually in millimeters. If you can't find such identifying characters don't disassemble the ocular in an attempt to find out. It's too easy to put an eyepiece back together with one of the elements reversed, and it will never work very well if you do. Also, the interior of an ocular is relia-

8.1 Range of focal lengths in a single type of eyepiece. Note that the longest (55 mm) ocular requires a 2" barrel to encompass its large field. (Courtesy Tele-Vue)

bly dust-free until you take it apart; after that, it will never be quite the same.

The two most popular designs among amateurs are the Kellner and the Orthoscopic. The former is probably the most popular on the market today, owing to its combination of wide field, long eye relief, negligible chromatism, and very low price. Its only flaw is a tendency to ghost. You can expect to pay upward of $18 for a good Kellner—somewhat more for an extra-wide-angle design. They usually come in focal lengths of 6, 12, 25, and 40 millimeters. The apparent field of a Kellner is usually 40 to 45 degrees, depending on design; 50 degrees is possible. The Kellner is best suited to short, fast reflectors, but it performs well with almost any objective.

The shortest Kellner does not give the maximum useful magnification for most telescopes; the shortest Orthoscopics will—coming in focal lengths as small as 3 millimeters. The Ortho has one of the best fields available: wide, flat, dark, and ghost-free. The apparent field is about 5 degrees narrower than the Kellner's, but at high powers this is not an important drawback. The eye relief is noticeably better, however, which is important in short focal lengths. Orthos are often available in *parfocal* sets; each ocular in the set is constructed so that you can change eyepieces without having to refocus the telescope—a small convenience. The Ortho is definitely superior to the Kellner, and costs about $10 to $20 more. At lower powers, a Kellner is probably the better buy; at high powers, the Ortho. In the middle range of focal lengths, ask your wallet what *it* thinks.

The Orthoscopic is not, however, the last word in either quality or price. the *Plössl* starts at around $50, and offers wide (50-degree), exceptionally sharp, dark, and flat fields, in focal lengths generally ranging from 7 to 26 millimeters. Somewhat less expensive than the Plössl is the *Erfle.* It has the widest apparent field of any commonly available eyepiece—65 degrees is typical. Focal lengths range from 7 millimeters (rare) to around 35 millimeters. They cost around $50.

Before you choose designs and focal lengths, a more fundamental decision is the eyepiece's diameter. Of the three available sizes, the most popular is 1.25 inches wide. Small catadioptrics and refractors often use a 0.925-inch barrel; very large telescopes, and many rich-field telescopes, use a 2-inch. A smaller size can always be adapted to a larger drawtube, but the reverse process is rarely satisfactory. The 2-inch barrel makes sense only for extremely wide-angle, long-focus eyepieces. Don't start with 2-inch oculars; the money would be better spent on high-quality 1.25-inch eyepieces. The 0.925-inch size is generally too small to use all of the light produced by any but small-aperture, long focal-length instruments.

The kind and number of eyepieces you buy at first will depend on the amount you have to spend. Unless you are determined to overspend, start off with a Kellner. If you can only afford one eyepiece, its focal length should be in the middle of the range—18 to 25 millimeters. Two eyepieces are the optimum for someone starting out; the urge to go to a higher power is as basic as curiosity itself. With two eyepieces, one of them can be a wide-angle. A 40-millimeter Kellner on a 6-inch, *f*/5 gives a 2.5-degree field, and is the equivalent of adding a giant finder to your telescope. Something around 9 millimeters makes a good higher-power choice.

In higher powers, buy the best you can afford. The Plössl eyepieces made by the Tele-Vue company are especially good. I recently bought their 7.2-millimeter, and developed a new respect for my old 6-inch Newtonian. Having an eyepiece of this quality offers an immeasurable boost to an observer's morale. Not only are the images in it simply superb; you will also find that *knowing* that you are getting the most out of your telescope gives you more confidence as an observer.

Your next eyepiece purchases should wait until you become familiar with your telescope and the oculars you already own. One shortcut toward increasing your range of powers is to buy a *barlow lens,* which is a negative (diverging) lens system placed in the optical path ahead of the eyepiece. It disperses the light slightly, like the amplifying secondary of a Cassegrain, so as to lengthen the effective focal length of the system. The increase in effective focal length varies: most barlows increase the power of an eyepiece by two times; some are adjustable, giving a range of amplifications. If you buy a barlow with a fixed amplification factor, don't buy eyepieces that simply duplicate the power available with the barlow and an eyepiece you already own. A 12-millimeter eyepiece, for instance, is redundant if you already own a 25-millimeter and a 2× barlow.

Because the barlow adds more glass to the light path, it becomes less useful at very high powers, where images are already dim. A poorly made barlow can be useless, so be prepared to spend at least $30. In addition to quality optics, the extra amount will also usually buy you a barlow with a field lens large enough to admit all the light coming from the objective. Bargain barlows (costing less than $25, usually), can have field lenses as small as 19 millimeters, usually too small for systems faster than *f*/8.

Mountings

9

A firmly mounted 2-inch telescope will always be a better instrument than a 6-inch on a shaky platform. The 2-inch will at least show you something; the 6-inch will just show a blur. A good mounting is vital. Not only must it be stable, but it should move freely and smoothly from one point to the next, stopping exactly where you point it. Stiff or wobbly motion that rebounds or slips when you take your hand away will turn your observing into a frustrating round of hide-and-seek. Other considerations in a mount include cost, ease of construction and use, and whether you want it to be permanently installed or portable; if portable, then weight and bulk become important. To someone of limited resources, buying a mount can seem a larger stumbling block than getting the telescope tube assembly itself. Fortunately, it is possible to build mounts with relatively simple materials and tools (Appendix 2 provides plans for a Dobsonian mount). By the time you need the kind of quality that only professional machining can offer, you may have saved up the money to go out and buy it.

The bare minimum requirements for a mounting are these: It should provide free motion in two perpendicular axes (one to track horizontally and the other vertically), so that the telescope can point to all regions of the sky. It should be made of materials strong and stiff enough to support the weight of the instrument, but also able to damp out small vibrations such as those caused by your knee tapping against the tube. Another essential feature is a *cradle* to support the telescope tube at its center of gravity. Some extremely cheap mountings support the tube from underneath: when you turn it skyward, the unbalanced tube pitches backward—into your eye. In most mounts, the cradle should also permit the tube to rotate around its optical axis; otherwise, the eyepiece can twist into awkward positions.

The mount should be balanced, so that it will remain in any position you set it in without requiring too strong a lock on either axis. In some designs, this requires *counterbalance* weights. The axes should turn smoothly. Ball bearings and Teflon are popular for this function, and inexpensive in some forms, but simpler materials such as pipe, wood, and felt also work well. In general, the larger the bearing, the smoother its motion. Finally, the mount should be as compact as possible. This is not for portability (though it helps that, too) so much as to ensure that the mount's weight stays as close to its center of gravity as possible. The farther a mount juts out from its central support, the more leverage it exerts, and the more sturdy its parts must be to bear its own weight.

TYPES OF MOUNTS

The simplest kind of mount is the *altazimuth.* The name is a contraction of altitude and azimuth. The azimuth axis points to the observer's zenith; the tube swivels around it horizontally, to all points of the horizon. The altitude axis lies level to the ground, and lets the tube swivel up and down. All points of the sky are accessible; motion is easy and natural. Tracking the rise and set of a star, however, is complicated: the observer must move both axes at once to keep an object in view as it arcs across the sky (Fig. 9.1).

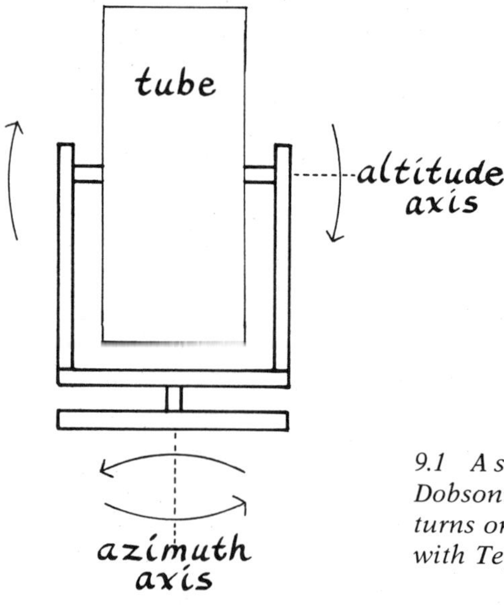

9.1 A simple altazimuth mount. In this Dobsonian mounting, the telescope turns on large bearing surfaces, lined with Teflon.

The *equatorial mount* simplifies tracking. The equatorial is basically an altazimuth tilted so its azimuth axis is parallel with the earth's rotational axis; it points toward the celestial pole. The axis parallel to the earth's axis is called the *polar axis.* As the earth turns, the observer need turn only the polar axis to track a star's apparent motion. The other axis is called the *declination axis.* It swivels the scope in declination only; rotating the polar axis moves the scope in right ascension. With an equatoral mount, the celestial coordinate system is built in. This permits the use of *setting circles,* which are circular scales attached to each axis. They are marked with hours and minutes of right ascension on the polar circle, and with degrees of declination on the declination circle. By rotating the axes so that pointers on the scales turn to an object's coordinates, you can "dial up" objects too dim to locate visually.

The equatorial mount is a necessity for some sophisticated uses, such as astrophotography (where a motor, called a *clock drive,* attached to the polar axis causes the telescope to follow an object over the course of a long exposure). Its greatest advantage for a beginner, however, is that it simplifies the tracking of objects. Since at high power a star can drift out of the field of view in a matter of seconds, simplified tracking is an important advantage.

The equatorial offers so many advantages over an altazimuth, it's hard to see why anyone would settle for the latter. The main problems with equatorial mounts, however, are that they are expensive to buy and difficult to build. Tilting the mount shifts its center of gravity, so that various parts, especially the axes and bearings, must be stronger. Other complications also arise: *slewing* (moving the telescope tube across the sky) from object to object can be maddeningly difficult near the pole, where the declination and polar axes are almost parallel.

There have been many different styles of equatorial mount developed over the past century and a half. The original, now called the *German equatorial,* is still among the most popular (Fig. 9.2). Its declination and polar

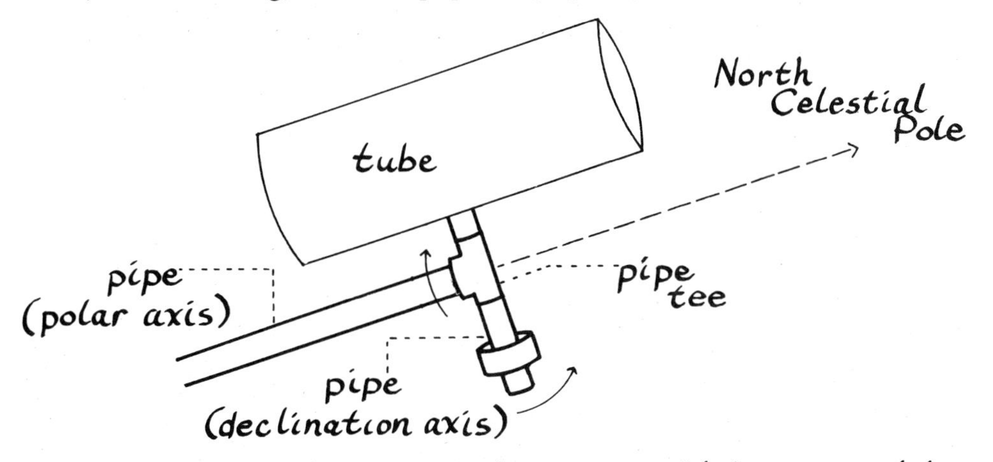

9.2 A German equatorial mount; notice that a counterweight is necessary to balance the off-center weight of the tube.

axes form a *T*, with the polar axis as the upright. The telescope rides at one end of the *T*'s crossbar; to balance the weight of the tube, a counterweight rides at the other end. The polar axis rests at an angle atop a tripod or vertical post (usually called the *pier*). The pier holds the axis assembly (often called the *equatorial head*), two or three feet off the ground. Commercially made mounts are almost always adjustable, so that the precise angle of the polar axis can be adjusted to match the altitude of the celestial pole above the observer's horizon. This angle is always equal to the observer's latitude.

The German equatorial is popular because it is versatile and relatively simple to build. It has two drawbacks, both of which stem from the need for a counterweight. The weight is heavy, of course, and requires a longer declination axis; both of these can cause greater vibration, as well as giving you fits when you try to stuff the thing into your car. You must also occasionally stop tracking an object and rotate the whole tube 180 degrees around the axis to keep the tube or counterweight from hitting the pier, or to raise the eyepiece to a comfortable height.

The second most popular kind of equatorial mount, the *fork mount* (Fig. 9.3), eliminates the counterweight; it is usually a compact structure, easily portable and also one of the easier mounts to use. Unfortunately, it does not work well with long tube assemblies; for short reflectors, however, and catadioptrics, it can be ideal. In the fork mount, the telescope rests between two arms (the "tines" of the fork). The declination axis shrinks to two stubby rods on either side of the tube; these fit into sockets at the ends of the fork.

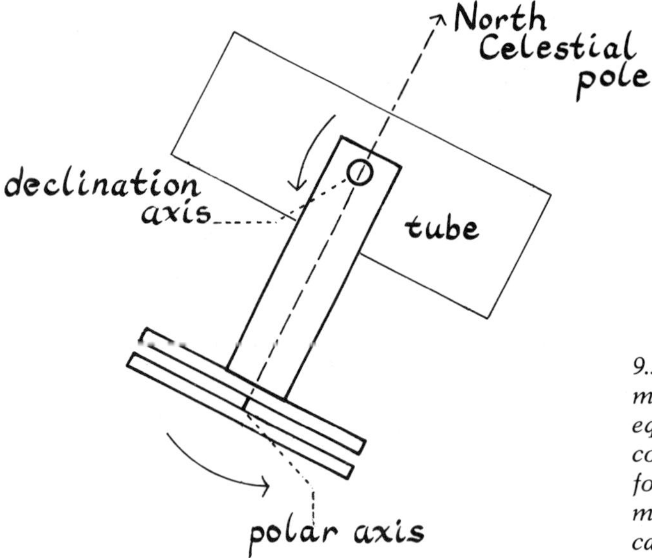

9.3 The polar disc fork mount is a popular type of equatorial mount; it needs no counterweight, which accounts for its popularity among manufacturers of portable catadioptrics.

The arms of the fork point at the pole, and the entire fork rotates around a polar axis, which can be either a long axle (the "handle" of the fork), or two circular plates with a ball bearing turntable between them. The latter design is called a polar disc fork and is the more portable of the two. The only structural problem with the fork is that the center of gravity can reach far out to the north of the mount.

There are at least a dozen other kinds of equatorial mounts in use by amateurs and professionals today, each with its own set of advantages and drawbacks; they are all either more difficult to build or transport than the German and the fork. The Bibliography lists several books that describe these various kinds of mountings, and the popular astronomy magazines frequently publish new designs as they come up. Would you believe that one used a bowling ball as its only moving part? The range of innovations seems unlimited, and you may find yourself sketching your own designs before too long. For now, you should probably limit yourself to one of the three designs described above, which account for almost every telescope sold today.

ALIGNING AN EQUATORIAL MOUNT

The convenience of an equatorial mount comes at a price. Unless it is a permanent installation, each time you set it up you must align its polar axis with the celestial pole. For most visual uses, this is a relatively straightforward matter. First, lock the telescope tube so that its optical axis is parallel to the mounting's polar axis. If you have setting circles, simply set the tube so that the declination pointer reads 90 degrees. If you don't have these circles, you can tell the tube is pointing correctly if the field does not move when you rotate the tube in right ascension: the stars in the field should revolve around the field's center. If the field itself slews in a small circle around the sky, move the tube toward the center of that circle until the slewing stops. With the tube locked in place, move the entire mount so that Polaris is centered in the finder; then, with a low-power eyepiece, set Polaris about 1 degree off-center in the field (at the edge of a low-power field). The north celestial pole is between Polaris and the Big Dipper, so move Polaris in the telescope field toward Cassiopeia.

For more advanced uses, especially astrophotography, more precise methods of alignment are necessary. Traditionally, this has involved up to an hour's worth of sighting on various stars, watching stars as they drift across the field, and similar finickinesses. More convenient methods have been developed in recent years. Most of these are based on the use of special eyepiece *reticles.* A reticle is a transparency inserted at the eyepiece focal plane; a coordinate grid inscribed on the transparency provides an overlay on the field. True Polaris North reticles mark the position of the pole relative to stars in the vicinity, including Polaris. They reduce the time required for accurate alignment to a few minutes, and are well worth the expense. As a beginner, of course, you won't have to worry about fine points of alignment.

Telescope Maintenance 10

COLLIMATION, CLEANING, AND SMALL REPAIRS

Regular maintenance is especially important for Newtonian telescopes, but refractors and catadioptrics also require periodic cleaning. Time your maintenance work to coincide with the full moon. Observing opportunities are poorest then. Such a schedule gives you a reliable once-a-month service interval, more than most telescopes need.

Collimation, or optical alignment, is necessary for Newtonians, which require readjustments from time to time, whenever the ordinary vibrations of use and travel jostle their optics. Collimation is much easier if you take a few preliminary steps. First, you need to set two reference marks permanently on your optics. Because these marks go in the center of each mirror, in the shadow of the secondary, they have no effect on the image. To mark the primary, place a self-adhesive, paper dot no more than ¼ inch across on the center of the mirror. The easiest way to find the center is by using a compass to mark out a circle the same diameter as the primary. Cut out the circle, and place it on the mirror, the hole left by the compass point marks the center.

Finding the center of the secondary is harder. Trace its outline on graph paper, and locate the center (at the intersection of the major and minor axes) with reference to the graph's grid. Punch out a hole at the center just large enough to admit the point of a marking pen; position the graph paper mask over the secondary, and mark the mirror's center with a single dot. Thin some fingernail polish, if necessary, and apply a dot of polish over the pen point mark (you might want to experiment on another surface until you get it right). Do not apply the polish with the mask still over the mirror; the fibrous edges of the holes in the mask will absorb polish, spreading it over the mirror. If the polish should spread slightly, don't worry—the sizes of these circles leave plenty of room for error.

Collimation is easier with a collimating eyepiece. This is a plug set in the drawtube, with a small hole centered on the optical axis. You can make your own out of a 35 millimeter film cannister. They fit snugly in a standard drawtube, and the cap has a small dot marking the center of its inner surface. Drill out that center with a 1/16-inch bit, and cut a larger hole in the cannister's bottom. Replace the cap, and set the cannister snugly in the drawtube.

First, adjust the secondary. Remove the primary from the tube, cell and all. Rack the drawtube out as far as it will go. Looking through the collimating eyepiece, you should see three circles:

1. The inner opening of the drawtube
2. Inside that, the circumference of the diagonal's angled silhouette
3. Inside that circle, reflected in the secondary, the circular opening of the far end of the tube

The process is easier if you take it in steps, two circles at a time. If the diagonal is not centered in the drawtube, the second circle will appear off-center in the first. Adjust the distance between spider and diagonal holder until the two circles are concentric. If the second circle is not round, rotate the entire diagonal holder until the secondary shows a circular silhouette. If these two steps are not sufficient to center circle two in circle one, check to see that the spider is installed correctly.

Once the diagonal is centered and its silhouette appears round, turn your attention to circle three, the reflected circular opening of the tube. If it is off-center, adjust the collimating screws on the top of the diagonal mount. You will probably do best at this stage to make adjustments by trial and error. Later, you will learn that tightening an adjustment screw causes the image to move toward that screw. For now, however, don't confuse yourself with too many facts. Simply work the adjustments until circle three is concentric within two.

Replace the primary in the tube. Looking through the collimator, you will see a confusing maze of reflections. Ignore everything but the two dots you placed on the optics, and the large, black spot that is the shadow of the secondary. Turn the adjusting screws on the primary cell until the large dot is centered in the secondary's shadow. Start with the adjusting screws on the primary as tight as possible, and proceed by loosening one screw at a time. When the large dot is centered in the secondary's shadow, the small dot on the secondary should be centered within the larger. If it is not, finetune the secondary until it is. When the dots line up, your system is collimated (Fig. 10.1).

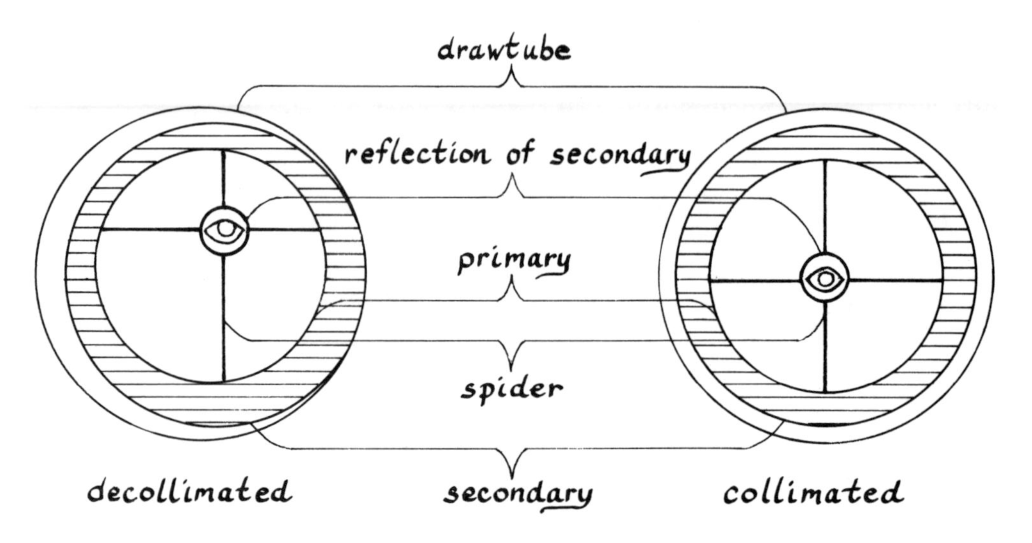

10.1 Collimated and decollimated optics.

The second major maintenance a telescope requires is cleaning of the optics. Because the surfaces involved are so delicate and easily damaged, the most important step in the procedure is to do it as seldom as possible. You will never see a lens or mirror that is entirely dust free. A little is always present, and you won't notice its effect. Should your objective ever grow positively fuzzy, however, you will need to clean it. If you find yourself needing to clean more often than once a year, take steps to protect your telescope from dust in the first place. You can prevent the premature formation of fuzz by keeping both ends of the tube capped whenever you're not actually observing. The best dustcaps I know are ordinary shower caps. Inexpensive, they also stuff conveniently in your pockets. Use your collimating sight to plug the drawtube; you'll always have it with you if you do.

There are two steps to cleaning a mirror, dry and wet. Dry cleaning requires a blower brush, which you can buy at a camera store for a dollar or two. The bristles of this brush should be camel's hair, and should not be trimmed ("natural end" in packaging lingo). Use this brush to stroke, *very gently,* across the surface of the mirror, stroking always in the same direction. If a piece of dust or grit resists, use the blower, but do not bear down. If it won't come loose, leave it. Clean the secondary, and your eyepieces too. You can use cleaning fluid, lens tissue (available from a camera store) and compressed air (such as "Omit!") on the eyepieces, but do not use them on any mirrors. Don't disassemble the eyepieces to clean them: they won't have any dust on the inside unless you do.

When you have finished the dry stage, assemble the following:

- Ivory soap or other *nondetergent* soap
- A *clean* glass, aluminum, or stainless steel bowl large enough to hold the mirror and several quarts of water
- A medium-sized box of surgical cotton, unopened
- Rubbing alcohol, unopened

Fill the bowl with hot water, and add liberal amounts of soap. Gently lower the mirror into the bowl, and let it soak for 10 minutes. Take a wad of the cotton and start swabbing in short, gentle strokes across the surface, stroking always in the same direction. Discard the wad after each stroke, to avoid scraping a piece of grit across the mirror. When you have cleaned the entire surface, run lukewarm water into the bowl, and lift the mirror out of the suds directly into the stream of water. The mirror will be slippery as a seal at this stage, so watch it. Keep the mirror in the rinse water constantly: if any water should evaporate, it will leave a residue.

When the soap is rinsed off completely, remove the mirror from the rinse and immediately pour rubbing alcohol across the surface of the mirror until it has sluiced all of the water off. This stage will sting like crazy, as the alcohol gets into all the cuts you didn't know you had on your hands. Once all of the water is off, set the mirror on a clean, lint-free towel. Rest it on one edge so that the alcohol drains off. It should be dry in about 10 minutes.

You will find that additional small maintenance chores arise from time to time. The most likely part to need attention is the major moving part, the drawtube. If it ever becomes stiff, it will present a nuisance, as the effort of focusing can leave your tube vibrating. Clean the gears with rubbing alcohol every year or so, relubricating with petroleum jelly. Do not use ordinary grease or oils for this; it will crawl right up the drawtube and onto your eyepiece. When cleaning optics, don't neglect your finder. Use the blower brush and lens paper on its objective and eye lens. Should your tube get cracked or dingy, remember that it's not too hard to remove the parts from it and give it a coat of paint (see *Sky & Telescope*, June 1982, for a clever method of refinishing fiberglass tubes). Try to give the entire instrument a thorough going-over every six months or so, using the opportunity to correct design flaws that have been bothering you, or to add options you've been considering. The chances for improvement are never-ending, and if you keep at it assiduously, you will find yourself in a few years with an increasingly versatile, useful telescope instead of one that snares you into an ongoing battle with frustration.

Accessories

11

One of the deepest pitfalls the beginning amateur can tumble into is overloading on gadgets. The problem for the beginner is twofold. First, until you have learned how to use a telescope, it's easy to feel that your equipment needs upgrading. Second, the beginner has a hard time telling necessities from luxuries from plain junk—all of which are offered on the open market. The cardinal rule is to keep it simple. Not only do you keep from wasting money; you also learn a little visual self-reliance. This chapter suggests optional equipment you are likely to find useful in your first year or so, emphasizing low-cost items whenever possible. After that, you're on your own.

OPTICS

Optical accessories can make an enormous difference in your observing. The most important accessory is a good pair of binoculars. Their wide, upright fields can make preliminary location of a target much easier. That ease can also take a lot of work out of stargazing. I find myself more likely to go out with only binoculars than with a telescope.

Binoculars are usually described by a pair of numbers, such as 10 × 35 or 6 × 20. The first number gives the magnifying power, and the second gives the aperture of the objectives, in millimeters. For use in astronomy, aperture is important, so the second number should be as large as possible. You will sometimes find large-aperture binoculars sold as "night glasses." Magnification is not as important, however. In fact, because you will be holding them above your head, with your neck craned back, for extended periods, high magnification will only exaggerate the inevitable shaking.

In binoculars for astronomy, 7 × 50's are a good choice, offering an optimal exit pupil, good light-grasp, and little weight. Because a hand-held instrument will shake, any higher magnification should be approached with caution—and a tripod. In recent years, several companies have started marketing extra-wide-aperture (80-millimeter) binoculars. Operating at the optimum low power (11×), such binoculars can give the best views obtainable of very dim, extended objects. They are also available in higher powers, usually 20× and 30×, but such instruments require a tripod; even the 11× binoculars work best if you have some kind of support for your arms, such as a beanbag chair.

Optical filters should be your next purchase. These come in three major varieties, all of which serve the same basic function. A filter cuts off some of the light before it reaches the eyepiece. With bright objects, such as the sun and moon, you need a filter to cut down all wavelengths of light. Other objects require selective filtering, so as to increase the *signal-to-noise ratio* in the gathered light. In a telescope, the signal is the light coming from the object you're interested in, and the noise is light from other sources. A typical example of a signal would be the reddish light from the Great Red Spot on Jupiter, which can be drowned out by "noise"—light from the bright, cream-colored planetary surface. Red light is the signal you want, and all other light is just distraction. It's difficult to increase the absolute strength of a signal, so rather than trying to brighten the red light in Jupiter's image, we use filters to increase the *relative* strength of the signal compared to the noise. The idea, in other words, is to increase the contrast in the image by filtering out the other light, darkening the background. The image filtered in this way is no brighter, but the detail you're interested in stands out more distinctly against its background.

There are three major types of optical filters. Solar filters are used only while observing the sun. They greatly reduce all wavelengths; their function is purely to protect the eyesight of the observer from the extremely dangerous solar rays. Light-pollution-rejection (LPR) filters filter out skyglow; they are one example of using filtration to boost the signal-to-noise ratio. By darkening bright skies, they increase the contrast between the sky background and faint, nebulous objects. Photovisual filters also adjust the signal-to-noise ratio in an image and are useful for increasing contrast in subtle features of planetary surfaces.

Solar filters come in several varieties. One of them is unsafe and can blind you. This kind is a small disc of darkened glass, set in a metal cell that screws into the barrel of an eyepiece. Among other failures, such filters have been known to shatter in use, subjecting the observer to the full force of concentrated solar radiation. If you have such a filter, destroy it now. Even if you can't afford an adequate solar filter, there are safer, more inexpensive ways of observing the sun, discussed in later chapters of this book.

Safe solar filters attach to the front of the tube, cutting down the light's intensity *before* it reaches the objective. Full-aperture filters come in a wide price range. The expensive ones are made of optical glass, with reflective coatings; inexpensive ones are made of one or two thicknesses of reflective Mylar film. (Mylar is a very thin, flexible plastic film, which can be coated with reflective substances, resists tearing, and, if properly installed, does not distort light passing through it.) The glass filters give a slightly superior image, but at about five times the price. Image quality shouldn't be a consideration to most amateurs in this case, since the majority of your solar observation will take place through an atmosphere so heated and turbulent that your telescope will not function at its limits of resolution anyway.

There are several brands of solar filters, of roughly equivalent quality. One I can recommend from personal experience is the "Solar Skreen" filter, sold by Roger Tuthill. Its images are excellent, and the price is right—around $35 for a 6-inch scope. It's available with a heavy plastic cell made to fit your tube, at extra cost. The cell is a convenience, but you could make one just as useful (that won't add so much weight to the front of your tube), from scrap Styrofoam in about 15 minutes. With or without the cell, the Mylar filter is a boon to amateurs. It makes possible direct, high-magnification solar observing at a fraction of the cost of the glass filters (Fig. 11.1).

Light-pollution-rejection (LPR) filters, also sold as "nebula" or "galaxy" filters, are a recent development, and a boon to amateurs working from suburban backyards. Although expensive (prices start around $45, and go above $100), they are such an effective weapon against light pollution that I recommend them strongly to anyone whose skies are frequently too bright to show fourth- or fifth-magnitude stars.

LPR filters are used in nighttime observing; they work by selectively blocking out only the specific wavelengths of light produced by the most

11.1 Mylar solar filter. The plastic cell fits over the front opening of the tube, and could as easily (and less expensively) be made of cardboard or styrofoam. The filter itself consists of two thin sheets of coated Mylar film.

common light polluters. Wavelengths in which most celestial objects radiate are almost unimpeded. The filter itself is simply a piece of high-quality glass that has been coated with numerous chemical layers, each of which filters out a specific region of the spectrum. Objects such as galaxies, nebulae, comets, and some kinds of star clusters benefit most from LPR filtering. Individual stars and planets, which concentrate their light into intense points, do not usually suffer so much from light pollution, and so benefit less dramatically from its filtration.

Most LPR filters screw directly into the barrel of an eyepiece, but they can also be hand-held between your eye and the ocular's eyelens (a convenience when changing eyepieces frequently). When comparing brands, select for durability over any other virtue. Durability is important in a much-used accessory that is likely to be dropped frequently. My impression has been that filters in which the coatings are protected inside a glass sandwich are more durable than the less expensive variety, made of a single layer of glass. The efficiency of the filter should be a secondary consideration, especially to a beginner. You can pay twice as much for a high-efficiency filter, but the difference between the contrast gain in the luxury model and the budget filter is much less than the difference between the budget filter and none at all.

The third variety of filters is the oldest. They were once simply called "filters," but now are distinguished from the newer types as (for lack of a better term) "photovisual filters." Like the LPR filters, these screw into the barrel of your eyepiece. Unlike them, they pass only a single, limited range of wavelengths (for example, red light), or, in the case of polarizing or neutral density filters, they cut all wavelengths equally. Since they are so simple, they are also much less expensive than other filters, going for around $7 apiece. Their primary use is in planetary observation.

A neutral density filter is fine for cutting glare in the image of a bright object, especially in a fast system such as an *f*/5. A set of two polarizing filters is even better for this. Turning one filter of the set changes the amount of light the two will pass, giving you an adjustable screen. To start, you may want to buy one or two colored filters, for observing Mars and Jupiter. They are extremely useful when you are searching for elusive details on the planets, but until you have developed a good eye for such things, the filter is likely to be nothing but a confusion and a bother. It's a good idea to master the skills before buying such an accessory, or you may come to use the filters as a crutch.

The best way to learn to use filters is by experimenting with a range of filters on whatever objects you are interested in. Atmospheric conditions change from night to night, and the conditions on the planets change as well. A few basic guides to how filters produce their effects will often be more useful than any hard-and-fast rule. Remember that the filter will brighten objects of its own color and dim all other light; light of the complementary color will dim the most. For example, the more red there is in the background, the more blue you want in a filter to dim it, and vice versa. Yellow, purple, and the ubiquitous shades of tan, however, resist this rule. There are often more components to the light that reflects from a planetary surface than we can see. Experiment, and you will soon achieve your own sense of what works.

The most important rule to remember in using filters is that there is more than one way to increase contrast. Your automatic tendency will be to try to brighten the object you are looking for, but there are times when the best course is actually to darken it. In general, try to brighten light objects that appear against dark backgrounds, but darken dim objects against bright backgrounds. A good example of the latter case is the Great Red Spot on Jupiter, which is darker than its background. A red filter will brighten the spot, but brightening such an object can actually decrease its visibility. A blue filter makes the spot much darker than its background, and it stands out much more clearly. Many of the dark, brownish or greenish markings on Mars follow the same rule: darken them with a filter of the complementary color to see them best. The polar caps of Mars are a good example of the need to brighten an object against a dark background. A blue filter will darken the reddish Martian disc, causing the caps to stand out more clearly.

You will learn to use filters more quickly if you don't burden yourself with a rainbow-full of colors. In my experience, the fainter the color of the filter, the more useful it will be: a deep, blood red filter needs a very bright object to show anything at all, but a pale pink filter will bring out a surprising amount of detail on, say, the surface of Jupiter. Start with two filters: one pale blue, the other salmon pink, and build from there. Remember that most of the commercial brands allow you to stack these two filters together, to make a purple filter. Your next purchase should probably be a pale yellow, or tan filter. You will probably find that those three will fulfill your needs for a long time.

MECHANICAL AIDS

Here is where the attractions of gadgetry lead most beginning amateurs astray. There is an entire galaxy of gizmos out there, ranging from computerized guiding systems, to clocks adjusted for telling sidereal time, electric focusers—the list gets longer every month. Although many of these devices are useful for particular applications, even more are of doubtful value. Until you have developed some sense of what your particular needs are, you should avoid almost all of them. If you start observing with a $2,000 computerized telescope, programmed to find the hundred most popular celestial objects for you, you certainly ought to get the jump on somebody who starts out searching by trial and error (that is, if you can get the thing set up). But when you go looking for that one-hundred-and-first object, you'll find only that you've never developed the skills necessary to know your way around the sky. As a general rule, if you resist the temptation to buy gadgets that promise instant gratification, your observing will be much more rewarding in the long run.

With that in mind, there is one device you may find that you want right away. This is an electric *clock drive.* A clock drive is simply a very slow electric motor, which is geared to turn the polar axis of an equatorial mount at precisely one revolution per day. A clock drive usually adds between $100 and $200 to the cost of an equatorial mount. It won't work with an altazimuth mounting, because these simpler mounts require motion in two axes to track the stars. Once you have the mount aligned on the north celestial pole and have found a means of supplying the drive with power (two challenges which, in my experience, outweigh whatever convenience the drive itself offers), your telescope will track a star across the sky, without your having to continually reaim it.

When purchasing a drive, you should buy one manufactured by the maker of the mounting so you will be certain that the drive will actually fit. Don't try to cut corners by setting a telescope on a driven mount intended for a smaller instrument. The added weight, especially if the telescope is used with a camera, or if it is out of balance, can burn out the motor.

The other important consideration in a clock drive is its gear system. A good drive uses a *worm-and-wheel* system. A worm gear is essentially a large threaded bolt with its head cut off. As the worm turns, the threads engage the teeth of an ordinary, wheel-shaped gear. With each complete rotation of the worm, the spiraling motion of the threads moves the wheel-gear ahead by one tooth: the worm-and-wheel combination moves very, *very* slowly. The main advantage in this system over ordinary *spur gears* is that the worm-and-wheel avoids *backlash.* Backlash is the small slippage that occurs in most gear systems, whenever one tooth disengages from another. Unless the gears are precisely machined, the gear being driven tends to slip slightly backward before the next tooth on the driving gear slips into place. This slight slippage, magnified at the eyepiece, can cause long time-exposures of the stars to blur.

As you can see, discussions of clock drives turn quickly into discussions of astrophotography, and that's where this chapter draws the line. Except for the simplest kinds of pictures, astrophotography requires so much complex, expensive equipment, it's just not a project for beginners. Not only is it complicated, but it's not observing: when the camera is looking through the eyepiece, you're not. Until you've logged quite a few hours at the telescope, keep the observing to yourself. If, later, you decide to extend your reach into space by taking advantage of the extra sensitivity of photographic film, you'll find an introductory project described in Appendix 4 to this book. Other books, for advanced amateurs, will take you from there.

CHARTS

A good set of celestial charts is a must. Quality in charts depends more than anything else on the amount of information they display and the clarity of the presentation. A chart showing stars to the ninth magnitude, for instance, must have a very large scale, or all of the stars shown will be packed too closely together to distinguish. A chart can show all of the celestial objects accessible to your telescope, but it should also show what kind of object each one is, and give a reference code so you can look up information about it in a catalog. The size of the chart is also important for other reasons. Is it too big to hold conveniently? Do the pages show such a small area of sky that you're constantly thumbing back and forth? The sets included in the Appendices 5 and 6 of this book are intended solely to allow you to start observing right away, with binoculars, or any other optical aid you now own. For serious observing, you must buy a separate set.

The *Norton's Star Atlas* is a good example of the right size and image scale for the beginning amateur. Its display of information is somewhat inconvenient, however. Celestial objects are all identified by the same mark (an asterisk), with numerical labels from a system of cataloging that has been obsolete for most of this century. This is a minor drawback, however, one that can be overcome with a little preparation before observing. The limiting magnitude of the Norton charts is 6.0, giving a good match with the skies as seen by the naked eye under good conditions; locating dim objects in dim starfields can be difficult, however. Despite their flaws, I still strongly recommend the Norton charts, for their convenient size and scale as well as for the wealth of additional information offered in a long introduction (which actually occupies the bulk of the volume). It is, as its full title says, a reference handbook in addition to an atlas.

Somewhat more advanced is the Tirion *Sky Atlas 2000.0*, a larger-format chart, with its coordinate grid adjusted to the precessional epoch of the year 2000. Because of precession, you may recall, the entire celestial coordinate system shifts across the celestial sphere. Charts compensate for this by specifying the year (*"epoch"*) for which they are printed. The Norton atlas, for instance, shows the coordinates for epoch 1950; the Tirion atlas is the first to anticipate the twenty-first century. Since the relative positions of the stars shift only very slowly, however, this is not an especially important consideration for the beginning amateur. The more important features of the Tirion atlas are its excellent graphic display, its large, clear scale, and the undistorted map projection used. The chart has two companion volumes, cataloging every star and other object shown, giving relevant information on each. The limiting magnitude of the charts is 8.0; it is available in field, desk, and deluxe editions. The field and desk editions are slightly smaller than the deluxe, which is a bit too large for convenient use at the telescope; the deluxe, however, shows different kinds of object in different colors, which is helpful.

It's a good idea to prepare for a night's observing by making a field copy of your charts. Photocopy the chart pages you will be using (provided, of course, that you don't violate copyright laws by running off multiple copies for your friends). The copy has the advantage of coming in a size that fits an ordinary clipboard. You can also scribble on them to your heart's content, and mark the particular objects you want to observe. Use a color that will stand out (if you will be using a red observing light, blue works best). You can also prepare a simple list, or stack of note cards, giving details about each object you mark. An observing list is a great help, giving information on the kind of object, its general appearance, and such information as the object's distance, age, and any other distinguishing characteristics.

How do you go about preparing such a list? There are numerous catalogs that give detailed information on the thousands of objects visible in an amateur telescope. The most famous of these collections is the *Messier catalog*, named for its compiler, the eighteenth-century French astronomer Charles Messier. The catalog includes 110 objects, among them most of the brightest nonstellar objects in the sky—a Greatest Hits of galaxies, star clusters, and nebulae that have been the first targets of amateurs observing the deep sky ever since. An excellent recent guide to the objects on this list is Mallas and Kreimer's *The Messier Catalog*, published by Sky Publishing. A much more inclusive list is the *New General Catalog* (*NGC* for short), which lists thousands of objects, from the largest and brightest open clusters to the dimmest and most distant galaxies. You will find a table of the *Messier Catalog*, and a selection of the brighter *NGC* objects, in the Appendix.

Guides to the plethora of objects visible in amateur scopes are almost as numerous as the objects themselves. By far the best—and arguably the best book of any sort written for amateur astronomers—is Burnham's *Celestial Handbook*. This exhaustive, three-volume set is a superb guide to objects outside the solar system. Dover offers the set in paperback, for around $8 per volume. The Webb Society of England has also recently released five hardcover volumes of detailed information on double stars, nebulae, and other classes of deep-sky objects. Handbooks such as Burnham's or the Webb Society's that can give detailed information on the full range of celestial objects are both rare and valuable. You will want one or both sets soon after you start observing.

Other such aids to observing include *ephemerises*. These are annual listings of special events upcoming in the skies. Eclipses, planetary movements, meteor showers, the return of periodic comets, times of sunrise and sunset, moonrise and moonset—all of these and much more will be covered in a good ephemeris. There are two of particular interest to amateurs. I prefer the *Observer's Handbook*. It's comprehensive, comprehensible, and offers new supplementary information each year on phenomena not ordinarily covered in other such sources. You can order from the publisher (see Bibliography) and several American distributors, who advertise in the popular astronomy magazines. The *Astronomical Calendar*, published annually by G. Ottewell, is perhaps the best for beginners, giving clear explanations of the phenomena involved in its calendars, with monthly star charts, diagrams of the positions of the bright planets, all illustrated in a dramatic, engaging style. The *Graphic Ephemeris* is a large, tall graph, intended to hang on a wall. It shows at a glance the times of events such as sunset or Jupiter's rising. The *Graphic Ephemeris* is also distributed by firms advertising in the astronomy magazines.

ADDITIONAL USEFUL EQUIPMENT

It's possible to write an entire book under this heading—hundreds of books, in fact. Each amateur soon develops his or her own list of inexpensive, clever adaptations to the conditions of observing. The following are a few of the more common.

Some source of dim light is a necessity for reading charts, choosing eyepieces, finding car keys in the grass, and so on. Red light, as I mentioned earlier, does not ruin the dark-adjustment of your eyes as quickly as white light. If you don't have a red sock handy, tape sheets of red cellophane on a flashlight, building it up in sufficient thickness so that the light is just bright enough to make out the print on a set of charts, and no brighter. More convenient than cellophane is a translucent, red plastic, sold at office- and art-supply stores as "blackout film." It comes with a self-adhesive backing and is easy to cut to size with a razor blade. Some people simply put their flashlights in brown paper bags.

Other things I generally tote around are mainly designed for comfort and convenience. A groundsheet is useful, not only to protect yourself and equipment from ground dampness, but also to catch the wingnuts you drop before they lose themselves in the grass. My present one is woven vinyl and has stood several years of hard use without a tear. A chair will seem a necessity, especially if you have a bad back or knees. A lightweight campstool, made of nylon mesh and aluminum tubing, weighs next to nothing and costs little more (under $5 at discount stores).

Other aids help ward off light. A black eyepatch, or a 2-foot square of black cloth, allows you to keep your free eye open while observing. (Observing with one eye squinted against the local light pollution strains the other eye.) Drug stores carry eyepatches for about a dollar, astronomy suppliers a good deal more. Also useful along these lines are rubber eyecaps for your oculars. These are simply sleeves that fit against your eye socket while you observe, blocking out stray light. They are especially useful with long-eye-relief oculars, but with any kind of eyepiece used in a brightly lit area, you will find that they aid your eye's light sensitivity immensely. Henzl and Edmund Scientific both carry these, for a dollar or so apiece. If you can't obtain them, a piece of black cloth draped over your head serves the same purpose; you may want to secure it under a hat or sweatband. (If the neighbors stare, gesture at them ominously, until they go away.)

Another protection against light pollution can also keep the wind from rattling your mount (and teeth): a sheet of nylon fabric stretched vertically between two tent poles, supported by a pair of guylines at each end. It is far more convenient to try to position yourself so that natural objects serve both purposes, but if you are setting up on a windswept pasture, a windscreen may be worth the trouble. Outdoor equipment stores sell collapsible tent poles and stakes; the nylon will be cheaper at a fabric shop. Finally, you will want a box to keep everything in: plastic fishing tackle boxes are ideal for holding oculars and flashlights, spare batteries and wing nuts, and a snack.

12

Using the Telescope

Telescopic observation is not a simple matter; it's a craft, an art, a skill learned over the years. You will probably find that your first experiences with a telescope (after the initial thrill subsides) are a little disappointing. You won't be able to see everything the photographs show. Experience helps, but in order to get that experience, you need to keep at it. This chapter points out some common pitfalls and offers a few suggestions, but the most important thing you can do is spend time at the eyepiece.

SELECTING AN OBSERVING SITE

Once you own a telescope, the first question you'll run into is where to use it. Access to the whole sky, calm, clear air, and darkness are the essentials of a good observing site. As with any ideal, the perfect site is impossible to find. The kind of compromise you make depends on where you live and what kind of transportation you have. In general, a pasture on a hilltop far away from any city will offer the best site. If that pasture is high in a mountain range, so much the better. The rest of us, however, have to take our sights where we see them. But even in the city, or a suburban back yard, there are ways to get the most out of a site.

Access to the sky should be your first concern. Trees and buildings can block an amazing amount of the heavens. What you have to do is ensure access to crucial parts of the sky. The bare minimum, in order to see every star in the sky at least once every 24 hours, is a strip of clear sky stretching from the north celestial pole to the south point on your horizon. Even if such a strip is only an hour wide, with patience you can see everything the universe has to offer as it passes over your "window." The only exception to this is in the case of special events, such as eclipses or comets, which are of limited duration and so occur only in a specific part of the sky. Generally, the most important criterion is an unobstructed south horizon. This is a necessity if you are to catch glimpses of the constellations on the borders of the southern circumpolar region.

The second most important consideration is the level of *light pollution* in your area. Light pollution is the stray light from streetlights, advertising, and so on, that drowns out dim stars as it fills the sky with a background glow. The reason for this is the *backscattering* of light from particles suspended in the air. Light pollution is especially pronounced near cities, where air pollutants increase the amount of backscattering even as brighter artificial lights increase the amount of light available to scatter. Light pollution has become a serious problem, not merely for amateurs but for professional astronomers as well. The great telescopes on Mount Wilson, Mount Palomar, Kitt Peak, and others of our major observatories are increasingly threatened.

Light pollution also has a more local aspect. Any light shining directly in your eyes prevents them from adapting to the dark. Choose a site with as few lights as possible shining onto it. In a crowded area, complete freedom from light is often impossible, but by positioning yourself strategically behind trees, chimneys, or phone poles, you can at least keep your eyes and the aperture of your telescope shaded. If you can't position yourself in available shadows, try rigging up a screen such as the windscreen described in chapter 11 (and if your back yard is really that well lit, start thinking about carrying your scope to darker pastures).

Finally, you need a site that minimizes atmospheric interference. Turbulence in the air above you causes the image to distort and shimmer, especially at high magnifications. The effect is the same as the heat shimmers you see while looking across a large parking lot on a hot day. Heated air has a different refractive index than the cooler air it disrupts as it rises. This varying refraction in the atmosphere is the reason stars twinkle, but in a telescope, the effect is less attractive. It shatters the image or causes it to crawl disconcertingly around the field.

Heat is the most common cause of turbulence. Avoid placing your telescope where its line of sight will be over objects that absorb heat during the day. Sheet metal roofs, pavement, rock, bare earth, and water all tend to lose heat steadily during the night, with metal losing it most quickly and water least quickly. Any object that generates its own heat, such as a house, will also cause trouble. A pasture covered with grass absorbs less heat, and a forest still less.

Avoid water for another kind of atmospheric interference: haze, caused by high humidity. In most of the United States on evenings after the temperature has been above 80 degrees Fahrenheit, the sky fills with water haze. Haze adds to backscattering of ground light or moonlight; faint stars disappear, and the images of brighter ones appear as fat, wet blobs. Such nights often indicate a steady atmosphere, and views of the planets and double stars can tolerate high magnifications. Observing sites near water are also subject to heavy dew, which is a major nuisance (see below), and, if that water is salty, you should avoid it for a special reason: salt will quickly tarnish the aluminum coating of a mirror. Only solar observers prefer lakeside observing, because water heats up slowly, reducing heat-caused turbulence during the morning hours.

FINDING CELESTIAL OBJECTS

Imagine: you are set up at your observing site. The sky, miraculously, is dark and clear. On your field notes, you read by dim red light your first entry. Your notes tell you the catalog number and name of an object, its right ascension and declination, its magnitude and angular diameter. These tell you where to look and what it should look like. Good notes might also include information such as an object's distance or age, and any special features that may or may not be visible, but help you to place what you're observing in the larger scheme of things.

It's easy to sight a bright celestial object with the finder: you point the telescope to a place among recognizable bright stars where the chart says an object should lie, and look for signs of the object in the finder. The usefulness of this technique depends on several conditions. First, you need recognizable stars nearby. This is one important reason to know the constellations well. You must also develop a sense for how objects move in the finder's field. In most finders, the field is upside down and reversed. To make a star move from the edge of the field to the center, you must move the scope *away* from that star's apparent direction from the center; the star will follow.

Recognizing a dim object in a finder is tricky. Many beginners expect an object to appear distinctly, even when it is dim and small. But an object at the limits of a finder's reach (as most objects will be) is almost always indistinct. You will notice it not as a clearly outlined, miniature galaxy, nebula, or cluster, but as a subtle *difference* in the blackness between the stars. Does something catch the corner of your eye, only to disappear when you focus on it? Center the fleeting mirage, then go to the low-power eyepiece in the main instrument. Chances are you've found your target—without ever seeing it distinctly.

To locate some objects, such as double or variable stars and very small planetary nebulae you need to know the star field around them. Here, a homemade *finder chart* will be helpful. Study charts or photographs to find the pattern of the brighter stars around the target. Don't try to include the dimmer stars; you need only draw enough stars to specify a particular pattern, unique to the region around your target. Holding that finder chart upside down, aim the scope in the approximate direction, and search for the pattern on the chart. You may find it helpful to defocus the finder slightly, so that the dimmer stars in the field disappear. If there is no distinct pattern of stars around your target, see if the chart shows one nearby. Once you have found such a neighboring pattern, you can use the next method to jump to the target itself.

The second method of locating objects is usually called "star hopping." On a chart, trace a line from your target through nearby stars to the nearest bright star. At the telescope, simply reverse the process: center on the bright star, then move the telescope a short distance to the next star on the line, then hop to the next, and so on, until you reach your target.

If the stars between your guide star and target are too widely spaced, you need to do some dead reckoning. To do this, you need to know how wide your true field is. If the next star in your line is 4 degrees away from the first, and your field is 2 degrees wide, move the telescope in the direction of the next star until the first star is at the edge of the field. Stop, and note any star at the field's opposite edge; move the scope until that star has crossed the field. The next star in your star-hop chain should now be on the edge of the field. You can move your scope many field diameters between guide stars in this way, although the farther you have to go without visible guides, the more likely you are to go astray. This is one important advantage to a wide-field scope or finder: you never need to go too many field diameters by this kind of dead reckoning (Fig. 12.1).

When all else fails, use setting circles if you have them. You should usually reserve these for a last resort, because in most cases the use of setting circles is time consuming, requiring that you read strings of numbers from

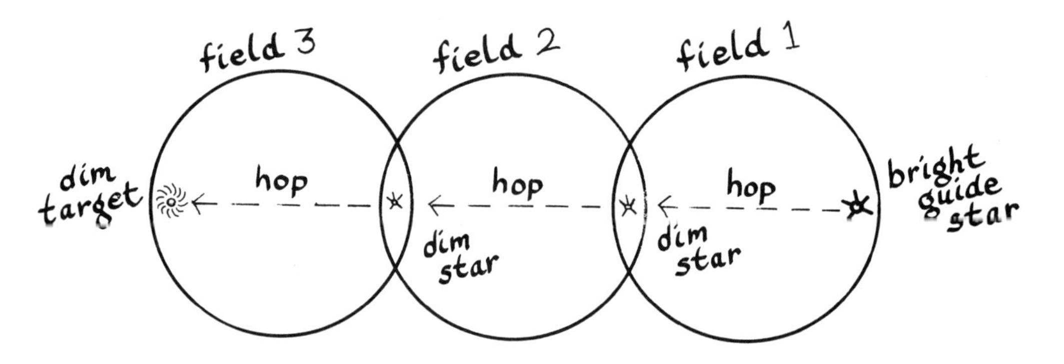

12.1 Star hopping. With a true field of known size, divide the angular distance between a dim target and a nearby, visible star by field size. Move the telescope from star to target one field at a time, using dim stars to mark the edge of the field.

your notes (not to mention requiring that you take some care with polar alignment when setting up). But if an object hides in a region with no guide stars to follow, the finder can't sweep it up, and you know you won't sleep peacefully unless you see it that night, use the circles. You can even rig up circles for use with an altazimuth mount; using them requires the help of a portable programmable computer, to convert from celestial to alt-az coordinates, but it can be done.

There are two ways of using setting circles. The more reliable method, especially with home-built circles and an indifferently aligned mount, is to go to the nearest bright star. Look up the coordinates of that star (you might want to keep handy a list of the coordinates of the 25 brightest stars). With the star centered in your field, set the circles to agree with its coordinates. Subtract the coordinates of your target from the coordinates of this star, or vice versa. The difference is the amount you have to move your telescope to reach your target. For example, say your guide star has coordinates 1800 hours, + 22 degrees, and your target's position is at 2030 hours, + 10 degrees. The target lies 2 hours 30 minutes east and 12 degrees south of the guide star. This method is adequate for most mounts and purposes. Its only limitation is that, unless you have a clock-driven right ascension circle, you need to move quickly once you've set the circle for the bright star's coordinates; otherwise, the motion of the earth quickly renders that setting inaccurate.

If you have an accurately aligned mount, with well-made setting circles, you can shortcut the above process by dialing direct, simply turning the scope to the position of the target. You will still have to set your circles at the beginning of the observing session. The simplest way of doing this is to center a bright star in a narrow-field ocular and turn the right ascension circle to agree with the right ascension of the star. (A properly aligned mount should not require adjustment of the declination circle; check it anyway, to make sure of your alignment.) If your mount has a clock drive, you simply engage the drive at this point, and your setting circles should remain accurate all night.

If you don't have a clock drive or if the drive does not turn the circle as it moves the polar axis, you will need to update the setting each time you use the circles. To update, subtract from the right ascension of the target object the hours, minutes, and seconds that have passed since the moment you set the right ascension circle. For example, to find an object at 12 hours right ascension, 2 hours after you set the circle, point the telescope to the 10-hour mark on the circle.

If your first experience using the telescope at night is confusing, you might want to get some practice during the day, when you can see more clearly both the telescope and the scenery around you. Practice finding distant earthbound objects by a variety of methods: sighting along the tube, using the finder scope, "star hopping" (use your imagination for this last). You will find it much easier to get a sense for the relationship between motion of the tube and the motion of objects in the field. Daytime is also the best time to collimate the telescope's finder. With a high-power eyepiece, center a distant object (more than ¼ mile away) in the field. Lock the tube firmly in place and then center that same object on the finder's cross hairs. You can also do this at night, using streetlights or the aircraft warning lights atop tall objects.

HOW TO SEE

No matter how much you spend on equipment, or how diligently you seek out the ideal observing site, unless you can *see* you will never realize your telescope's potential. Seeing (this term also refers to atmospheric turbulence, explained in the next section) involves both psychology and physiology. The physiological factors revolve primarily around the dark-adaptation of your eyes, but conditions such as their blood supply, muscle tension, and other imponderables also play their part.

The eye adjusts to low light levels in two ways. First, the iris expands to its maximum diameter, over a period of several seconds. Next, the chemical balance in the retina changes. This process takes as long as half an hour: your eyes become more sensitive, increasing their light-grasp by several magnitudes. This sensitivity can be destroyed by a split-second exposure to bright light. Once you have been in the dark for half an hour or so, shield your eyes from all bright lights. Dim, red light, however, does little harm.

The blood supply to the eyes is also important. Nicotine and alcohol both restrict your circulation. Even changes in your blood pressure can make a difference; you will see more while comfortably seated than while crouching. Squinting or straining the eyes in any way—even the eye you aren't using—has the same effect. Try to observe with both eyes relaxed and open. Strain can come from tension in the muscles that focus your eyes. That tension is at a minimum when you focus on distant objects, so focus your telescope by drawing the eyepiece slowly out until the image clears. The moment it does, stop moving the focus: your eyes will then be looking at the image as if it were at infinity and will be at their most relaxed. Cold and fatigue will also cut way back on your sensitivity.

But seeing is more than the physical registration of photons. The most important part of the process goes on inside your mind. Relax, and you will be in the best frame of mind for observing. Slow down. The stars aren't going anywhere, at least not at a rate you won't be able to follow. The best way to start observing is to put your dustcaps back on the scope, stretch out on the grass, and just look up at the stars. Take the full half-hour your eyes need to adjust. You have the time to do this. One hour of high-quality observing, with your mind more in tune with the slow motion of the sky, will be worth a whole night of unproductive struggle.

The observer's greatest emotional enemy is expectations. Expect nothing. If you observe without expectations, you can never experience the grotesque sense of being disappointed by the universe. Your mind's eye will bring no prejudices to the task. You will see what is actually there—not the shadow of what you had hoped to see. This is a difficult discipline, but it is the essential of observation (not only in astronomy, but in any endeavor). You won't succeed at it every night, not even long after you are familiar with the feeling. Some nights one just can't take things as they come. Accepting the inevitability of those nights will make them rarer. One way to recover the proper receptivity, should you lose it, is to put your charts and notes aside and sweep at random through the sky. Not knowing what you'll find, you'll find yourself seeing more.

Along with the proper attitude, there are some tricks you can use. Most important is *averted vision,* a technique that takes advantage of the varying sensitivity of your retina. The retina is most sensitive away from its center. When trying to get the most detail out of a dim object, don't look directly at it but slightly to one side. This goes against the grain at first, but this technique can make a completely invisible object pop into view.

Staring for a long time at any object is tiring. Break off every few minutes and let your eyes relax at infinity. Switching eyes will also help. Tapping the tube gently will help to make small details such as sunspots visible, especially if your optics aren't as clean as they might be: dust spots shake relative to the background, but the image detail won't. Any rapid change in the appearance of an object helps. If you have a light-pollution-rejection filter, flick it in front of the eyepiece for a moment; emission nebulae in the field will brighten suddenly, and the variation will call your attention to them.

NUISANCES, REAL AND IMAGINED

You already know about light pollution, but we have neglected the largest natural cause of light scattering: the moon. Moonlight is as much of an impediment to observing as clouds. The −11-magnitude light of the full moon can wash out all but the brightest objects; even in its partial phases the moon severely restricts observing. Dim or extended objects such as galaxies and nebulae practically vanish while the moon is above the horizon. When you plan your observing, remember the phase of the moon, and note when it will rise and set.

Haze and turbulence can have their local causes, but large-scale weather systems also disrupt the atmosphere. *Transparency* and *seeing* are the names given to the qualities affected by haze and turbulence, respectively. Haze reduces transparency; turbulence degrades seeing. Transparency is important for dim, extended objects, such as galaxies and nebulae; good seeing is essential for high magnification, as with double stars and planets. Fortunately, weather conditions that cause haze frequently give good seeing, and turbulent skies can be fabulously transparent (Fig. 12.2).

Haze is easy to recognize: the stars appear as fat blobs, and dim stars disappear. Forget most kinds of deep-sky observing on such a night. Turbulence is apparent to the naked eye in the twinkling of bright stars. Through the telescope, you will most likely see it as "boiling" in an extended image,

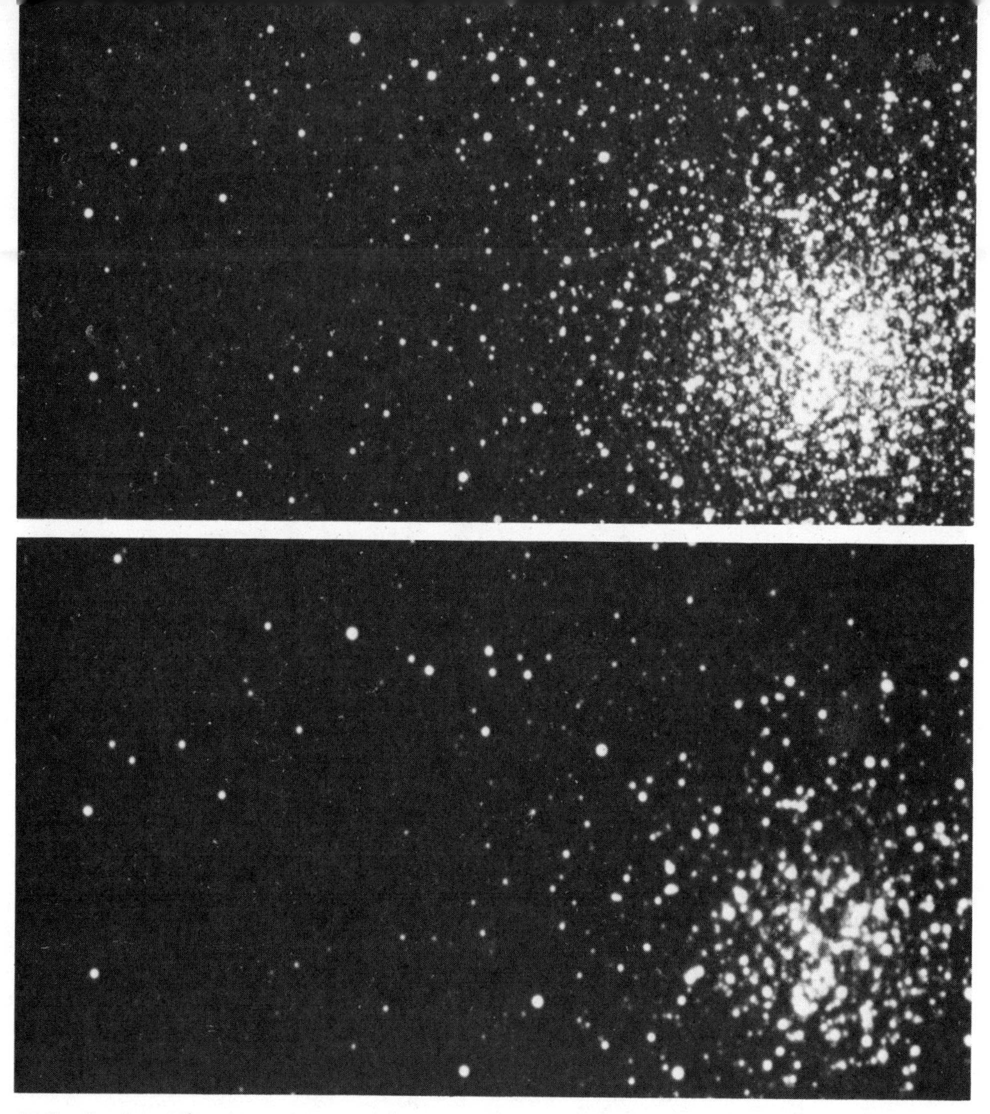

12.2 Seeing. These two images of a typical extended object (M 13, the Great Globular Cluster in Hercules) show the effect produced on a good image (top) by atmospheric instability (bottom). (Lick Observatory photograph)

or as it disrupts or entirely destroys diffraction rings around medium-bright stars. Seeing is commonly ranked on a scale of one to ten, with ten as the best possible, one the worst. You may want to keep records of the kinds of objects or details visible under different conditions. An observer's log can give you a sense of what to pursue on a given night.

Even more useful than the ability to recognize haze and turbulence is knowing when to expect them. Haze builds whenever a high-pressure system lingers in warm weather. The weather will be oppressive, with high humidity. The winds are light and from the south (more or less). While the high persists, skies clear shortly after sunset. If the air cools into the sixties, haze can diminish after midnight. At these times, the atmosphere is as still as lead crystal, and the seeing can be superb, although heavy dew is then a problem. Strengthening winds can help to clear the haze away and usually signal the approach of a front. The passage of a cold front, with a shift in the wind into the north and an accompanying drop in the humidity, clears the haze away. As the wind shifts, however, the mixing of cold and warm air causes turbulence. The sky clears, but the seeing deteriorates. Poor seeing is generally a cold-weather phenomenon; haze is a summertime problem. Spring and fall, in most parts of the country, are just plain cloudy.

The best combination of seeing and transparency will come a night or two after a front passes through, just before the passage of the next high-pressure cell. This is especially likely if the weather is cool, the humidity low, and the winds light; look for conditions to improve after midnight when heat radiation from the ground diminishes and the cooler air has lost most of its humidity as dew. You can anticipate such conditions by watching the weather map in your newspaper. Throughout the United States, weather systems tend to move from west to east. Should a strong high-pressure system develop west of you, try to estimate its time of approach and clear your calendar for that evening. The same kind of foresight can also spare you the disappointment of a rained-out or clouded-out evening.

Two other atmospheric phenomena to watch out for are dew and local winds. Unlike the weather, you can do something about these. Dew is a problem on clear, humid nights. As the temperature drops, many objects, especially metal and glass, cool faster than the air around them. When they cool below a certain temperature (which varies with the air-temperature and humidity), moisture condenses on them. Exposed glass will mist over with a fine haze of water droplets. Sometimes, the first warning you have of this is that you seem to be going blind; you see less and less through the finder or scope. The worst thing you can do at this point is to try to wipe the dew away. The most reliable way to remove or prevent dew is to warm the affected part. An eyepiece held inside your shirt, or wrapped in a warm bandanna, will clear magically.

There are several methods of applying heat to larger surfaces. Taping a flashlight bulb near the surface, and wiring that to a battery, will generally provide enough heat to prevent dew on most nights. A reflector's primary is usually warmed slightly by heat radiating from the ground, and inside its tube it tends to radiate heat more slowly than it would if exposed. This is fortunate, because attempts to heat it can cause other problems. The low-tech solution for a dewing refractor's objective is an improved dewcap. Dewcaps on refractors are never long enough: they should be twice as long as they are wide; a cone (not a straight tube) of black construction paper taped to the present dewcap should increase the level of protection. The high-tech solution is a hand-held hair dryer, powered by battery or extension cord. At this point, however, you have trespassed over the line into serious gadgetry.

Local winds are a problem primarily because they shake the telescope. Using a building or hedge as a windscreen is the best solution, or you can use the screen described earlier (although it's a nuisance to carry). The most local wind problem you can have is inside the tube of a reflector. *Tube currents* tend to form early in the evening, as heat from the objective, the ground, or even the observer's body rises through the open tube as if it were a chimney. These currents tend to hug the walls, so a tube of adequate size will keep them from encroaching on the light path. You can prevent their formation by setting up over grass rather than stone, pavement, or bare earth, which radiate strongly in the evening. After midnight, the problem becomes less severe.

Thermal adjustments in the mirror itself, as it cools to the surrounding temperature, will change the mirror's figure enough to harm the image. Allow time for the primary to cool before you start observing. Set up the telescope at sunset; in most cases, a mirror requires an hour to reach complete thermal equilibrium.

THE DIEHARD OBSERVER

If you have an ordinarily busy life, you may find it hard to devote as many evenings to observing as you would like. The good nights—clear sky, no moon, no other commitments—may be so rare that you hate to let one pass. Such an attitude will get you out under the sky more often, but it has its risks: fatigue and frostbite. Learn to recognize when you are too tired, especially if you will be driving home from your site. An unconquerable quiver in your neck muscles, facial tics around the eyes, and dimming vision are good signs it's time to hang it up. Don't ignore them. The pleasure has gone from your observing by then, and the next object on your list isn't going anywhere for many millennia to come. Save a little energy for packing up.

Cold-weather observing will bring on fatigue even faster, and in the cold, exhaustion can be deadly. Learn to recognize the symptoms of hypothermia. These include fatigue, confusion, dimming of vision, muscle stiffness, and uncontrollable shivering—a deadly combination that takes away your judgment, leaving you helpless to take steps to protect yourself. It's possible to freeze to death from this imbalance in the body's heating system, even on a temperate day; imagine how quickly it can act on a sub-zero night. Recognizing frostbite can be difficult, since it requires a visual inspection under bright light. Play it safe and call it quits when any extremity has been numb for fifteen minutes or more. (The recommended treatment for frostbit digits seems to change from year to year. This year, they're calling for bathing in cool water. Look up the latest in a first-aid manual *before* you go out in cold weather; the same for hypothermia.) If you don't already know it, learn the wind-chill table. A light breeze can drop the chilling power of a sub-zero night by several dozen degrees.

You can protect yourself from many cold-weather threats by common sense and proper clothing. Common sense should tell you (a) when to quit, and (b) when not to start. Although I have read one report of an amateur observing through an Antarctic winter (long nights, but he found his eye kept freezing to the ocular), you wouldn't catch me out there. But if you find the transparent winter skies draw you outside on a night when common sense says no!, adopt the second line of defense.

Overdressing is the key to protection, and overdressing in layers is the secret to doing it inexpensively. The total thickness of clothing between you and the night air is all that matters. The U.S. Army calculates that an immobile human (sleeping or stargazing) needs 2 inches of clothing on a 20-degree-Fahrenheit night; 2.5 inches at 0 degrees; 3 inches at −20 degrees—more, or course, if there's a wind. So layer up bulk wherever you can. Avoid sweating: getting wet is a surefire way to bring on hypothermia. You can reduce the risk of a chilling sweat by using materials that wick away moisture from inner layers. Wool is good for this, if a bit itchy; synthetics, such as polypropylene, are also good. Avoid all-cotton materials; they don't insulate when wet.

So what would you wear on a typical winter night? Heavy wool pants are fine, but layering thermal long underwear, sweat pants, and large overalls is even better. Insulated boots will keep your feet warm, but a pair of galoshes over four progressively larger pairs of thick wool socks will also work. Cover your torso with (again) thermal longies, a wool shirt, a light sweater, a large, bulky sweater, and a big jacket. Thin liner gloves inside large mittens are a good idea; you can take off the cumbersome outer shell to change eyepieces or focus and still hold some heat next to your skin. Don't neglect your head; it radiates as much as half your body heat. Precautions against cold are also in order on spring and fall evenings, and if you live in the northern tier of the United States you will want to take along at least a sweater on most summer nights as well. A 1.5-inch layer of clothing is necessary to stay comfortable on a 40-degree-Fahrenheit night.

EXPECTATIONS, AGAIN

One piece of advice from this chapter bears repeating: the hazards of expectations. Approaching the sky with an open mind is important, not only to sharpening your perceptions but to your whole enjoyment of astronomy. In an era dominated by enormous, enormously sophisticated research instruments—when long-exposure, full-color photographs taken with the world's largest telescopes are widely available, and cinematic special effects can take an audience into black holes and distant galaxies—an amateur needs to keep a sense of proportion. If you expect to see a technicolor light show through your telescope, watch stars evolve for your enjoyment, or see a comet every night, you will be disappointed. The sky isn't an entertainment put on by human beings for humans to enjoy. It's the world we live in. Approach it with a certain respect and with the knowledge that the world doesn't give up any mysteries worth knowing without the active participation of your mind. The view through the air above is a spectacle, but its brightness and coloring, its activity or profundity, depend on the kind of intelligence and imagination, knowledge, and awe you bring to it. Expect nothing. You will be rewarded beyond your dreams.

The Solar System: Aspects, Operations, and Origins

13

ASPECTS

As you learn the constellations, sometimes you will find an uncharted, bright star. If you watch from night to night, you will notice it move among the fixed stars. Its path will follow the *ecliptic*, the path of the earth's orbit around the sun. (The planets of the solar system orbit the sun on approximately the same plane as the earth, and so they too appear in the sky on or near the ecliptic.) This uncharted "star" is, of course, a planet. From its motion comes its name, from the Greek word *planetes*, for "wanderer."

Each of the bright planets has a unique appearance, although its position and brightness constantly change. The innermost planet, Mercury, is the most elusive of the ancient five (Mercury, Venus, Earth, Mars, and Jupiter). It appears as a dim, ashen star, close to the western horizon at sunset or in the eastern sky at dawn. Venus, too, is a morning or evening star, but it is far brighter than Mercury. It is the brightest of all the planets, with a maximum light of magnitude −4.4.

About once a year, each of the planets disappears for a time behind the sun. The time between these disappearances, when a planet is readily visible, is called an *apparition*. Because they orbit between the earth and the sun, Mercury and Venus are called *inferior planets*. An apparition of an inferior planet begins with *inferior conjunction*, when the planet is between the earth and the sun and usually is invisible in the solar glare. A few weeks later the planet appears west of the sun and shines as a morning star before sunrise. For weeks it climbs to its maximum elevation above the eastern horizon. When it reaches its greatest angular distance from the sun, it is at *greatest western elongation.* As the planet continues around its orbit, this angle shrinks, and the planet appears to sink back toward the horizon. At *superior conjunction,* it is behind the sun, invisible from the earth, and at its greatest distance from us. Next it appears in our sky to the east of the sun, rising as an *evening star* before the sunset. Again it climbs to its greatest angular separation from the sun, at *greatest eastern elongation.* From there, it moves back in toward inferior conjunction again, sinking toward the western horizon.

Both of the inferior planets show phases, like the moon's. At superior conjunction, the face nearest us is fully sunlit. The planet appears as a circular disc; this is *full phase.* Full gives way to *gibbous phase*—between full and *half phase;* the latter occurs at greatest elongation. Half phase narrows to *crescent phase* as the planet approaches inferior conjunction. Then the planet, like the new moon, shows us its dark side. After inferior conjunction, the crescent phase swells to half again, and then grows through gibbous phase to full (Fig. 13.1).

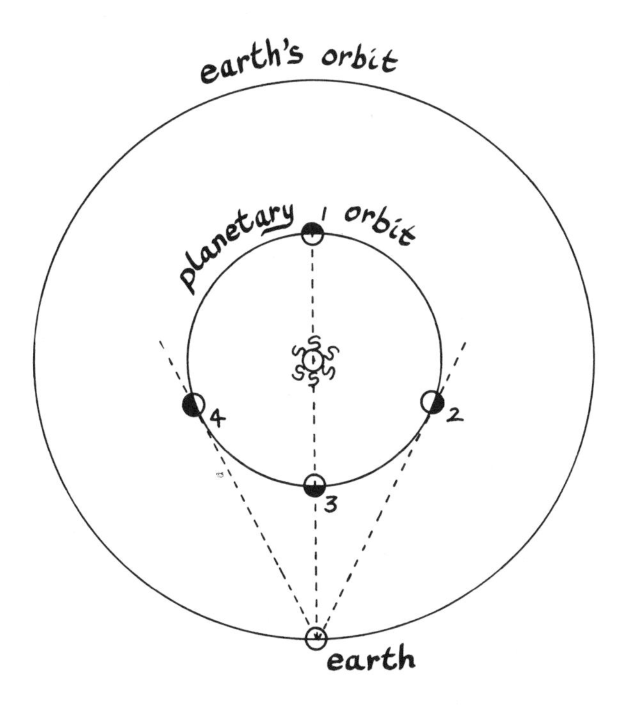

13.1 Aspects of an inferior planet. (1) *Superior conjunction; planet shows a fully illuminated disc.* (2) *Greatest western elongation; planet shows a 50% phase and rises before dawn.* (3) *Inferior conjunction; planet shows an unlit disc.* (4) *Greatest eastern elongation; planet shows a 50% phase and sets after the sun.*

Mars, Jupiter, and Saturn are the *superior planets* visible to the naked eye (Uranus, Neptune, and Pluto are also superior planets, too dim to see). Their orbits lie outside Earth's, and their aspects and apparitions differ from those of Mercury and Venus.

An apparition of a superior planet begins, like an inferior one, at *conjunction,* when it is on the far side of the sun. After conjunction, a superior planet rises in the dawn, just before the sun. As the weeks pass, the planet moves farther west of the sun, and rises earlier each morning. This westward motion, like that of the stars, is owing to the earth's orbital motion. The planet's own orbital motion carries it toward the east, but earth's greater orbital speed makes this a losing race. So the motion of a planet is actually two motions in one: relative to the background stars, the planet moves generally to the east; relative to the sun, however, it moves inexorably into the west.

As this motion continues, the planet reaches a point where it rises at midnight. At this time, the angle between sun, earth, and planet is 90 degrees, and the planet is at *western quadrature.* The planet continues to move west, rising before midnight; it becomes an evening star at this time. Eventually there comes a night when the planet is on the eastern horizon at the moment of sunset, and remains in the sky all night. This is *opposition;* the planet reaches its maximum brightness and apparent diameter, and is closest to the earth (Fig. 13.2).

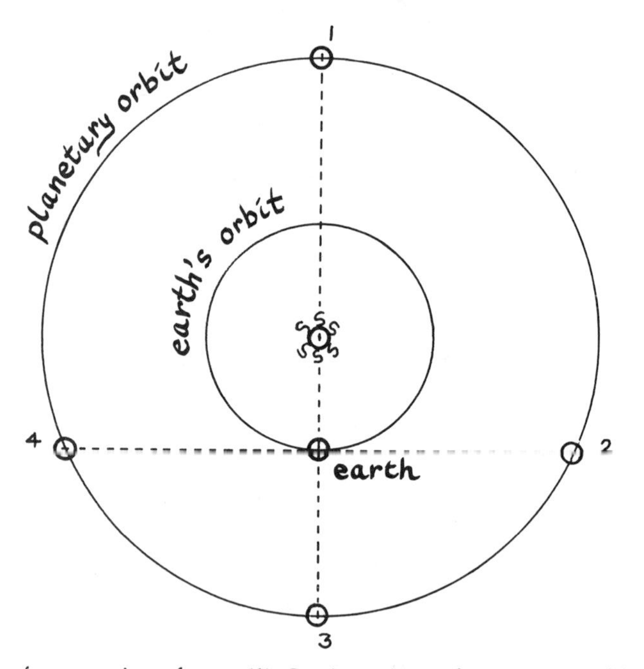

13.2 Aspects of a superior planet. (1) *Conjunction; planet not visible, rises and sets with the sun.* (2) *Western quadrature; planet rises at midnight; on meridian at sunrise.* (3) *Opposition; planet visible all night.* (4) *Eastern quadrature; planet on meridian at sunset; sets at midnight.*

Around opposition, the superior planets move strangely: they stop going west across the sky and for a time move backward, toward the east. Figure 13.3 shows this *retrograde motion* to be an optical illusion, the effect of the difference between the orbital speeds of earth and the planet. As the earth catches up to a superior planet, earth's higher orbital speed makes the slower outer planet seem to drift backward against the distant stars. The point on a superior planet's orbit where it starts and stops its retrograde motion are called its *stationary points;* its ordinary, forward motion is also called *prograde motion.*

After opposition, the planet rises before sunset. Each evening finds it higher above the eastern horizon; each day it appears farther to the west in the sky. In a few months, sunset finds the planet on the meridian: it is at *eastern quadrature,* halfway between opposition and conjunction. Eventually, the planet falls into the sunset and fades into the glare of the sun. On the sun's far side, it is in *conjunction* once more, setting with the sun and rising at dawn. It is at its greatest distance from the earth and appears dimmest and smallest. For a few weeks, it is usually completely hidden by the solar glare. The outermost planets, Uranus, Neptune, and Pluto, also follow this pattern; they are rarely (Uranus) or never (Neptune and Pluto) seen with the naked eye.

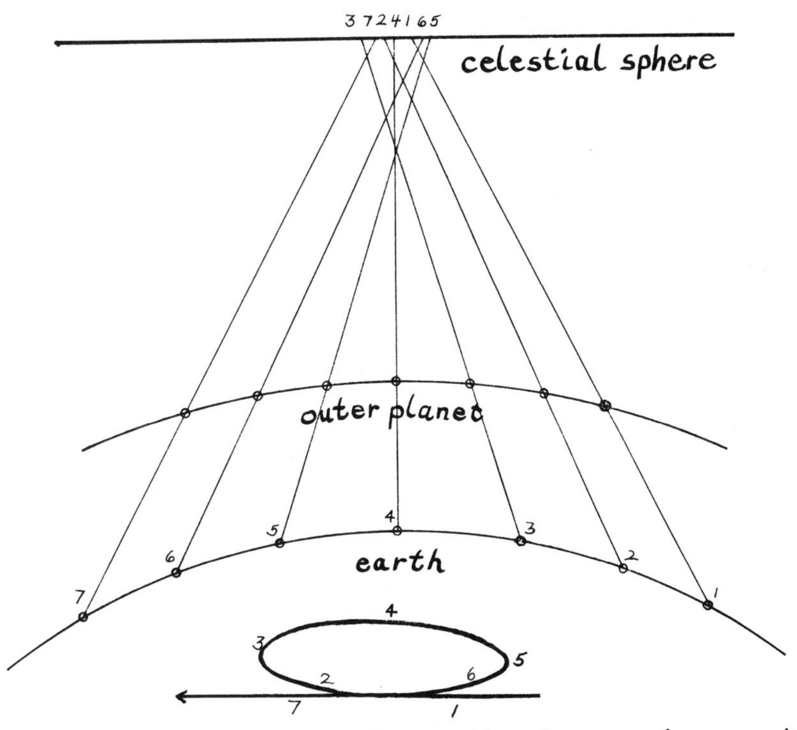

13.3 Retrograde motion. As earth catches up with and passes a slower-moving outer planet, the line of sight from earth to planet describes a loop (below) *on the celestial sphere.*

Each planet has its distinct appearance. Mercury is a dim, ashen gleam, hard to glimpse in the twilight sky. Venus is so bright you may easily mistake it for an aircraft with its landing lights turned on (and it is often the source of "sightings" of UFOs—unidentified flying objects). Mars is a vivid red glow, ranging from a dim +1 to a brilliant −2 magnitude. Jupiter is a steady, creamy white, and its magnitude varies less, from −2 at opposition to −1 at conjunction. Saturn is dimmer, and its brightness varies least, shining with a yellow light of magnitude 1.

The planets' orbits do not lie exactly on the ecliptic. Each orbit tilts a few degrees and thus cuts through the ecliptic at two points on opposite sides of the celestial sphere. Half of each planet's orbit lies north of the ecliptic on the celestial sphere; the other half lies to its south. The point where a planet passes from north to south of the ecliptic is called the *descending node*; the point of passage from south to north is the *ascending node*. At these nodes, the planet will appear exactly on the ecliptic. The angle of intersection is always extremely shallow, so a planet will be very close to the ecliptic for several weeks or months on either side of a node. Only Pluto ever strays more than a few degrees from the ecliptic.

When an inferior planet passes through a node at the time of inferior conjunction, it will pass directly between the earth and the sun. Such an event is called a *transit* of the sun, during which the planet appears as a small dark dot against the solar surface. Mercury transits about a dozen times each century; Venus, far less often.

Any two bodies sharing the same right ascension are in *conjunction*. A planet can be in conjunction with the sun, and any two or more planets can also be in conjunction with each other. When this happens, they rise and set at the same time, and appear close together in the sky. Usually, the two will be separated by several degrees on a north-south line. An *occultation* occurs when one body (the moon, a planet, an asteroid) passes between the earth and a more distant body (most often a star, but occasionally another planet). When this happens, the nearer body blocks our view of the farther. An *appulse* is a close apparent approach of two bodies, when two planets appear within a few degrees of each other, but are not actually in conjunction. These events are often spectacular to watch, with or without a telescope.

OPERATIONS

The orbital speeds of the planets differ according to some regular rules, first deduced by the seventeenth-century astronomer Kepler. The two of Kepler's laws you are most likely to notice are:

- Each orbit is not a perfect circle, but an ellipse, with the sun at one focus of the ellipse
- The farther a planet lies from the sun, the slower it moves on its orbit

The elliptical orbit brings each planet closest to the sun at one point, called *perihelion*. The farthest point is *aphelion*. (These terms can be modified to apply to the orbit of any body around another; satellites of the earth, for instance, have *perigee* and *apogee*. The *gee* ending comes from *geo*, for "earth"; *helion*, of course, means "sun." The same rules apply to their orbits as well.) When a planet is near perihelion, it moves more quickly than when at aphelion. The changing speed is enough to make a visible difference in the apparent motion of some planets. The shape of an orbit is always an ellipse, but ellipses can be almost perfectly round, or very elongated. The amount of elongation is called *eccentricity*. An ellipse that is perfectly round has an eccentricity equal to zero; a very elongated ellipse has an eccentricity approaching 1.0 (orbits with eccentricities greater than or equal to 1.0 are open-ended; some especially fast comets follow such orbits, never to return). Most of the planets have orbits of very low eccentricity.

As each planet completes one orbit around the sun, it goes through its own year or *sidereal period*. Planet watchers also count another kind of year, the time between oppositions, called the *synodic period* (Fig. 13.4). The sidereal periods of the planets grow longer with increasing distance from the sun. Mercury's is 88 days; Jupiter's is a dozen earth years. The synodic periods vary, generally decreasing with increasing distance from the sun. Pluto's synodic period is barely longer than a terrestrial year; Jupiter's is 13 months, Mars's can vary, averaging around 2 years.

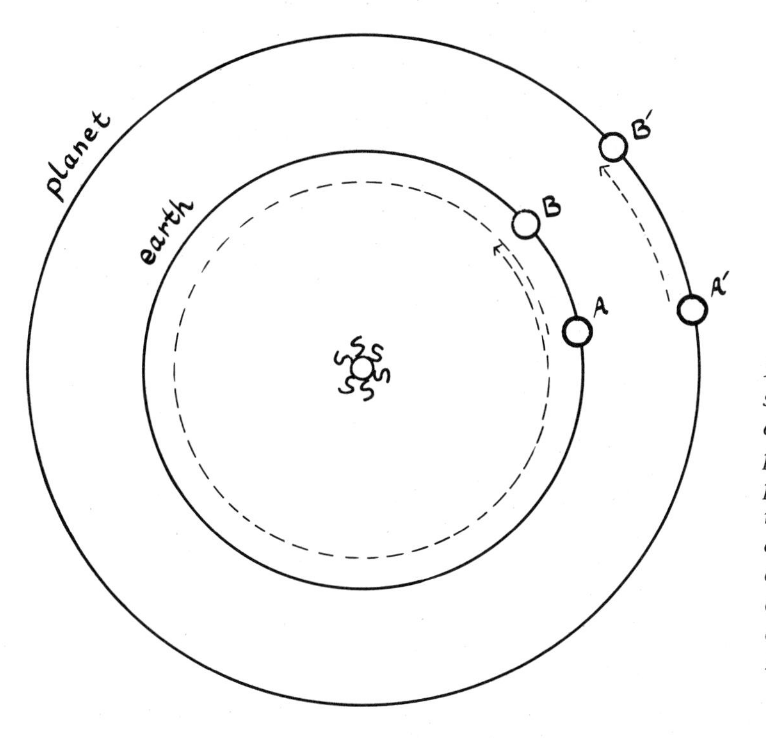

13.4 Synodic period of a superior planet. After one opposition (A–A') *the outer planet moves along its orbit to point* B'. *To catch up at the next opposition, earth must complete more than one full orbit, moving past* A *to* B. *The outer planet's sidereal period is one complete revolution, from* A' *to* A' *again.*

ORIGINS

A census of the solar system today turns up the following:

- Nine planets: Mercury, Venus, Earth, Mars, Jupiter, Saturn, Uranus, Neptune, and Pluto. They fall into two distinct categories. The inner four are small, dense, rocky worlds, the *terrestrial planets.* The next four are giant spheres of gaseous slush, the *Jovian planets* (Jovian for Jupiter), also called the *gas giants.* Pluto, a small, frozen world, is in a class of its own.
- Several dozen moons: one for Earth, two for Mars, sixteen (and counting) for Jupiter, fifteen (and counting) for Saturn, five for Uranus, two for Neptune, and one, the newly discovered Charon, for Pluto.
- Thousands of *asteroids:* small fragments of rock and metal following orbits concentrated between Mars and Jupiter in the *asteroid belt,* but ranging in to the orbit of Mercury and out beyond Saturn. The asteroid belt marks the boundary between the *inner solar system* and the *outer solar system.*
- Millions of comets, swooping on highly elongated orbits into the inner solar system from beyond the outer planets. Some graze the sun; others never come within the orbit of Mars. Some return periodically; others visit once, then vanish forever into the vast *Oort cloud* of comets that marks the boundary of the Solar System.

Before we look at each of these worlds, both in the detail possible only with the great research telescopes and interplanetary probes, and in the more direct way possible only with an amateur telescope, it will be helpful to understand how the solar system came into being.

You can see solar systems forming elsewhere in the sky, in stellar nurseries such as the great nebula in Orion, or the Lagoon Nebula in Sagittarius (Fig. 13.5). They are bred in dense clouds of gas and dust, composed primarily of hydrogen, enriched by traces of other elements blown into space by dying stars. "Dense" may seem an exaggeration here; such clouds are thinner than the most rarefied vacuum attainable on earth. There is enough material in them, however, to cause even denser knots to form. Molecules of gas and dust, drawn together by the mutual attraction of their gravity, clump closely enough to overcome the cloud's own turbulence. Once such a lump forms, it attracts more matter, slowly, over eons, until a *proto-stellar nebula* forms. At the center of this nebula, a star is forming.

The cloud is collapsing toward its center, drawn by the ever increasing accumulation of matter there. But as it collapses it also spins. Along the equator of that spin, centrifugal force resists the inward pull of gravity, and matter falls in toward the center more slowly: the nebula flattens into a

13.5 The Lagoon Nebula in Sagittarius. (Official U.S. Navy photograph)

spinning disc. It is densest at its center, where 99% of its matter eventually falls. But out along the disc, other condensations form: *planetismals,* clumps of the matter in the nebula. They also spin, as they revolve around the nebula's center. Just where along the disc such clumps occur, how many and how large they are, depend on many factors still poorly understood. It appears that many combinations are possible.

At some point in the evolution of the system, the new star at the center flickers into life and slowly heats up. Eventually, it burns through the thick clouds that shroud it, flooding the disc with light and heat, blowing away the gas not yet gathered into the star or planetismals. These too lose their veils and stand revealed as spheres of gas, cold miniatures of the new star. Heavier elements, most commonly iron, settle at the center of each planetismal. Hydrogen, helium, and other light, gaseous elements hover at their surfaces. Water is abundant throughout the nebula, as are methane and ammonia.

On the planets nearest the sun, the light elements are heated so much that they quickly reach escape velocity. They evaporate, leaving heavier gases, like water vapor, around a sphere of light minerals with an iron center. Two other processes abet this heating and eventually strip the inner bodies of their atmosphere. One is the decay of radioactive forms of light elements, such as aluminum. Another is intense bombardment, as the smaller planetismals remaining in the disc collide with the larger. The surfaces of the planets heat up, boiling away the heavier gases. For a time, these small, rocky worlds lie naked to space. Soon, however, the heated minerals start to *outgas,* spilling gases like oxygen, nitrogen, and carbon dioxide from their interiors. Water vapor follows, and, if the conditions are right, condense into a rain that falls for millennia.

These are the terrestrial worlds. Mercury, a small world harshly heated, was too small and hot to retain the second atmosphere its rocks outgassed. Venus, heavier and not so hot, lost its first atmosphere, but kept the second. But this second atmosphere set off a disaster in the form of a runaway *greenhouse effect.* Carbon dioxide in a planetary atmosphere traps heat from the sun, turning the planet into a giant solar collector. Oceans, such as Earth's, have a terrific capacity to absorb carbon dioxide. Venus lacked enough water to form oceans, and so all of the carbon dioxide that outgassed from its rocks stayed free and absorbed the heat of the nearby sun. The result is an atmosphere many times denser than ours, with a temperature of 630 degrees Celsius (1170 degrees Fahrenheit), thick with sulphuric acid strong enough to etch metal.

Earth escaped the fate of Venus. At just the right distance from the sun, it kept its plentiful water, and the water remained liquid, forming oceans to absorb the large quantities of carbon dioxide in its atmosphere. In these benign conditions, life formed, generating oxygen. Oxygen made a radiation shield high in the atmosphere and sheltered the abundant, constantly evolving life below.

Mars suffered another fate. It was too small to retain its lighter gases, and so cold that its abundant water (it has more than Earth) froze permanently in its soil. It has only a thin atmosphere of carbon dioxide, so cold it freezes at the poles as dry ice. Eroded channels that look like river beds suggest that Mars may once have had abundant surface water, but that water remains locked below the surface as permafrost. Long-term climatic change could even release that water, and Mars may even support life. But today, the weather forecast is bleak; at noon, on the equator, the ground temperature may rise to the mid-30s Fahrenheit (about 0 degrees Celsius).

The asteroid belt is an enigma. Why did so much matter fail to condense into a single large planetismal? One theory holds that the gravitational pull of Jupiter stirred up the asteroids' orbits, preventing the formation of any larger body. If they have never formed a larger planet, the asteroids may store remnants of the solar system's original material, unaltered by the forces of planetary formation. The gas giants, too, tell us much about the early solar system. Here, far from the light and heat of the sun, the high masses of these planets were sufficient to hold their light gases. Their atmospheres are a sample of the solar nebula, only slightly changed by heat and chemical reactions.

The Jovian worlds are all atmosphere. If you were to try to land on the surface of Jupiter, for instance, the gases around you would get progressively thicker, until you arrived at a region too thick to penetrate. There is no distinct surface to land on. At the cores of the gas giants, the immense atmospheric pressure does strange things. Hydrogen becomes a solid metal, capable of conducting electricity; as the planet spins, the metallic hydrogen becomes a giant dynamo. This gives the gas giants extremely powerful magnetic fields, which cause great displays of aurorae in their atmospheres. Jupiter's mass is so great that it radiates more heat than it absorbs. This heat is generated by gravitational pressure, not the thermonuclear forces that drive the sun. If Jupiter had been much more massive, however, the solar system would have had not one, but two suns.

All of the gas giants share Jupiter's general form. Their masses are less; their surface temperatures are lower, too, and often their densities. Saturn, for instance, has an average density less than that of water; could you find an ocean of water large enough, Saturn would float in it. The atmospheric winds of the other gas giants tend to be less violent than Jupiter, and their electromagnetic fields are probably weaker. Most of the Jovian worlds share another intriguing quality, a high rate of rotation. Jupiter and Saturn spin so fast that centrifugal force* makes them visibly fatter at their equators.

Two other intriguing similarities of the Jovian planets are the rings that girdle at least three of these worlds, and their large flocks of moons. For centuries, Saturn was thought to be unique in the solar system for its set of thin, icy rings orbiting above its equator. In the past decade, however, astronomers have found rings around Jupiter and Uranus, and their presence is suspected around Neptune as well. Swarms of moons surround Jupiter and Saturn, and may soon be found around Uranus and Neptune.

Beyond Neptune, our knowledge evaporates. We know little about Pluto—its size is uncertain, and we have only the vaguest information about its composition. Its origins are fertile ground for speculation. We do know that it is much smaller than a Jovian planet, and seems to be much less dense than a terrestrial planet. More planets may lurk out in the darkness, but we could be decades stumbling on them, if they do exist. Space is too large, and sunlight reflects but dimly off surfaces so far from the sun, for any purposeful search to pay off. Searches from orbiting infrared telescopes hold the most promise, but observing time on such instruments will be scarce and expensive for many years to come—a major hindrance to a time-consuming search.

*Any physics major will tell you that "centrifugal force" is a "fictitious force." What we feel as an outward tug when we whirl a bucket on a rope or make a tight turn in a car is actually not an *outward* pull. As Newton once remarked, objects in motion tend to go in the same direction, unless some force turns them aside. When you swing a bucket, your arm keeps forcing the bucket to move in a circle. The bucket would much rather go off in a straight line tangent to that circle. The difference between that tangent and the circle creates the force we feel as "centrifugal." If the sun's gravity were turned off tomorrow, the planets would not move directly away from the sun. They would travel on straight lines tangent to their former orbits.

Beyond the planets lies the *Oort cloud,* where comets circle in slow orbits around the sun. These are the parts of the solar nebula least touched by the passage of time; heat has not melted them, nor does light brighter than starlight reach them. Only after long eons does the sun pass close to another star, and the orbits of the comets are disturbed, sending millions diving through the inner solar system at the rate of thousands a year. When the disturbance passes, most of these comets are swept out of the inner solar system by the whipcrack effect of a close passage by Jupiter. The number of surviving comets drops to the present dozen or so per year. Most of these are comets that were captured by the gravitational pull of the sun or Jupiter and swung into various elliptical orbits, taking 3 years to several centuries to complete one revolution. Others flash by once and return to the Oort cloud; some may even achieve escape velocity and leave the solar system forever, passing beyond the two light-year distance where the sun's gravitational field merges into those of other stars.

With this introduction to the solar system's past and present form and function, you can go on to the next chapters, which give detailed information about the planets. You will notice that all of the remaining chapters of this book are divided into two sections. The first describes the celestial object's physical characteristics. Many of the phenomena discussed under this heading are not directly visible to amateur observers. Your knowledge of them, however, will contribute to a better understanding of the features that you can see. The second section of each chapter gives information on how and what to observe. In the later chapters, this section concludes with a list of typical objects selected for the beginner.

The Moon

14

ASPECTS

An amateur astronomer needs a good understanding of the moon's motions, because the presence or absence of the moon in the night sky dominates your observing plans. The moon behaves more like an inferior than a superior planet, showing phases. Unlike an inferior planet, however, the moon can appear in opposition to the sun. The phases of the moon have their own traditional names (Fig. 14.1). At *new moon,* it is in conjunction with the sun, and the face turned toward the earth is unlit. At *first quarter,* the moon appears on the meridian at dusk, with its western half lit. *Full moon* occurs at opposition, the fully lit face of the moon rising at sunset and setting at dawn. At *last quarter,* the moon rises at midnight and transits at dawn, the eastern half of its face appearing lit.

The circuit from new moon to new moon is called a *lunation* and requires about one month to complete (Fig. 14.2). One lunation takes 29.5 days; this is also called the *synodic month.* Over the course of one lunation, the earth moves in its orbit, changing the angles between earth, moon, and sun. Because the time of new moon depends on this geometry, the synodic month is therefore slightly longer than the 27.32-day *sidereal month.* This is the length of one lunar orbit measured with respect to the stars. The difference is similar to that between a solar and sidereal day, and occurs for the same reason. The synodic month is the one we notice, because it is linked to the moon's phases.

From night to night, the moon's orbit takes it east across the celestial sphere. It covers about 6 hours of right ascension in a week, just under one

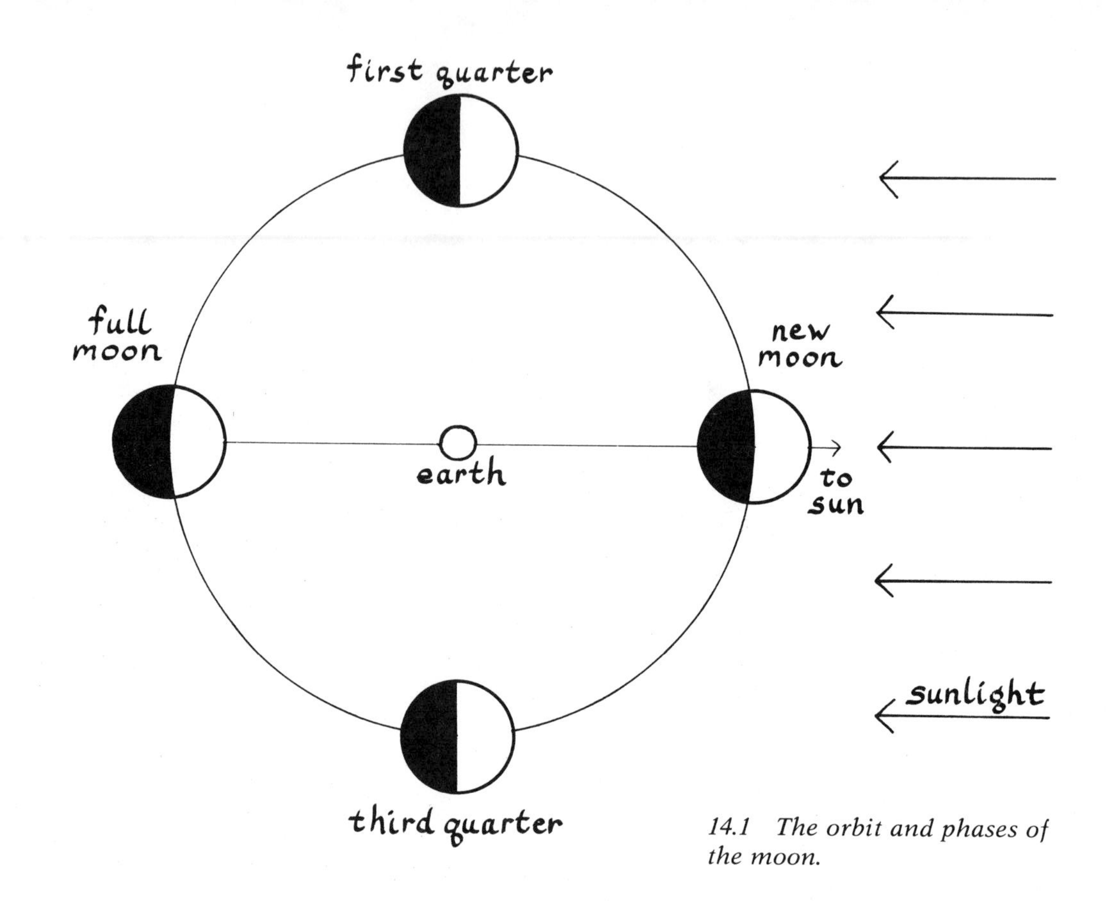

14.1 The orbit and phases of the moon.

hour of right ascension every night. The moon's orbit is inclined 5 degrees from the ecliptic. This is why eclipses don't occur at each new and full moon. That 5 degrees is ten times the moon's angular diameter, and gives it plenty of leeway to miss an eclipse. Straying 5 degrees from the ecliptic, (which, you'll recall, is titled at 23 degrees) the moon can range 28 degrees north and south in declination. The moon's phases appear in different parts of the sky at different times of year. In the summer, for instance, full moon appears in the southern constellations of the zodiac: Scorpius, Sagittarius, and Capricornus. Winter full moons ride high in the northerly constellations of Taurus, Gemini, and Cancer.

The moon's eastward motion causes moonrise and moonset to occur about an hour later each night. The precise interval between risings, however, varies widely. The angle at which the moon's orbit meets the eastern horizon determines how far below the horizon the moon travels along its orbit from one night to the next. If the angle is steep, the moon dives deeply below the horizon and will take longer to rise from one night to the next. If the angle is shallow, the moon can seem to rise at almost the same time for several successive nights. You are most likely to notice this effect around the autumnal equinox, when during the full moon its orbit slants very grad-

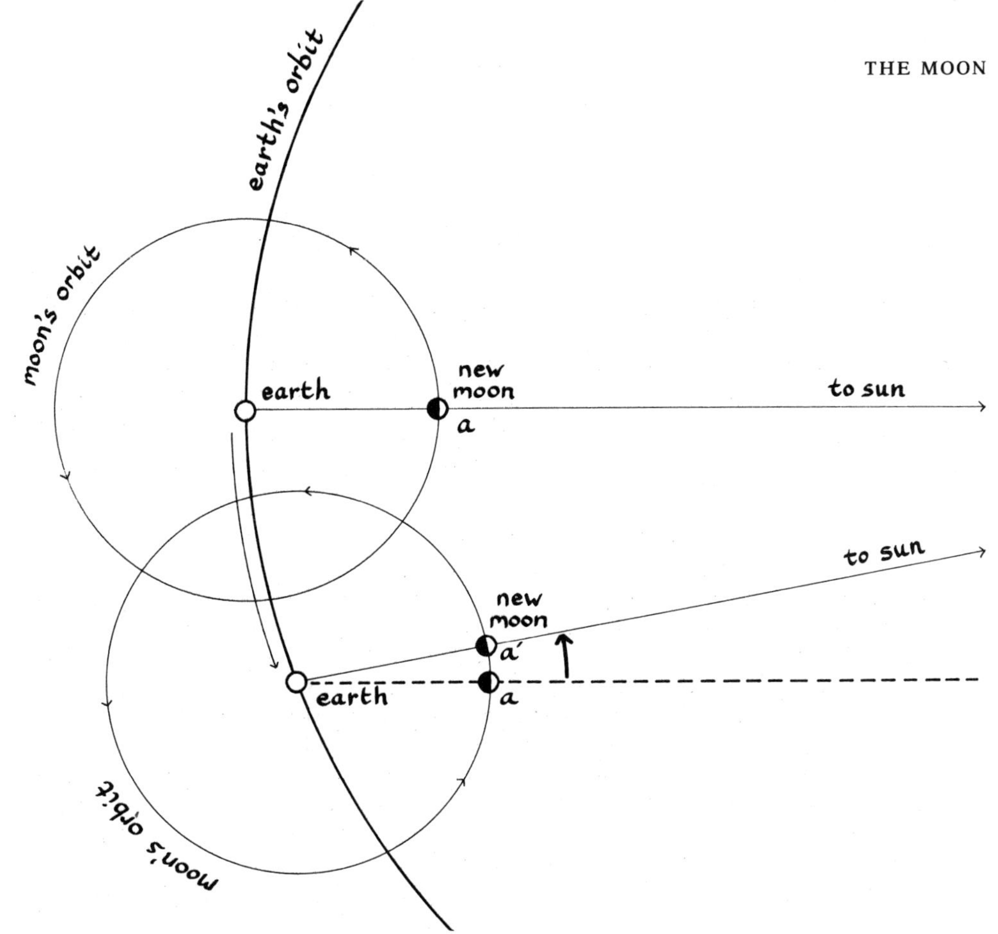

14.2 The synodic and sidereal months. Over the course of one month, the earth moves along its orbit, changing the angle between sun, earth, and the distant stars. A sidereal month measures the moon's orbit with respect to the stars; points (a) *mark the beginning and end of one sidereal month. To go from new moon to new moon, however, the moon must move beyond the second point* (a) *to* (a′), *making the synodic month longer than the sidereal.*

ually below the eastern horizon (Fig. 14.3). The *Harvest Moon* rises around sunset for several days, giving farmers a light to get their crops in.

The moon's orbit also precesses around the ecliptic, shifting the nodes between the ecliptic and its orbit in a slow, 18-year precessional cycle. Only if the moon passes through one of the nodes when it is at full or new phase will the moon pass between the earth and sun, causing a solar eclipse, or directly into the earth's shadow, causing a lunar eclipse. The precessional cycling of the lunar orbit causes eclipses to occur in a regular pattern, called the *Saros cycle.* Certain alignments of sun, moon, and earth recur regularly

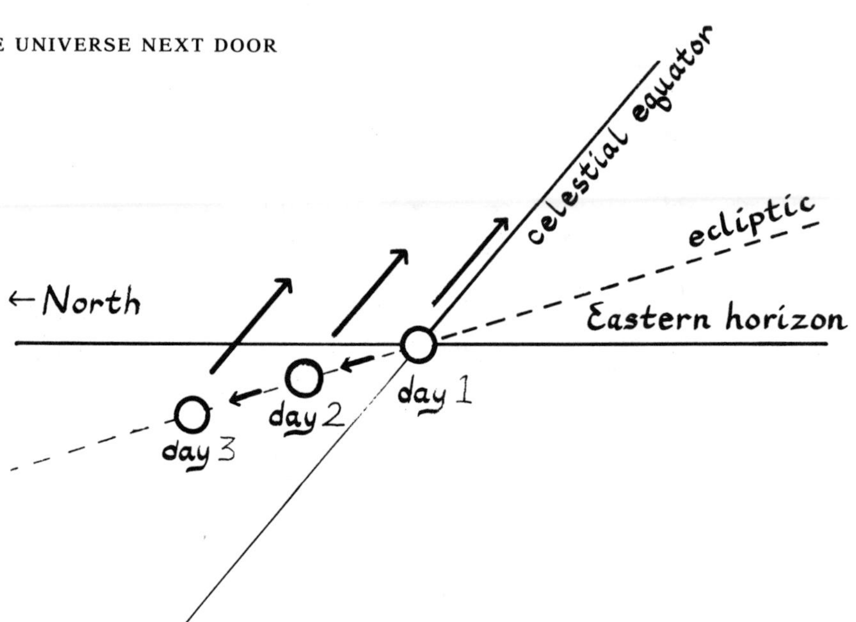

14.3 The Harvest Moon, showing the position of the moon relative to the eastern horizon, at sunset on three successive evenings around full moon. The shallow angle between the ecliptic and the horizon at this time causes the moon's eastward motion to carry it only slightly below the horizon from day to day. The moon rises only slightly later each evening.

over the course of many cycles, causing eclipses to recur in the same region of the sky and earth.

The moon always keeps the same face turned toward the earth. It does this not because, as you might think, it does not rotate, but because it rotates once each month. Try this for yourself by walking around a table. If you keep your face toward the table as you circle it, you will notice that you also rotate once for every revolution. This situation is not uncommon in the solar system, and is called a *spin-orbital lock.*

The earth-moon lock is not absolutely rigid. The moon rotates at a perfectly steady speed, but because its orbit is not a perfect circle, its orbital speed varies. Near perigee, for instance, the moon travels more quickly along its orbit, and its rotational speed lags slightly behind the amount required to keep its trailing edge turned away from us. We see as much as 7 degrees 45 minutes of lunar longitude into the far side. The opposite occurs near apogee. This slight apparent wobbling of the moon is called *libration,* and occurs in latitude as well as longitude. Libration in latitude is caused by the moon's axial tilt (similar to the earth's axial tilt, which causes the seasons). The moon's axis inclines to its orbit by 41 minutes, which allows us to see 6.68 degrees of lunar latitude beyond the moon's north and south poles over the course of each month. Altogether, libration makes 60% of the moon's total surface visible from the earth.

You probably know that the moon raises tides on the earth by the tug of its gravity on the ocean's waters, but exactly how this works is worth knowing. Tides are common throughout the universe; they have helped to shape the surfaces of other bodies in the solar system and occur on the surfaces of stars as well, since the gases the stars are made of act like fluid. The moon's gravity pulls on all of the earth, pulling harder on the nearer side. Water, being fluid, resists the tug least, and reaches toward the moon; this is a high tide. But there are two high tides, one on the side of the earth nearest the moon and one on the opposite side. Where does the opposite high tide come from?

The mystery clears up when we consider what actually happens to two bodies, such as the earth and moon, in orbit. Notice: not "one body in orbit around the other" but "two bodies in orbit." The moon does not orbit the earth, nor do the planets orbit the sun. Rather, two bodies in orbit revolve around their common center of gravity, or *barycenter.* Imagine a dumbbell being twirled like a baton: the point it twirls about is the dumbbell's center of gravity. If one end of the dumbbell is much heavier than the other, the center of gravity shifts toward the heavier end. But even if one end is 1,000 times heavier than the other, the center of gravity will still not be exactly at the center of the heavier mass but displaced slightly toward the light end. As the dumbbell twirls, the heavier end will not rotate in place, but will perform a tight orbit around the dumbbell's center of gravity.

This is the situation with the earth and moon. The earth is the heavier end of the dumbbell, and as the earth-moon system spins around its barycenter, the earth performs a tight circle of its own. The side away from the moon is on the outside of that orbit. The centrifugal effect generated by the earth's motion around the barycenter causes the ocean on the far side to bulge away from the barycenter, raising the second high tide. (The water of the ocean tends to keep moving in a straight line, not in the circular path of the earth-moon rotation, and so the ocean bulges.) The crust of the earth itself rises slightly, too little to notice without sensitive instruments but perhaps enough to play a role in triggering earthquakes.

The moon, too, has tides. On the moon, however, the only large fluid mass is its molten interior. Tidal stress on this mass of liquid rock may explain the large quantities of lava that have poured out of the moon's interior on its earthward side (it is these lava fields that form "the man in the moon"). Tidal motion causes friction: friction between sea and sea floor, between atmosphere and surface—even between the sections of the earth's crust. This friction has slowed the earth's rotation by more than one-third since life first appeared on its surface. The energy taken from earth's orbit in this process must go somewhere. Some of it dissipates as heat, but some also transfers to the moon, boosting it into a higher orbit. As these effects go on, the moon will orbit the earth more slowly (in obedience to Kepler's laws), its orbital distance will increase, and the earth will slowly spin down into a mutual orbital lock. Earth and moon will someday stare fixedly at each other, across a widening space.

Physical Properties

Orbital radius: 384,404 kilometers (237,000 miles)
Diameter: 3,476 kilometers (2,145 miles)
Mass: 7.35×10^{21} kilograms (8.08×10^{18} tons)

The moon's mass is less than one eightieth (1.23%) of the earth's. Its diameter, however, is fully a quarter of earth's 12,756 kilometers. That large fraction is enough to make the moon unusual in the solar system. Except for the Pluto-Charon pair, our moon is larger, relative to the planet it circles, than any other satellite in the solar system.

The moon is made of rocks essentially similar to those found on earth's surface. Some slight chemical differences tell of an early history different from our planet's. Like most of the terrestrial planets, the moon's interior is layered, like an onion. The outer shell, the *crust,* is about 60 kilometers (37 miles) thick. It is cold, and relatively stable. Lightweight rocks, immediately familiar to terrestrial geologists, compose the bulk of the crust. Deeper, in the *mantle,* heavier rock, apparently similar to the constituents of earth's own molten mantle, extends to a depth of 800 kilometers (about 500 miles). Moonquakes do occur, and tend to be centered deep in the mantle. The moon's core is a mystery to us. If it is like the earth's it is molten iron. It probably still retains some of the terrific heat in which it formed.

The moon's history is typical of the inner worlds. About 4.6 billion years ago, the moon was born as a planetismal condensing out of the solar nebula. We still do not know whether it formed as part of the earth and then split off, or condensed nearby the earth, or formed elsewhere and was later captured in earth's gravitational field. Like the rest of the terrestrial planets, it underwent early bombardment and melting, which caused large lava flows still visible today with the naked eye.

After the heaviest bombardment ended, approximately 4 billion years ago, radioactive elements began to melt the moon's interior, and lava flowed out through the shattered surface, filling in the largest scars of the bombardment. The floods of molten rock lasted half a billion years. In the 3.5 billion years since these seas cooled and hardened, the moon has been almost unchanged, down to the very pebbles strewn across its plains. Each year, slightly more than 100 rocks weighing from a few ounces to a ton strike the surface. Dust and ashes rain in steadily, as they do on all the worlds, but the changes these small impacts wreak on the lunar surface are miniscule, compared to the whole.

One lunar day is the same length as a synodic month. The sun lights any given point on the surface for over 14 earth days, and then a darkness colder than any Antarctic night descends for another 14 days. Because there is no atmosphere, the temperature range on the surface is extreme: in full sun, rocks bake in over 100 degrees Celsius (212 degrees Fahrenheit); in shadow (either nightside or behind any boulder), the temperature dives to −173 degrees Celsius (−279 degrees Fahrenheit).

Without any air or water, with a crust that never shifts as earth's does, the surface of the moon is changeless. A rock that fell atop a boulder 3 billion years ago still balances there today; the tracks of the astronauts, their lunar landers and rovers, will remain sharp and clear in the lunar dust for millions of years. Since it is changeless, the moon's surface is a clear visual record of its history (Fig. 14.4). The casual visual observer can easily gauge the relative ages of impact craters, lava flows, and mountains by their relative positions and the amount of impact damage they have sustained. For instance, a small, sharp-edged crater breaking up the outline of a larger one must have formed later; any surface free from large craters must have solidified late in the moon's chaotic early history. As we observe the moon through its phases, we can see direct evidence of world-forming forces, silent now for billions of years.

14.4 Craters Theophillus, Cyrillus, and Catharina. Younger craters break the outlines of older. (Lick Observatory photograph)

14.5 The northern limb of the moon as it rounds toward first quarter. This is similar to the view at moderate powers in a 6" telescope. (Official U.S. Navy photograph)

A variety of features confronts the telescopic observer. Looking at the disc, the largest phenomena you will observe are the disc itself and its division (except at full or new) into a sunlit and a shadowed area (Fig. 14.5). The edge of any celestial body is called its *limb* and is usually identified by a compass direction, for example, "north limb." Any rotating body, such as a planet, has its own set of poles, labeled north and south like earth's; the north pole is usually the one lying on the north side of the ecliptic. A potentially confusing convention has long governed the orientation of lunar maps; they depict the moon upside down and backward, as it appears in a telescope. To match such a lunar chart with the naked eye view of the moon, you must hold the chart upside down. More recent charts often print the moon right-side up; know which you're dealing with before you go out to observe.

The line between the sunlit and dark areas of a planet's surface is called the *terminator*. Between new and full, the terminator moves east across the lunar disc. The region near the terminator is marked by long, dramatic shadows, or bright specks of mountain peaks catching the first rays of day.

The largest features visible are the *maria* (Latin for "seas"; the singular is *Mare*). There are 14 maria on the nearside, and each marks the spot where long ago a large asteroid crashed into the lunar crust (Fig. 14.6). Maria have smooth surfaces because of the lava that filled in the impact scar during the second great melting. Most of them are round, like giant craters. Some are indented by smaller scars, called *Sinii* ("bays"; sing. *Sinus*). The maria are the youngest regions of the lunar surface, but they are not entirely free from the scars of later impacts. Craters pock their surfaces.

The *walled plains* and *craters* are the next features you will see. Countless numbers of these circular pits, from microscopic to 150 miles across,

14.6 Central region of the moon's limb, before first quarter, showing maria as dark circular regions. (Official U.S. Navy photograph)

cover the surface. Because no internal forces change the lunar surface, and no wind or water erode it, once a crater forms, it remains. In a given region, as time passes, more and more craters appear. Because the rate of cratering is the same over the entire lunar surface, the number of craters in a particular region will tell you how much time has passed since that region formed. The more cratered the surface, the longer it has been exposed to the rain of meteorites.

Walled plains and craters are distinguished by their size and structure (Fig. 14.7). Walled plains are the largest craters. They are comparatively rare and most show signs of great age. Their walls have crumbled, and in places are completely broken by later, smaller craters. Craters are smaller, more recent and numerous. Their walls rise high above the surrounding plains. The crater floor, however, is usually sunken below "sea level," making the wall's inner face steeper and higher.

Some crater walls rise almost 2 miles above their floors, falling in steep terraces to a smooth, circular plain dozens of miles across. In the middle of this plain, often a single mountain peak will rise thousands of feet above the level waste. The shape of a crater is the frozen "splash" of a meteorite in rock. Below 10 miles in diameter, small craters and *craterlets* proliferate. These usually lack the dramatic structure of their larger kin, appearing mainly as shallow bowls, with rims pitched higher than the countryside. Unusual shapes, representing oblique angles of impact, occasionally appear in this size-range.

The third major class of surface feature is the *mountain range* (Fig. 14.8). On earth, mountains form where vast areas of the crust collide. Lunar mountain ranges usually occur in telltale arcs, which mark them as rem-

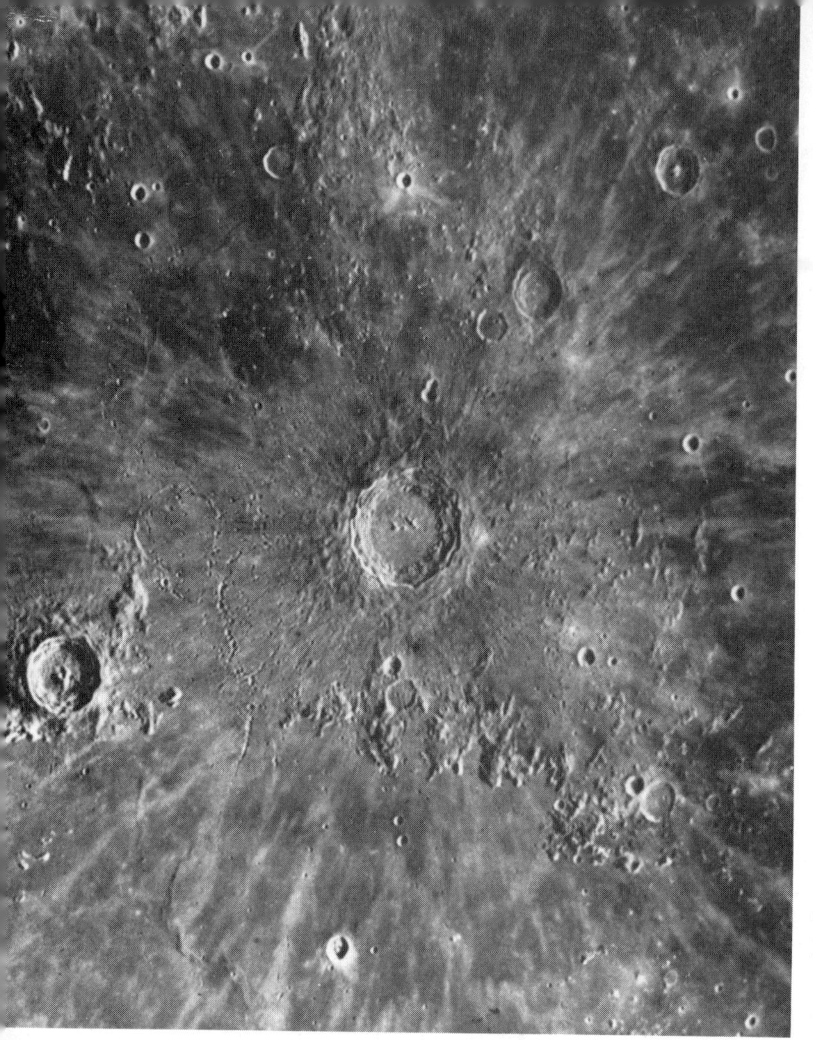

14.7 Crater Copernicus
(Lick Observatory photograph)

nants of old impact scars. They are the ramparts splashed up by the formation of the maria. The lava flows that filled the seas left only these isolated arcs, standing 5 miles above the surrounding lava plain. Isolated peaks also occur, standing above the maria like the great buttes and granite towers of the Southwest—only several times their height.

Another kind of terrain covers large areas of the surface, especially in the moon's southern hemisphere. These are the lunar *highlands*, a chaotic, ancient land of craters piled upon craters (Fig. 14.9). Here the oldest surface rocks are found. Spared the cataclysmic impacts that formed the maria, the highlands have survived from the end of the first major melting. Here, in the myriad craters of every size, from the great ruin of walled plain Clavius, to the bright, fresh ring of crater Tycho, you can see the result of 4 billion years of uninterrupted impact.

14.8 The Lunar Alps (Lick Observatory photograph)

The southern highlands also contain a fascinating array of other, smaller features found over the entire surface (Fig. 14.10). Valleys—too sharp and straight to be the work of water—cut deep gorges across the rugged terrain. Smaller, sometimes meandering crevasses, called *rilles,* are apparently scars of ancient lava flows. Great cracks and sharp, straight cliffs cut across the maria, evidently caused by contraction as the hot lava cooled and shrank. Finally, around the time of full moon you will notice bright, linear markings radiating from some of the large, fresh craters. Tycho, especially, shows a burst of these *rays,* some of which stretch clear across the lunar disc. These are apparently light-colored debris pitched out of the moon's crust by the impact that formed the crater at their center.

The rays are about the only features that show best under the vertical illumination cast by the sun on the moon's surface at full phase. Most other

14.9 Craters Clavius (top center) *and Tycho* (below and right) (Lick Observatory photograph)

features show best when their shadows lend contrast to the soft grays of the rocky highlands and lava plains. You will find that craters and mountains, valleys, cliffs, and rilles show best when the terminator lies near them. In the low light, their shadows stretch for miles around them, and mountain peaks stand brightly limned against the dark.

OBSERVING

Every month, the first challenge to the lunar observer is finding the new moon. There is actually some competition among amateurs for the record sighting, which is now around 14 hours after new. To come close to this figure, you need an absolutely flat western horizon and an elevated observing site. Binoculars help, too. Pay attention to the angle between the moon's or-

14.10 The Aridaeus rille (Lick Observatory photograph)

bit and the ecliptic, and the ecliptic and the equator. Your best chances are when both of these angles are high. At the equinox, when the moon has passed through the ascending node of its orbit, sunset will find the moon placed directly above the setting sun, at the highest possible altitude.

The very young moon is a beautiful sight, especially when, as often happens, the dark portion of the sphere is lit by the phenomenon long known as "the old moon in the new moon's arms." You may have noticed this yourself before now, during the early phases of a lunation—the thin sliver of sunlit moon encloses a pearly gray, sometimes coppery circle, faintly lit, and blotched dimly by the maria. The source of this illumination is earthshine: sunlight glancing off the clouds and ice-fields of the earth.

After the challenge of the youngest moon, lunar observation requires a telescope. Little extra equipment is necessary, but some extras you may want include a set of filters and a detailed lunar map. A good lunar chart is a must for extensive lunar observation. The Bibliography lists several sources for maps. The considerations in choosing lunar charts are similar to those for star charts: convenient size and scale, clarity and completeness of display, and so forth. I use two most often. One is a small set of photographs in Menzel's *Field Guide*. They are convenient to use at the telescope, and the photographic format matches the view in the eyepiece better than some hand-drawn charts (although for many observers this is a matter of taste). I also refer very often to a very large wall chart (almost life-sized, it seems at times), published by the Hammond company.

Filtering for lunar observation is primarily a matter of cutting glare. Neutral density and polarizing filters installed at the eyepiece are both useful. Mylar solar filters often come in two layers; you can use one of them alone as a moon filter, especially when observing with a fast system around the time of full moon. Colored filters can enhance whatever hues there are (and they are there, as the Apollo program photographs demonstrate) on the largely monochrome lunar surface. Some amateurs use red filters to look for "transient lunar phenomena." These are rare reports of fugitive gleams, most likely caused by outgassing from the lunar interior, but thought by some to represent volcanic activity.

The photographs with this section (Figs. 14.11, 14.12, 14.13, 14.14) have been labeled with some of the most prominent features, those that typify the major classes outlined above, and those that are unique on the lunar surface. What these photographs show are merely the highlights—distinctive features you can use as jumping-off points for extended trips into the countryside. You can devote many moons to acquiring a fuller knowledge of what lies a quarter of a million miles overhead, by observing directly, and reading further. The Bibliography lists a few texts devoted entirely to the moon, for amateur observation or for a purely descriptive interest. The observer's most important single text to study, however, is the moon itself. How you proceed depends on your own sense of where your interests lie, how your memory works, and how much energy you wish to devote to the purpose.

Patrick Moore, the dean of amateur astronomers, gave over a year to his first intensive study of the moon, memorizing every crater, rille, and peak he could see through a 3-inch telescope. He drew sketches prepared in outline from standard charts and photographs, and then filled in at the telescope, as a way of imprinting what he saw upon his memory. You may find a similar technique useful. Other methods involve observing the changing appearance of given features at different sun angles (and sketching them, if you're so inclined, although verbal notes will also do); extensive measurements of the heights of lunar features (for this, you may wish to look up precise formulae and methods—see C. P. Sherrod's text, in the Bibliography); and extensive crater counts, leading to comparative age determinations.

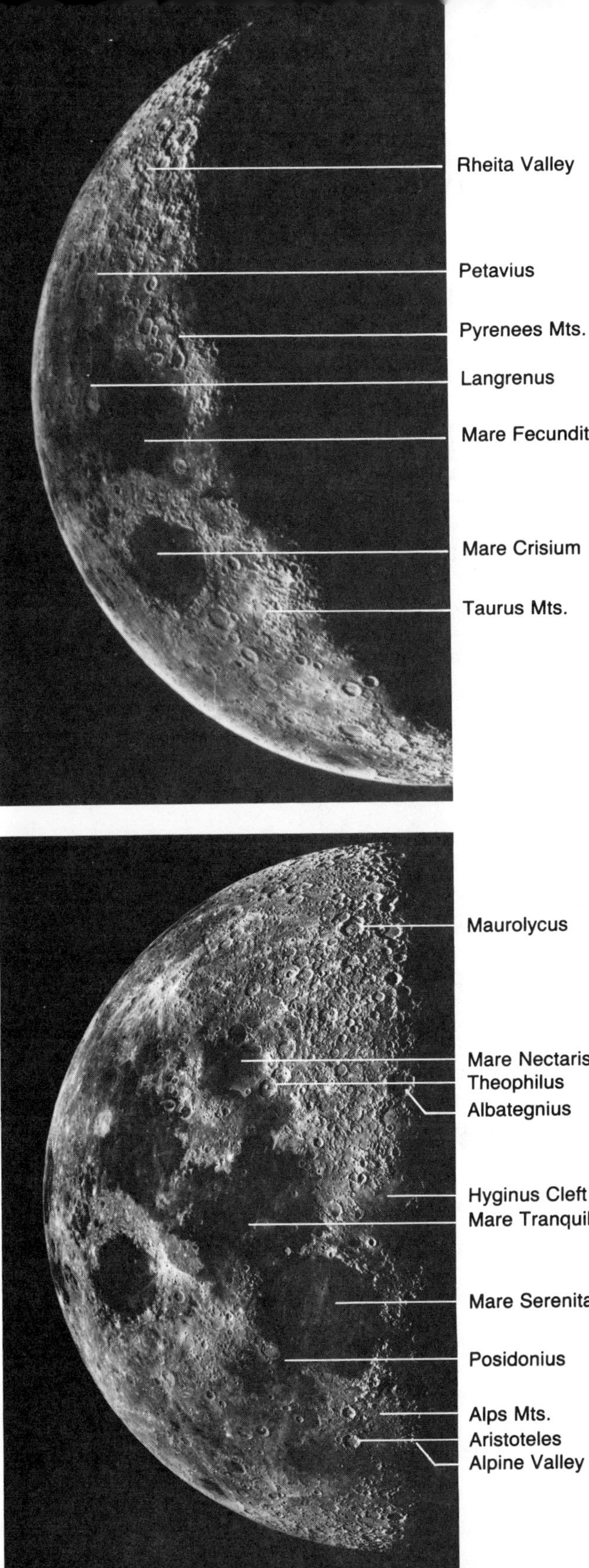

14.11 The moon: 4 days old. South is up. (Lick Observatory photograph)

14.12 The moon: 7 days old. South is up. (Lick Observatory photograph)

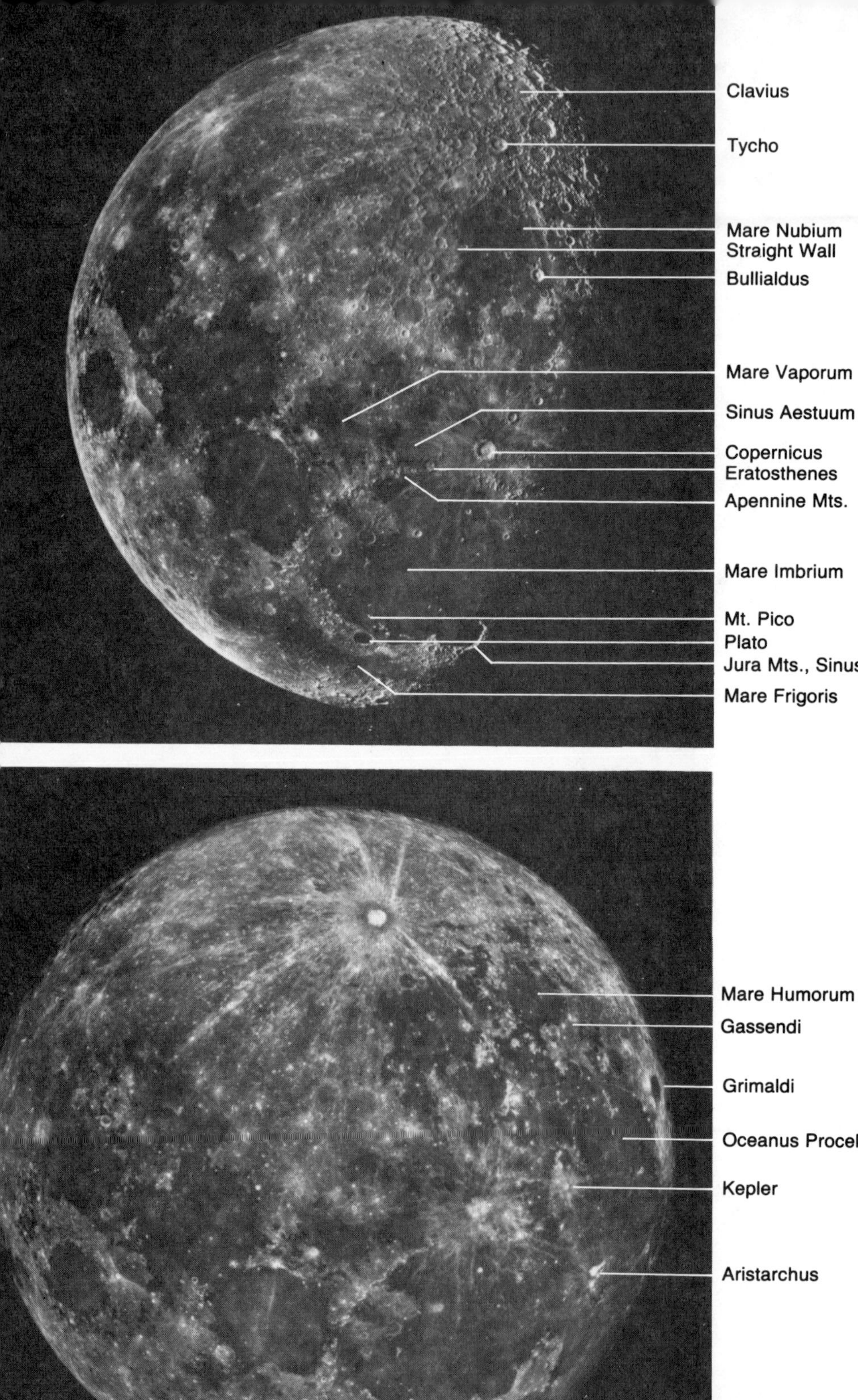

14.13 The moon: 10 days old. South is up. (Lick Observatory photograph)

14.14 The moon: full phase. South is up. (Lick Observatory photograph)

If none of these methods appeals to you, involving as they do an extra burden of equipment, both material and mental, you may simply want to wander at will among the scenery. This last method may actually require the most discipline. To keep at it, steadily, until the lunar landscape becomes familiar to you, over a variety of powers and phases, without some specific task to keep you engaged, you need an open, receptive mind, immunized against boredom by a speculative interest in everything. The ability to ask yourself questions—How old is that crater? Why is this rille broken? What do those ridges on the maria floor mean? What is the cause of this light stippling?—about everything you see, posing those questions in terms of what you already know about the moon and going to more advanced texts to seek their answers is the key here. The longer you continue such a process, observing, questioning, and learning, the more you will come to understand the moon—intimately, with the intuitive feel for the subject that books alone cannot supply.

Occultations and Eclipses

15

The moon is centrally involved in a peculiar kind of celestial event, in which one body passes between an observer and a more distant body. These events are peculiar, in part, for being instantaneous: much of what we see in the sky is (in human terms, at least) eternal. These events happen in a flash, and are often visible only from remote, restricted regions of the earth. This elusiveness makes them uniquely fascinating to many observers, who have formed a gypsy subculture among amateurs, pursuing their interest to the corners of the globe. The two most common types of this event are *lunar occultations* and *eclipses.* You are probably familiar with the latter, which are strikingly apparent when and wherever they occur. An occultation, on the other hand, usually requires a telescope to appreciate. It is the passage of the moon, as it moves along its orbit, in front of a star or planet.

OCCULTATIONS

Occultations of stars are the most common, and because a number of bright stars (Regulus, Spica, Antares) are within 5 degrees of the ecliptic, occultations bright enough to observe easily are frequent. More interesting are the moon's occasional passages through one of the star clusters such as Praesepe or the Hyades that dot its path. Occultations of planets, when the planetary satellites, rings, disc, and so forth disappear abruptly behind the moon's rugged limb, offer a fascinating view of a planet appearing to swoop in on the lunar terrain. Most exciting of all are *grazing occultations,* when the moon's northern or southern limb brushes by a star or planet. The bright, distant star or planet blinks in and out among the mountain peaks silhouetted on the moon's extreme edge.

All occultations are visible only over a limited area of earth's surface, called an *occultation path;* the path for a grazing occultation is especially limited. The astronomy magazines publish advance notices of such events, with times and maps for each; *Sky & Telescope* runs a complete forecast of occultations for the coming year each January; the *Observer's Handbook* does the same in each edition.

In the past, the precise timing of an occultation has been valuable for refining measurements of the lunar orbit. Hundreds of amateurs would turn out along the path of such an event, tuned to radio time signals with stopwatches, tape recorders, and clipboards to mark the exact disappearance of the star. Now, with laser ranging devices in place on the lunar surface, the moon's position is available to within a matter of inches, without waiting for an occultation. But the pleasure in watching an occultation can be simply in watching the unusual encounter of two far-distant worlds, of being able to watch the slow turning of the solar system's wheels. Especially pleasing is the disappearance of a double star. As the brighter star of the pair disappears, a dim, close, companion star can become momentarily visible, and then it too is gone.

Whether you watch for pleasure or for science, there are a number of things to look out for. First, you need to know the path of the occultation. Ordinary occultations will usually be visible either north or south of a line arcing hundreds of miles across the earth; in a grazing occulation, visibility is restricted to within a few miles of such a line. The visibility of a given occultation also depends on whether the limb at which it occurs is light or dark: a dark-limb occultation will be easier to see than one that happens along a sunlit edge of the disc. An occultation will occur in several stages. For a star, there are two: *immersion,* the moment the star disappears behind the moon; and *emersion,* the instant of reappearance. Planets in occultation progress through four stages: *first contact,* when the moon encroaches on the near edge of the planetary disc; *second contact,* when the planet is completely immersed; *third contact,* when the leading edge of the planet reappears; *fourth contact,* when the planet is entirely clear of the moon.

A particular kind of occultation has an important scientific value. The planets, and also the asteroids, can also occult distant stars. When they do, astronomers pay attention; these are rare and valuable opportunities to learn more about the atmospheres, satellites, and physical characteristics of the distant members of our solar system. The rings of Uranus were discovered during such an event, as the light of the occulted star winked off repeatedly *before* and *after* the planet itself dimmed it for an instant. The stunned observers, 8 miles up aboard NASA's Kuiper Airborne Observatory, were astonished to find that they had made the first discovery of planetary rings since Galileo noticed "handles" around Saturn. Even more recently, an occultation of a star by an asteroid has suggested that asteroids may travel in close pairs, or have moons. In the near future, watch for announcements of upcoming asteroidal occultations. The *Observer's Handbook* for 1983 had an excellent, detailed discussion of them. In addition to forecasting forthcoming occultations, the annual *Handbook* also lists addresses where you may send any results of your own observations. Because the occulting bodies are small, the paths from which these events are visible are also small. If you find yourself near an occultation path, make the effort to observe. You may have the rare privilege of discovering a new body in our solar system.

The key to such a discovery is to watch closely for at least several minutes before and after the published time of occultation. Calculation of an orbit, especially an asteroid's, is usually so inexact as to leave some doubt as to the time of the occultation itself. But you should also allow yourself enough margin to be watching if any secondary object—a close companion such as a planetary ring or an asteroidal moon—occults the star ahead of or behind the primary. Because such a companion may be to one side of the primary in its orbit, you should also observe even if you are not exactly on the predicted occultation path. In such a case, *any* sign of occultation, no matter how brief, is important: either the primary's orbit needs revising or you've found an unknown companion.

The whole subject of asteroidal companions is still so controversial that any datum is valuable; if an opportunity comes up, try not to miss it. The width of an asteroidal occultation path is so small that it's not unlikely that you could be the only one to report it. It will be helpful to enlist a more experienced observer when timing such an event; if you can't, the experience you gather now, before the opportunity comes up, will be invaluable—so start practicing on the next lunar/stellar occultation visible from your area. To learn more about the observation of occultations, you may want to contact the International Occultation Timing Organization, at P.O. Box 596, Tinley Park, IL 60477. This organization publishes newsletters and bulletins, giving instructions and detailed predictions for occultation-observers.

ECLIPSES

Eclipses are less fraught with individual responsibility (you're never alone at an eclipse), and a good deal more exciting. There are, of course, two kinds of eclipses, lunar and solar. An eclipse is simply a straight alignment of three bodies, in which the middle blocks the line of sight from the first to the third. When these three are the sun, moon, and earth, the eclipsing body will cast a shadow over the face of the third. In a *lunar eclipse,* the shadow of the earth covers the lunar disc, darkening it considerably but not completely. In a *solar eclipse,* the moon's shadow passes over the face of the earth, darkening a narrow path as the moon passes in its orbit high over earth's turning surface.

Lunar eclipses occur only at full moon, when the sun and moon are on opposite sides of the earth. Solar eclipses happen only when the moon is new, lying between earth and the sun. This does not mean that such eclipses happen twice a month, however. As mentioned before, the moon's orbit does not lie exactly on the ecliptic; only when the moon is on a node of its orbit at new or full, lying on the ecliptic (which is how that great circle on the sky got its name), can an eclipse occur. Such alignments happen usually twice a year at new moon, and about twice a year at full, giving an average two solar and two lunar eclipses per year.

Eclipses of either sort can be *total eclipses* or *partial eclipses.* In a total lunar eclipse, the entire face of the moon is obscured by the darker, central region of the earth's shadow; in a partial eclipse, part of the moon's face remains unobscured. Likewise for a solar eclipse, in which the moon can cover all or only part of the sun's disc.

Lunar eclipses are visible from a much larger area of the earth's surface than solar eclipses, so they are visible from your area far more often. A lunar eclipse is more likely to be total as well. At the moon's orbit, the earth's shadow is almost 6,000 miles across, considerably wider than the moon. The moon can miss dead center of the earth's shadow by several thousand miles and still be entirely obscured. The requirements for a total solar eclipse are more precise. The moon and sun have almost the same angular diameter: if the moon is only slightly off-center on the solar disc, the eclipse will be only partial. Moreover, the shadow cast by the moon in a solar eclipse is only a few dozen miles wide, although it may be travel thousands of miles across the earth. You must be within that shadow to see a solar eclipse as total. A lunar eclipse is visible anywhere on earth that the moon happens to be visible at that time.

In each kind of eclipse, the amount of darkening depends on the special characteristics of shadows cast by the extended surface of the sun. Only a point source of light casts a simple shadow. When the light source has an extensive width, it is possible for an eclipsing body to block only part of the disc, cutting off some, but not all of the light. The outer region of the shadow, which is only partially darkened, is called the *penumbra.* When the moon passes through earth's penumbra, only a very slight darkening of the lunar disc occurs—so slight as to escape notice by the casual observer. Only when the moon passes into the smaller central cone of the fully darkened shadow, called the *umbra,* does the moon darken dramatically.

In a solar eclipse, it is easier to tell if you are in the moon's penumbra; a moon-shaped chunk seems to have been bitten out of the sun. The area of earth's surface that lies within the moon's penumbra during a solar eclipse is always vastly larger than the small strip shaded by the umbra. The precise timing and location of any eclipse is determined by the Saros cycle, mentioned above. The orbits of the earth and moon are known accurately enough to allow predictions of eclipses centuries in advance (Fig. 15.1).

Each kind of eclipse has its distinctive phenomena. A lunar eclipse starts with *first penumbral contact* as the leading edge of the moon's disc enters the earth's penumbra. The moon dims only slightly in this stage, until

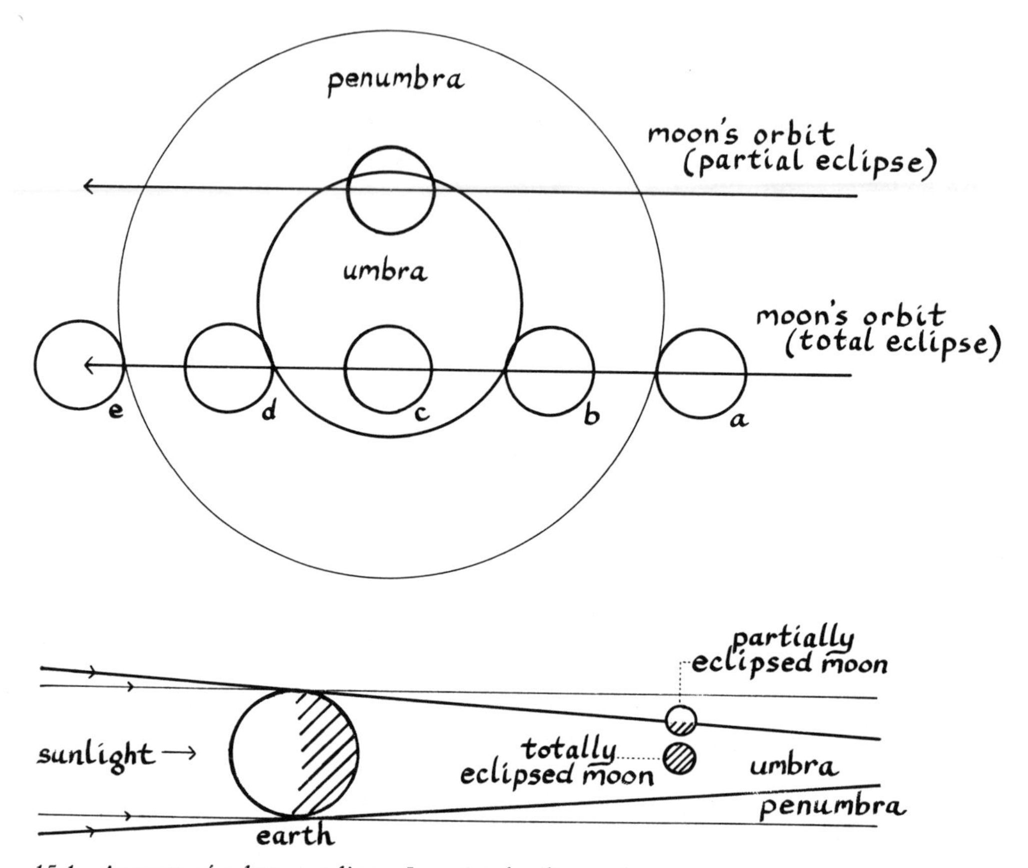

15.1 Aspects of a lunar eclipse. In a total eclipse, the moon passes entirely into the umbra of earth's shadow. (a) *first penumbral contact, beginning of eclipse;* (b) *first umbral contact, beginning of visible eclipse;* (c) *mid-eclipse;* (d) *fourth umbral contact, end of visible eclipse;* (e) *fourth penumbral contact, end of eclipse. In a partial eclipse, the moon skims the northern (shown) or southern edge of the umbra, darkening its southern or northern hemispheres, respectively. The bottom of the diagram shows a side view (not to scale).*

first umbral contact. You will notice the penumbral phase of the eclipse usually by a subtle change in the quality of the moon's illumination, rather than by a vivid change in its brightness. The shaded portion of the surface may seem dull, ashen or leaden. You won't likely see it as distinctly darkened until it actually enters the umbra. At that point, somewhere on the leading (eastern) limb of the moon, a crescent shadow with an arc roughly three times the moon's radius will bite in. For about an hour (unless the eclipse is partial), the shadow will pass from east to west across the lunar disc, until, at *second umbral contact* the whole will be obscured. During this process, you can watch through a telescope as individual craters fall into shadow; some amateurs keep precise timings of these events, and relay them to *Sky & Telescope.* A lunar eclipse is generally a naked-eye event, however. They are best enjoyed from a blanket spread on the ground, preferably with a friend or two.

When the eclipse reaches *totality,* between second and third umbral contact, the moon lies entirely within earth's umbra. The moon will not disappear entirely, because even at its darkest, earth's shadow contains some sunlight, refracted around its disc by the atmosphere. This light is generally reddish, like a dim, dark sunset, and for the same reason: only the longest wavelengths of light escape the scattering dust and moisture of our air. There is a great deal of variation, though, in the dimness and color of a totally eclipsed moon. The exact appearance depends on the state of earth's atmosphere. During the total eclipse of July 1982, for instance, the moon was so dark it was difficult to find in the sky. A large cloud of volcanic ash in the air, launched by a volcano in Mexico, may have been responsible. Ordinary climatic change can cause the darkness and color of an eclipse to range from bright, brick red to dull gray, and occasionally to blue.

As long as there are great unpopulated waste areas of earth's surface, solar eclipses will statistically tend to occur far from civilization. If you are not the wandering kind, this can limit the number of total solar eclipses you see to none. There will be none visible from the North American continent during the rest of the 1980s. The spectacular variety of phenomena visible during a solar eclipse, however, has convinced many amateurs to pack up their telescopes and travel across the earth in their pursuit.

The progress of a solar eclipse is much like that of a lunar. At first contact, the moon's disc notches the western limb of the sun, moving inexorably across it until only a thin crescent remains. If you are off the path of totality, the solar crescent will reach a minimum width, and then widen. Even this much is a wonderfully eerie experience, as the quality of light around you changes. The shadows of trees are studded with dozens of crescent images of the sun, formed by tiny holes among the leaves that act as pinhole cameras (for more on this, see below).

If the eclipse is total, in the last minutes before second contact, events happen quickly. From the west, the moon's shadow rushes upon you, preceded by the *shadow bands,* rippling waves of light and dark. At the last moment before totality, *the Diamond Ring effect* flashes from the eastern limb of the merging solar and lunar discs as the last slice of sun passes behind the moon (Fig. 15.2). Then *Baily's Beads,* glimpses of the sun's red outer layer, blaze briefly through the valleys and high mountain passes of the moon's limb: second contact.

The sky is dark. The brighter stars appear; planets string out along the ecliptic. The dark lunar disc is surrounded by the thin red halo of the sun's atmosphere. The sky for several degrees around is lit by the silver iridescence of the sun's corona. At times of heightened solar activity, the corona will appear almost like the lines of force around a bar magnet; at others, it is

15.2 The sun in eclipse, showing solar corona. (Lick Observatory photograph)

shrunken, almost featureless. An active sun will also show loops and spikes of brilliant red hedging the moon's limb; these are solar prominences, world-sized surges of solar matter leaping tens of thousands of miles into space. The air grows chill, and a hush settles (broken only by whirr and clack of motorized cameras); birds twitter uneasily; night-calling insects may start tuning up. Too late: third contact is upon them, and the western limb is brightening: Baily's Beads, the Diamond Ring, the shadow bands pass over the terrain, and the shadow races away into the east.

The length of totality in a solar eclipse depends not only on how accurately the moon passes across the center of the solar disc, but the moon's distance from the earth at the time. The closer the moon is to perigee, the larger the angular diameter of its disc. If the size of the disc is the same or only slightly greater than the sun's, totality can be very brief. When at apogee, the moon's apparent width is smaller than the sun's, leaving a ring of sunlight visible around its rim, even at totality. Such an *annular eclipse* leaves sufficient sunlight to rob the event of most of its drama: the solar corona never appears, and the landscape never darkens completely. Close to perigee, totality can last as long as 7½ minutes. Because the moon and earth are both moving during this time, the moon's umbra sweeps in a curving path across the earth, at speeds over 1,000 miles per hour. The width of the path depends entirely on the moon's distance; its length depends on other factors, primarily the straightness of the sun-moon-earth lineup. At its widest, the shadow is over 150 miles across; most are about half this.

If you are lucky enough to be anywhere near the path of totality, watch the astronomy magazines. For several months before the event, they will publish detailed, updated information on times and locations, and usually give a summary of where the clearest skies are likely to be found. If you can't move toward the path of totality, you can still put on a good show for yourself and the neighbors by setting up a number of devices to provide safe viewing for small groups. To avoid causing serious harm to your own or others' eyesight, be sure to read the sections on solar observing and filters.

It is especially important to protect your eyes during an eclipse, because the low light levels can fool people. In the dim light, the eye loses its natural impulse to squint and blink, but there is still enough harmful invisible radiation to damage your eyesight permanently, especially if you stare (as there is a powerful tendency to do) for a minute or so. Smoked glass or sunglasses do not provide sufficient protection. Welder's glasses, or several thicknesses of exposed, developed, photographic film, are better, but still carry some risk. A Mylar solar filter is best for direct viewing. Just hold it between your eyes and the sun, and you will be able to see plenty, at no risk. A telescope with an adequate solar filter is good for eclipses as well as ordinary solar observation, showing clearly the rugged limb of the moon silhouetted against the spotted solar disc. If you are on the path of totality, however, you may not want to be looking down into an eyepiece as so much happens around you.

A very simple, effective, and inexpensive means of allowing several people to view the eclipse is a *pinhole camera* (also called a *camera obscura*). Take a large cardboard box and cut away one of the long sides. In the middle of one end of the box, cut a hole an inch or so across, and tape a sheet of aluminum foil over it. With a sharp pin, poke a clean hole in the foil about 1 millimeter across, spinning the needle to round the hole. Line the opposite end of the box with white paper, and you have your camera. Take it out and test it in the sun before the eclipse: you may need to try different widths of pinhole to achieve the best effect. With the pinhole aimed at the sun, the box should project an image of the sun on the white paper. As the moon passes over the disc, the crescent sun will show plainly in the box. For a better view, complete with sunspots and details of the lunar limb, rig a projection screen to your telescope. Chapter 25 tells how to make one from inexpensive materials.

Mercury

16

PHYSICAL CHARACTERISTICS

The innermost planet, airless Mercury is burned by the sun. It takes its name from the winged messenger of the gods because it is also the fastest of the planets. Mercury's sidereal period is a mere 87.97 days. Its average orbital distance from the sun is 57.9 million kilometers (0.39 astronomical units*), but the high eccentricity (.206) of its course takes it as far from the sun as 70 million kilometers at aphelion, and as close as 46 million kilometers at perihelion. Mercury rotates on its axis every 58.65 days, exactly two-thirds the time it takes to complete one sidereal revolution. Mercury's orbit is also inclined unusually far from the ecliptic, ranging as much as 7 degrees away to north or south.

The smallest of the terrestrial planets, Mercury's equatorial radius is barely a third of Earth's; its mass is a mere twentieth. The low mass and small radius combine to give an object on Mercury only a third the weight it would have on Earth. The average density of the planet is 5.5 grams per cubic centimeter (water, by comparison has a density of 1 gram per cubic centimeter). This is in the upper range of the terrestrial planets' densities, exceeded only by Earth's.

The high density indicates that Mercury probably has a sizable iron core, like Earth's. Large quantities of metals at a planet's core are thought

*An astronomical unit is the average distance from earth to sun, approximately 150 million kilometers or 93 million miles.

to act like giant magnets, surrounding the planet with a magnetic field. A planetary magnetic field, like Earth's, attracts a compass needle to its poles, and it can also trap much of the sun's more harmful radiation before it reaches the ground. But Mercury's large iron core has not produced a strong magnetic field. The planet's low rotational speed has apparently kept the iron from acting as an internal dynamo. The result is only a weak magnetism; its most noticeable affect is to trap an extremely thin cloud of charged hydrogen atoms around the planet. This is all that Mercury can claim as an atmosphere, and it is so thin as to be on the limits of detection.

Although internally much like Earth, Mercury has a surface that looks more like the moon. With its minimal atmosphere, it has had little protection from the continual infall of meteors. Neither has erosion worn away the scars of early impact, as has happened on Venus, Earth, and to some extent on Mars. The major difference from the moon is that mountains and välleys, ridges and crater walls are shallower, owing to the higher surface gravity. A slightly worn appearance of the walls of its many craters, and an even more chaotic, baked and cracked look to the floors of its impact basins, indicate that this is a world that has suffered billions of years of high heat. At perihelion, the Mercurian surface receives ten times as much solar energy as the moon, raising its temperatures to as much as 700 degrees Kelvin (467 degrees Celsius—see Appendix 3 for notes on the Kelvin scale); during the long nights, the surface cools terrifically, dropping below 100 degrees Kelvin (−173 degrees Celsius).

This extreme cycle of heating and cooling has subjected the surface rock to repeated expansion and contraction. The entire crust apparently slumped and tightened early in its history, until, like a stretched-out sweater, it no longer fits. The *lobate scarps*—low, rippling cliffs that cross the Mercurian plains—are the signs of this process. That they sometimes cut through craters, riving their walls in two, indicates that the planet-wide contraction took place after the early period that cratered the planet's surface. Mercury escaped some effects of meteoritic infall, however. Its higher surface gravity seems to have kept the debris thrown out of each impact crater (such debris is commonly called *ejecta*) close to the original site. Ray systems found on Mercury are more compact than lunar rays.

From evidence gathered by the *Mariner* probes, a picture of Mercury's origins and history emerges. After Mercury condensed from the solar nebula, its heavier elements, such as iron, settled to the planet's center, leaving the lighter, silicate materials at the surface to form the crustal rocks. Then came the planetismals falling in upon the surface, carving it up into the wilderness of craters that remains today. After the worst of the bombardment was done, lava upwelling in the larger impact scars smoothed much of the surface. These volcanic plains preserve the evidence of the next major event, the global shrinking and cracking of the crust into the lobate scarps. One of the last events of the bombardment changed the planet's surface forever. About 3.7 billion years ago, something very large hit the young world amidships, leaving the *Caloris Basin,* an impact crater some 1,400 kilometers across. So great was the force of this collision that on the far side of the planet the terrain heaved upward, forming a jumbled badlands. More volcanic upflow followed this last shock, covering areas of the surface with rock essentially similar to the material already on the surface. To this day, this similarity deprives Mercury of high contrast between its mountains and its plains.

OBSERVATION

Because Mercury's orbital speed is so high, the planet's synodic period is almost 30 days longer than its sidereal year. And thereon hangs a tale, one that involves a series of mathematical coincidences. You will recall that the planet's day is 58.65 days long. This figure is exactly two thirds of its 88-day sidereal year. The synodic year is 116 days; this happens to be the interval between its greatest elongations from the sun, the times when the planet is best-placed for observation. This period also happens to be exactly two Mercurian days. The upshot of all this is that Mercury always presents one face to Earth at eastern elongations, and another face at western elongations. For many years, observers assumed from this apparent sameness that Mercury and the sun were in a simple spin-orbital lock. Older textbooks still give 88 days as the rotational as well as the orbital period of Mercury. Only in 1965, when astronomers at the Radio/Radar Observatory at Arecibo, Puerto Rico, were able to bounce radar signals from Mercury's disc, was the true 59-day period discovered, and the reason for the mistake discovered.

Astronomers quickly deduced the reason for this apparent coincidence: the tidal force of the sun has influenced Mercury's rotation, just as Earth's gravity influenced the moon's. But Mercury did not settle down to a one-to-one ratio of orbital-to-rotational periods, as our moon did. Mercury's rotation has adjusted so that the same face meets the sun at perihelion three times in every two orbits, a 2-to-3 orbital-to-rotational ratio. Such an even fraction occurs frequently among satellites of the larger planets. The phenomenon is generally known as *orbital resonance* or *orbital harmonics.* The cause for the resonance in Mercury's case is apparently the buried mass of the meteorite that made the Caloris Basin. The off-center weight is pulled by the sun, and has acted as a drag on the planet's rotation.

For the amateur observer, the most important effect of Mercury's orbital phenomena is that you will find the planet well-placed for evening observation every 114 days. To find Mercury in your skies, consult one of the astronomy magazines or an ephemeris. They will generally present the information in some form such as "Elongation: 7° E." This means that the planet appears in the sky 7 degrees east of the sun. It will follow the sun as it moves across the sky, and is therefore visible after sunset as an evening star. This is an *evening elongation.* A *W* in such a notation indicates elongation west of the sun, causing the planet to rise before the sun in a *morning elongation.* As an inferior planet, Mercury never wanders far across the sky from the sun. When greatest elongation occurs near the time of Mercury's aphelion, its angular separation from the sun can be as much as 27 degrees; observers call this a *favorable elongation*. An unfavorable elongation occurs near Mercury's perihelion; the planet can fail to rise more than 18 degrees from the sun before sinking back toward conjunction.

Mercury's altitude above the horizon is always somewhat less than its angular distance from the sun. This is because the ecliptic always meets the horizon at a slant. At certain times of year, this slant is steep, at others shallow. If the angle is steep, Mercury will appear higher in the sky than it will when the ecliptic slopes shallowly. A regular pattern arises from this rule. Morning elongations are favorable when they occur in the fall, and unfavorable when they come in the spring. Evening elongations are best in the spring, and worst in the fall.

Precious little of Mercury's surface is visible even in the largest telescopes. The feature of its disc you will notice first is its moonlike phases. (Figure 16.1 shows the relationship between orbital aspect and phase for both of the inferior planets.) Mercury is brightest when 80% of the visible surface is illuminated. This occurs 1 to 3 weeks before greatest eastern elongation, and 4 to 6 weeks after greatest western elongation. At that time, its brightness reaches magnitude −1.8. It is still low in the sky, and that sky is usually well lit by a sun only a few degrees below the horizon, so even at greatest brightness Mercury is an extremely elusive object; tradition has it

16.1 Favorable and unfavorable evening elongations of Mercury. On spring evenings, the angle between the ecliptic and the horizon is steep; Mercury appears high above the horizon. In the fall, the angle is shallow, and Mercury's altitude is less, even though its elongation from the sun may be the same. For morning elongations, spring is unfavorable, and fall is favorable.

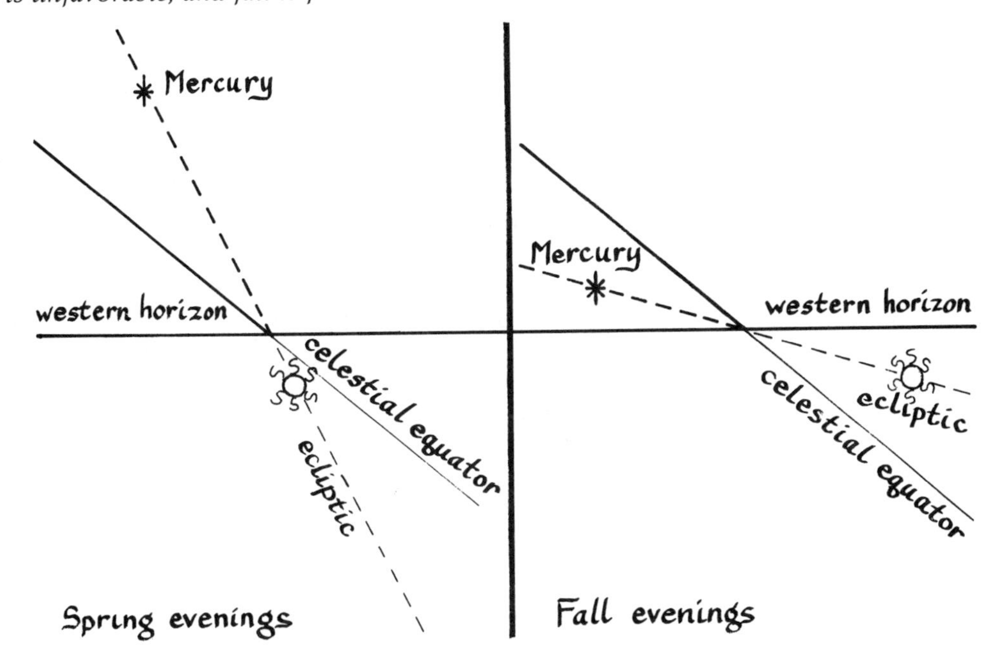

that Copernicus, the great sixteenth-century astronomer who determined that the earth revolves around the sun, never saw Mercury.

Adding to the difficulty of observing Mercury are its small size and relatively great distance. At closest approach the planet is still 80 million kilometers from Earth, and at elongation the distance is even greater. The planet's small diameter presents a disc from five to nine arc-seconds across, which is visible only through the thick layer of Earth's atmosphere close to the horizon. Such a small image needs magnifying, yet these are precisely the conditions at which magnification suffers most from atmospheric turbulence. This is especially true for evening apparitions, when the air you look through has been left in turmoil by a full day's heating; at dawn, the air is usually calmer. For a 6-inch reflector, powers in the 150× to 200× range give the best results, and under good conditions will show some very faint streaking or mottling on the disc. Don't expect to see anything more than the phase the first time you look, however. Even that much will be hard to see if the atmosphere is unsteady, or if your optics are poorly aligned.

If you detect a distinct crescent, look at the southern horn; does it seem blunt? This was one of the first anomalies to be detected on Mercury, and is apparently a large, relatively dark feature on the surface, near the south pole. Features smaller than this are subject to a great deal of distortion—both by turbulence, and by the mind's eye, which tends to arrange random markings into ordered geometrical forms, and by the eye itself, which can "see" mottling on a surface that is actually unmarked. Elaborate maps of the Mercurian surface, drawn by skilled observers with powerful refractors, were found by the *Mariner 10* probe to be figments of the observer's eyes and minds.

Two projects for the amateur with more modest equipment and expectations include finding Mercury by daylight and observing a *transit of Mercury* across the disc of the sun. Daylight observation has several advantages. The daylit sky diminishes the glare of the planet's illuminated disc, allowing some of its lower-contrast features to emerge. You can also follow the planet up to (and at times all the way through) conjunction. Locating the planet, especially when it is close to the sun, can be very dangerous. If your mount suddenly slips, you could find yourself staring at the sun. Blindness can follow shortly upon such an accident, so don't attempt this until you're confident of your touch with the telescope.

Because the daylight sky is almost as bright as the planet itself, finding Mercury by day is a challenge. Setting circles are very helpful, if you can look up in an ephemeris the coordinates for the sun and Mercury. Aim the telescope at the sun by turning it until the shadow of its tube contracts into a circle. *Keep the telescope capped during this procedure.* Set the right ascension circle to the sun's right ascension. Then move the tube by the amount indicated by the difference between the sun's coordinates and Mercury's. For instance, if the sun's position is right ascension 00 hours, declination 0 degrees, and Mercury's is 0130 hours, +9 degrees, you would move the tube east 1 hour and 30 minutes, and north 9 degrees.

Without setting circles, you can still sweep up Mercury by estimating the distance east or west of the sun, and then offsetting north or south. When the planet is east of the sun, you can let the Earth's rotation do part of this for you. After centering on the sun, wait as many hours, minutes, and seconds as the difference in right ascension between the two. After that interval has passed, raise or lower the tube by the difference in declination between it and Mercury. You should find Mercury drifting through the field of your lowest-power eyepiece; if not, sweeping in very short arcs around that point should pick it up.

A light red filter will make the planet stand out from a daylit sky. When observing in the daytime, keep your scope protected from heat and away from heated ground. You should also wait until the planet is close to the meridian (which will always be within two hours of noon), to keep it above low-lying haze. The closer the planet gets to the sun, the more difficult (and potentially dangerous) it is to observe, but the view of the crescent phase as the planet nears inferior conjunction is worth the attempt.

Transits of Mercury occur, like solar eclipses, when the planet reaches inferior conjunction at one of the nodes of its orbit. Unfortunately, the next one of these visible from North America isn't scheduled until November 1999. Mercury in transit appears as a dark, round spot, much more sharply defined than the typical sunspot. Some distortions occur around the planet's limb; it may even appear to trail a small teardrop of darkness behind it. Until recently, such effects were thought to confirm the presence of a Mercurian atmosphere. We now know that they are optical illusions. Even though a transit offers no further opportunity to observe phenomena, atmospheric or other, on Mercury's surface, these events do offer the best view you can get of the size of Mercury's disc at inferior conjunction, when it comes closer to Earth and looks larger than at any other point in its orbit.

Venus

17

PHYSICAL CHARACTERISTICS

Second of the terrestrial planets from the sun, Venus has long been called Earth's twin: its mass is 81% of Earth's; its equatorial radius, 91%. If you were standing on its surface, you would have 88% of your present weight. There, unfortunately, the resemblance stops. Standing on the surface, you would be crushed by the atmosphere, which is 90 times denser than Earth's. Venusian "air" is a fog of carbon dioxide and droplets of sulfuric acid. The scenery around you would be a low, rolling plain strewn with sagging outcroppings of rock, glowing fitfully red in the shadows. Temperatures hover in the low 700s Kelvin, hundreds of degrees hotter than the average surface temperature of Mercury. Above you, the sky would be a dim boiling of thick, yellow cloud, its darkness broken incessantly by bolts of lightning, The sun is barely visible as a diffuse glow that crawls from *west* to *east* across the sky. This backward motion is the result of Venus's unusual retrograde rotation. At 243.01 days, Venus takes 20 days longer to complete one axial turn than it takes for its entire 224.7-day sidereal year.

How could such a promising planet have turned out so badly?

The answer: the *greenhouse effect.* This effect has some relevance to our own situation on Earth. A good many compounds, among them glass and the gas carbon dioxide, are opaque to infrared radiation. The glass in a greenhouse keeps the sun's infrared rays from entering. But when the visible light that penetrates the glass strikes the objects inside, they absorb energy, and warm up. Warm objects emit infrared radiation. This infrared cannot escape through the glass, so the interior of the greenhouse gets much hotter

than it would if infrared could pass freely in and out through the glass. The greenhouse effect turns the building into a giant storage battery for heat.

The greenhouse effect can also occur on a planet-wide scale. On Venus, the enormous quantities of carbon dioxide in the Venusian atmosphere have turned the planet into a global greenhouse, raising the temperature by about 500 degrees Kelvin over the temperature it would attain without an atmosphere. If Venus's atmosphere were similar to Earth's, its average surface temperature would be very like our own, and perhaps adequate to sustain life. If Earth's atmosphere had as much carbon dioxide as Venus's, however, its temperature would skyrocket.

Carbon dioxide is a natural byproduct of many chemical reactions, especially biological ones. We exhale it when we breathe; the burning of petroleum, coal, and wood creates millions of tons of carbon dioxide each year. Growing plants absorb carbon dioxide. In recent years, staggering tracts of the great rain forests of the tropics have been cut down. The average temperature of the northern hemisphere has already risen by a degree or so; increases of three to ten degrees are forecast for the next century. This small rise may be nothing to worry about. But consider Venus.

Where did all of Venus's carbon dioxide come from? It was cooked out of its rocks. On Venus, as the temperature rose, more carbon dioxide cooked out, raising the temperature further, producing still more carbon dioxide. If this vicious circle were to start on Earth, as the temperature rose, the ice caps would melt, increasing the ratio of ocean to ice. Seas absorb more solar heat than ice, further accelerating the rise in temperature. The cycle would get out of hand even faster. No one knows how much of a push it takes to disturb the temperature balance of a planet like Earth and set off a runaway greenhouse effect. We only know, from the example of Venus, that it has already happened in our solar system at least once.

This disaster has left Venus shrouded in a beautiful, impenetrable veil of cloud, which hid the nature of its surface for centuries. Recently, radar investigations from Earth and from the *Pioneer Venus* orbiter have allowed us to map the planet's solid surface. After the spectacular terrain of the moon and Mercury, Venus's topography seems subdued. The thick atmosphere has protected it from the later stages of meteoritic bombardment and has eroded the signs of the earlier, catastrophic infall that formed the great impact basins of other worlds.

The surface of the planet is surprisingly level; 60% of the total keeps to within 500 meters of the average surface level (the equivalent, for a waterless world, of sea level). There are mountains, however. Because the highest altitude any peak can attain is determined by a planet's surface gravity and the strength of the materials it's made of, Venusian mountains rise as high as the highest peaks on Earth. The highest peaks are in a range called *Maxwell Montes,* which is on a high plateau called *Ishtar Terra.* They reach to 12 kilometers (7 miles) above the norm—higher than Mount Everest's altitude above sea level.

If Venus had seas, it would have two continents. Ishtar Terra is one, occupying an area in the high northern latitudes about the size of Alaska. *Aphrodite Terra,* a long, equatorial highland, runs east-west for over 10,000 kilometers (6,200 miles). It bears in its southern regions a large, circular scar, called *Artemis Chasma,* one of the few obvious impact scars on the surface. Other highlands and low plains, named for the goddesses and legendary or historical women of Earth, account for the rest of the known surface. The highlands average a few hundred kilometers across, scattered widely over the surface. The lowlands are larger, so a Venusian ocean would be extensive, dotted with archipelagoes, with a shoreline as rugged as the coast of Maine.

The presence of mountains, and continentlike masses standing out above the normal level, suggests that Venus, like Earth, has had a history of crustal movement. But our present state of knowledge is so rudimentary, compared to what we know about every other planet between Mercury and Saturn, that we can say little more for sure. A planned instrumented probe, the Venus Orbiting Imaging Radar, should tell us much more, if funding for the mission clears Congress (it was denied funding once, but has since been revived in a new version, using a probe with one-quarter the resolution of the previous one). In light of the many similarities between that world and our own, some of which hold important implications for our future survival, we can only hope that the American program of planetary exploration recovers its vitality soon.

OBSERVATION

Venus is altogether a more tractable object to observe than Mercury (Fig. 17.1). It remains visible for a full seven months at each apparition, and can achieve an elongation as much as 47 degrees from the sun. It reaches a piercing magnitude −4.4 at maximum brightness (which occurs 35 days before greatest eastern elongation, and 35 days after greatest western elongation). As with Mercury, different seasons are more favorable than others, for the same reasons. Evening elongations are best in the spring, and morning elongations in the fall. The synodic period, 584 days, includes a long spell around superior conjunction. Elongations occur in quick succession before and after inferior conjunctions. The best possible combination of orbital geometries occurs every eight years at eastern elongations; the next of these will occur in 1988, when Venus will dominate the evening skies of spring, remaining visible throughout the evening.

So bright is Venus when near its maximum brilliancy that telescopic observation becomes a challenge. Like Mercury, it can be seen better against a light sky, when the contrast between the planet and background reduces glare. An Orthoscopic or Plössl eyepiece will do best, especially when the sky is dark. Around inferior conjunction, when the phase is a distinct crescent and the disc is largest, a very slight magnification will reveal the sharp, moonlike crescent. Even at these times, however, magnifications of around

17.1 Venus (Hale Observatories)

150× will reduce glare, so use as much magnification as the atmosphere will permit. Filters are also useful: neutral density to reduce glare, red to aid in finding in a daylit sky. Blue filters seem to show more detail along the terminator, and at the tips (also called *cusps*) of the crescent.

Venus's dense atmosphere diffuses sunlight into the night side of the disc. Around inferior conjunction, the tips of the crescent stretch beyond the poles of the planet, becoming far longer and sharper than the horns of the crescent moon ever become. Such atmospheric effects can at times illuminate the entire dark portion of the disc with a very faint glow called the *ashen light.* The effect is extremely delicate, especially compared to the brilliance of the sunlit crescent, and takes an experienced eye to appreciate. The appearance is similar to the "old moon in the new moon's arms," only much fainter.

The same atmospheric refraction is responsible for *Schröter's effect,* which distorts the terminator. This is most noticeable at half-phase, when the planet is at greatest elongation. At that time, Venus ought to show a terminator perfectly straight across the disc, but infiltration of sunlight into the night side makes the lit half of the disc seem to bulge farther into the unlit hemisphere than it should. At eastern elongations, the planet reaches apparent half-phase a week or so early, and at western elongations, a week or so late.

Other markings reported on the Venusian disc are necessarily short lived, and can only be the result of large-scale turbulence among the upper banks of cloud. These are most apparent in ultraviolet light; amateurs have long found that blue filters make some mottlings appear on the sunlit disc. Cloud patterns observed in the ultraviolet from the *Pioneer* orbiter appear as parabolic arcs, looking like the wake of a great ship sailing east along the equator (Fig. 17.2). These are the marks of global wind currents, formed as air heated in the long Venusian day rises at the equator and sweeps toward the poles, while cool polar air heads south along the surface. At the same time, centrifugal force sends the upper atmosphere racing west at 363 kilometers (225 miles) per hour. The combination of the two forces creates banks of cloud sweeping diagonally across the northern and southern hemispheres.

If you should see markings in the atmosphere, watch for them to change rapidly from night to night. If they remain fairly constant, chances are that some optical effect is really responsible. Some features do appear in Venus's upper atmosphere, but they are subtle, and years may pass before you can be at all certain of your ability to distinguish them from optical effects.

Venus is capable of transiting the sun, but such events are very rare. There have been none in this century, nor will there be; they occur in pairs, and the next is scheduled for 7 June 2004, followed by another on 5 June 2012.

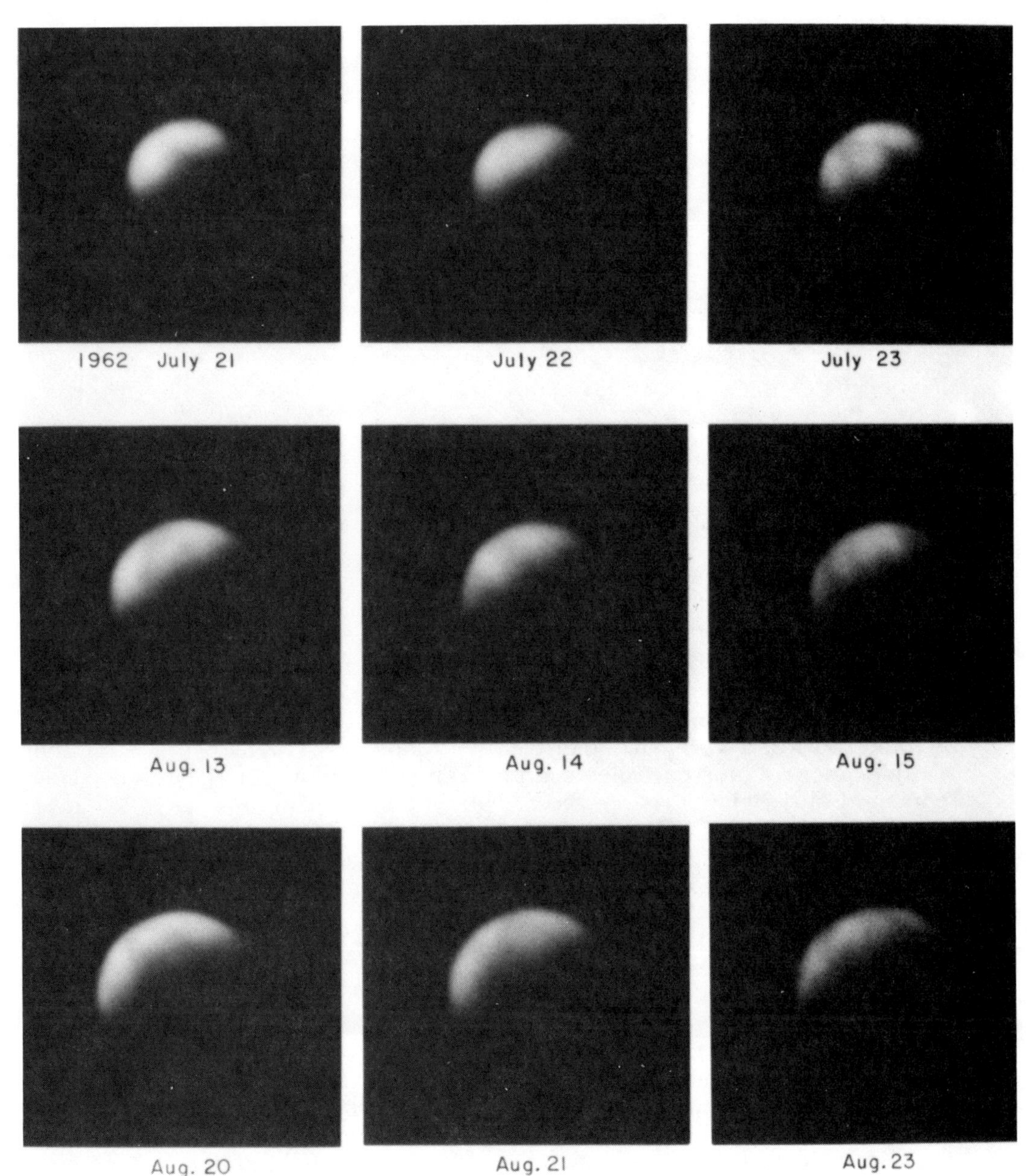

17.2 Venus: a series of images in ultraviolet light, showing cloud markings. (Lick Observatory photograph)

18

Earth

Our planet's motion through the solar system carries us through a variety of phenomena often visible in our skies (Fig. 18.1). A *solar wind* interacts with the earth in a shimmering display called the *aurora.* The ancient trails of comets, littered with ash and gravel, send showers of *meteors* into our upper atmosphere. Finer particles scattered over the plane of the solar system gleam faintly, giving rise to the *zodiacal light.*

The aurora is caused by the interaction of the solar wind with the earth's magnetic field. The solar wind is a spray of atomic particles, primarily protons and electrons, that streams continually into space from the sun. Protons and electrons carry electromagnetic charges, and are therefore attracted or repelled by magnetism.

The earth, like Mercury, has a core of iron. Current theories hold that fluid motions in the earth's iron center cause the core to act as an electrical generator, setting up an enormous, powerful electromagnetic field around our planet. If you think of the earth as having a giant bar magnet running through it, you have the picture. The earth's magnetic field extends far above the atmosphere. The region enclosed by this field is called the *magnetosphere.* It is a teardrop-shaped bubble in the solar wind, presenting a smooth, curved front to the wind, and trailing off behind in a region called the *magnetotail.* The boundary of the magnetosphere is called the *magnetopause.* The magnetosphere provides an effective shield against the solar wind, preventing its charged particles from bombarding earth with harmful radiation.

18.1 Earth and moon. Voyager I returned this image of our twin planet on 18 September 1977, when it was outward bound for Jupiter, at a distance of 11.66 million kilometers. From Voyager's perspective, the moon was beyond the earth. (NASA)

At times, however, the solar wind strengthens and presses against the magnetopause. The pressure distorts the magnetic field, initiating a complex (and as yet poorly understood) mechanism. From a relatively calm region in the magnetotail, charged particles come spiraling in toward the earth. These particles, since they are electrically charged, are affected by magnetism. The earth's magnetic field directs them downward toward the poles, as they spiral around the lines of force in the earth's magnetic field. It is only at the magnetic poles that these lines of force reach the earth's surface. There, they penetrate the earth vertically, directing the charged particles down from space and into the upper atmosphere. Here, at altitudes around 1,000 kilometers, the particles start to collide with molecules of air and set up a complicated process that sets a variety of charged particles into rapid motion. The end result is something very like the device that makes the picture tube of a television glow. In this case, however, the television screen is the earth's atmosphere.

At 75 to 150 kilometers above the earth's surface, air molecules energized by high-speed particles glow brightly, exactly as the gas in a neon tube glows when electricity flows through it. The sky glows in great, shimmering curtains of light that hover over the horizon, or in bursts of rays called *coronae* that appear to radiate from overhead. These auroral displays can be colored—usually red, or a ghostly, greenish gray, though white and yellow lights also occur, depending on the atmospheric gas involved. Hydrogen, for instance, typically glows red; oxygen glows green. These displays often flicker, the entire sky shifting dizzyingly. Or they can hover insistently, in eerily regular patterns (Fig. 18.2).

The visibility of aurorae depends most strongly on your latitude; they are more common the closer you are to the pole. The aurora, also known as the *aurora borealis,* or northern lights, occurs almost continually in a halo hundreds of miles in diameter, around the Arctic Circle. Occasionally this halo widens, and we have reports of aurorae from the northern United States and occasionally as far south as Arizona. Such events seem to depend on the strength of the solar wind. Strong displays usually occur about 2 days after a *solar flare*; weaker displays occur 3 to 5 days after the passage of other disturbances across the solar disc. Solar flares occur most often at 11-year intervals of the solar cycle (see chapter 25). Very strong auroral displays tend to occur in the two or three years following a peak in this cycle. The cycle last peaked in 1979–80; it will be at its weakest in the mid-80s. In the northern tier of the United States, you can see as many as ten aurorae a year; in the southern tier, you may wait several years between displays.

A display will most often become visible low in the north, rising in rays that look like a distant searchlight. In a good show, these horizon lights will brighten and climb, and can not only reach the zenith but cover the entire sky. A very active aurora can be awesome: the entire sky pulses in a giddy array of patterns, often to an eerily insistent rhythm. Waves of light march along the horizon, or cluster in a burst of rays. At times, swirling spiral pat-

18.2 Aurora borealis. This curtain formation is one of the most common forms the aurora takes. (Official U.S. Navy photograph)

terns roll across the sky; these are apparently the visible sign of the lines of force in the earth's magnetic field.

METEORS

Much more predictable than aurorae are meteor showers. The meteors that fall as "shooting stars" are mostly very small, fluffy bits of dust—about the size and appearance of the ash on a cigar. They are the debris that flakes from the head of a comet as it travels its elongated path through the inner solar system. In some cases, they are all that remains of a comet that long ago melted away. In either case, the trail of a comet is marked by a scattering of small particles, traveling around the sun at the same speed and over the same path as the parent body that sloughed them off millennia ago. Several dozen of these trails cross the orbit of the earth, and once a year, as the earth passes through them, millions are caught in our gravity and plunge into our atmosphere. So high is their speed that air friction heats them to their melting point, and they evaporate long before reaching the ground. The air itself is heated to an auroralike glow along their paths, forming a trail (or *train*) that glows bright gold or green for a second, fading quickly to a faint, gray wisp that quickly dissipates.

Meteors occupy an anomalous position in astronomy, which the terminology attached to them suggests. While it's still in space the particle is called a *meteoroid.* A meteor that has survived its blazing traverse of the atmosphere and reached earth's surface as a solid body is called a *meteorite.* The term *meteor* is reserved not for the particle itself but for the particle in the act of falling. A meteor is a very short-lived phenomenon indeed.

Planetary scientists eagerly pursue meteorites. An intact fragment is a tremendous windfall to those interested in the physical history of the solar system. The scientific value of these rocks, which can be metallic (predominantly iron), stony, or a mixture of the two, is one reason to watch for meteors. When a particularly bright meteor, usually called a *fireball,* crosses the sky, a record of its apparent path across the constellations, combined with observations from only a few other locations, can lead meteorite hunters to its point of impact.

Meteors can be divided into two categories: *shower meteors,* which are associated with particular comets and fall in an organized pattern; and *sporadic meteors,* which fall constantly to earth, at the rate of millions per day, in a continuation of the process of accretion by which the planets formed. On a typical night, you can see seven sporadic meteors per hour. During a strong meteor shower, you can see as many as one per minute. During rare meteor storms, tens of thousands of meteors each hour etch the sky.

You can tell a meteor shower from the continual infall of sporadics by tracing the meteors' paths back across the sky. A whole-sky map, such as the ones published monthly by the astronomy magazines, is ideal for this. When a meteor passes, mark its path on the chart. During a shower, the majority of these marks will all seem to radiate from a single point on the celestial sphere. This point is called the *radiant* of the shower. It marks the position of the stream of meteoroids from which the meteors came. The meteor trails seem to radiate from a single point on the sky for the same reason that railroad tracks seem to converge to a single point in the distance. The tracks are actually parallel, as are the paths of meteors in a shower. Our perspective makes their tracks appear to converge to a single point.

Over a period of several nights, the position of the radiant will shift, as the earth's orbit carries us into and past the swarm. Most swarms are so small that the shower starts, reaches its maximum intensity, and dies away over a few nights; some showers can last, although at a low level of intensity, for weeks. In either case, the shower is usually named for the constellation in which the radiant appears during maximum intensity (the few exceptions to this rule, such as the *Giacobinids,* are named for the shower's parent comet). The *Perseids,* for instance, are a highly active shower that occurs each year around 12 August; their radiant appears in the constellation Perseus. No matter where the radiant lies, however, meteors will be visible over most of the sky. Their trails may start and end some distance from the radiant; only by extending the line of their path backward do they seem to converge on one point.

To observe meteors, it helps to observe during a shower and to know where the radiant lies. You need little special equipment. Telescopes and binoculars restrict your field of view too much to be useful, but you may want a chart, red flashlight, pencil, and a blanket to lie on. Low horizons are also helpful, because they allow you to see more of the sky, but since meteors are best seen overhead, they are not absolutely necessary. Other conditions affecting the intensity and visibility of a meteor shower are familiar to you already: the dimmer the moon, and the less haze and light pollution, the more meteors you will see.

Meteor showers all intensify after midnight. After midnight, you are on the side of the earth facing forward in its orbit. Meteors swept up by the earth at that time meet us head on instead of having to catch up from behind. More meteors fall, and those that fall fall harder and burn brighter, in the hours between midnight and dawn. For this reason, phases of the moon between new and first quarter do not disrupt the best meteor watching, but a full or last-quarter moon will interfere with viewing.

You should also be aware of daily and annual variations in shower strength. The astronomy magazines and ephemerises usually give predictions for the maximum of each shower. Showers also vary from year to year, intensifying as the comet that spawns them nears the earth. The Perseid shower, for instance, seems to have peaked in the past several years, with a strong display of over 100 meteors per hour in 1981. The *Leonid* shower, radiating from Leo each 17 November, is a spectacular example of this. Every 33 years, the earth passes through an extraordinarily dense section of the swarm: tens of thousands of meteors fall each hour. This last occurred in 1966, and the next such display should come in 1999. These dates are not infallible, however; astronomers who eagerly anticipated the 1899 display saw only a dozen or so per hour!

The meteors in each shower tend to have a characteristic color, brightness, and speed of descent. Some actually make noise, sizzling or (rarely) booming as they pass. Some have trains that persist for several seconds, a gray cloud like a jet's contrail; others disappear as soon as the flash fades. All of these phenomena depend on the speed of the swarm, and to some extent on the composition of the meteoroids involved. You may want to take note of unusual variations within or among showers—for your own curiosity, or to pass on to other amateurs or professional astronomers interested in meteors. Shower watching is a good activity for a club, because teamwork can allow for more complete coverage of the sky. With one or two people devoted entirely to record keeping, the others can keep their eyes on the sky, calling out meteors' paths as they appear.

If you don't have such help, you will quickly find that it's hard to keep your eye on the entire sky at once. If you watch one particular area of the sky, and keep watching it, you will see as many meteors as possible for one individual. The radiant is not the best place to look. Meteors appearing at or close to the radiant are coming straight at you. The meteor's path appears foreshortened, and can be hard to tell from a star. Look about 45 degrees away from the radiant. Meteors there move across your line of sight about as far as they fall in toward you, stretching their trails far across the sky. Farther from the radiant, the trails are even longer, but the supply of meteors is spread more thinly across the widening area of the celestial sphere so your odds of seeing a meteor decrease.

If you do keep records of the paths of meteors, what can you do with them? This is one area in which amateurs can easily make observations that are of interest to scientists. You can record simply the raw number of meteors you see each hour. This figure, along with the time of observation and sky conditions (measured best by the magnitude of the faintest visible stars), can help to compose a picture of the shower's strength. *Sky & Telescope* collects such information and gives more detailed instructions for such a project. The Perseid shower is an especially good shower to start with. Next August, you might want to try it.

To help you get started with meteor watching, the following table lists the major showers.

Shower	***Date***[1]	***Radiant***[2]	***Rate***[3]	***Speed***[4]	***Duration***[5]	***Comments***
Quadrantids	1/3-4	1528+50	40	41	1.1	very brief; bright, blue trains
Lyrids	4/21	1816+34	15	48	2	ancient, dying shower
eta Aquarids	5/3-4	2224 00	20	65	3	possibly from Halley's comet
S. delta Aquarids	7/28	2236−17	20	41	7	slower, yellower than Perseids
Perseids	8/11	0304+58	50	60	4.5	lasts several weeks at low strength; best of year
Orionids	10/21	0620+15	25	66	2	fast, bright
Leonids	11/16	1008+22	15[6]	71	?	unpredictable, very fast, sometimes extremely rich
Geminids	12/13	0732+32	50	35	2.5	year's second best; brief

[1] month/days
[2] position at maximum, in hours and minutes right ascension; degrees declination
[3] the average number of meteors seen by one observer in 1 hour
[4] in kilometers per second
[5] in days, from one-fourth strength to maximum
[6] varies from nil to thousands; peaks every 33 years

THE ZODIACAL LIGHT

The *zodiacal light* is a dim glow in the sky, at times as bright as dimmer patches of the Milky Way. It usually appears as a cone of light tapering upward from the eastern horizon before dawn, or from the west after sunset, and is most visible around the equinoxes, when the ecliptic rises more nearly perpendicular from the horizon. This very faint gleam follows the path of the ecliptic, at times forming a band across the sky, thickening and brightening at the point opposite the sun. The source of this light is the sun, whose rays are backscattered by fine dust that extends throughout the solar system. It is exquisitely difficult to see. Try it, any night you are far away from city lights and the air is clear. The particles that gleam so dimly are all that remains of the nebula from which our solar system condensed, more than 4 billion years ago.

Mars

19

PHYSICAL CHARACTERISTICS

Long associated with the fire and bloodshed of war, Mars has been the scene in modern times of equally violent controversy among astronomers arguing the planet's suitability for the evolution of life. In the past hundred years, dispute over the existence of the Martian canals (they don't) has given way to more sophisticated discussions over the results of the *Viking* landers' search for the presence of microbes in the Martian soil (they probably don't exist either). Thanks to the *Viking* and *Mariner* expeditions, our knowledge of the planet has increased enormously, revealing a world marked by air and water erosion, with great polar ice fields and volcanoes larger than any on the earth, where cyclones swirl through the upper atmosphere, snow accumulates in light dustings on the terrain, and clouds wreathe the peaks of mountains. Compared to the moon, Mercury, or Venus, Mars is the most hospitable planet we have found. But its smaller size, greater distance from the sun, and different geological history have left it with an atmosphere too thin to support life, no free surface water, and bitter cold.

Mars completes an orbit every 687 days, on a path that is the third most eccentric in the solar system. Mars's distance from the sun varies from 206.7 million kilometers at perihelion, to 248.6 million at aphelion. Its distance from the earth at opposition can range from 55 million kilometers at *perihelion oppositions* to more than 96 million kilometers at *aphelion oppositions.* These are vital terms to observers of Mars; the planet's disc is so small, yet so rich in detail, that the difference between the view at perihelic and aphelic oppositions is like viewing two entirely different worlds.

A superior planet, Mars's orbital relations with Earth are different from the inferior planets', displaying retrograde motion, and coming to opposition once each synodic year. Because Mars's orbital speed is not much less than Earth's, each time the two planets approach opposition, Earth must struggle to catch Mars from behind. As the earth draws closer, we see Mars from close up, increasing the apparent speed of Mars's retrograde motion. The result of these two factors makes Mars's motion across the celestial sphere more dramatic than any other planet's. It evades opposition longer than any other world, lingering near conjunction with the sun for almost a year, and swooping wildly across the sky when near opposition. Its orbital eccentricity further complicates matters. At perihelic oppositions, Mars is moving fastest along its orbit, and Earth takes longer to catch up with it then. When opposition occurs near Martian aphelion, Earth catches up more quickly. The synodic period therefore varies: two successive perihelic oppositions will take place 800 days apart; aphelic oppositions, 765 days apart. The average interval between oppositions is about 780 days.

The Martian day is only 37 minutes longer than ours, and the planet's axis is tilted 24 degrees from its orbit, giving it days and seasons like our own. The direction of the polar tilt is somewhat different, however: the pole star in the Martian skies is Deneb. With a mass one-tenth of Earth's and an equatorial diameter half as large, Mars's density is only 3.94 grams per cubic centimeter, closer to the moon's than to the rest of the terrestrial worlds'. Standing on the Martian surface, you would weigh barely a third your present weight. But you would have more on your mind than your weight: the temperature, for instance. On a cold winter night at around 48 degrees north latitude, the temperature falls to −190 degrees Fahrenheit (150 degrees Kelvin). On a summer's noon, however, at 22 degrees north latitude, the temperature can rise to about ten below (250 degrees Kelvin). These were the extremes of temperature recorded by the two *Viking* landers in their first years of operation. At the poles and equators, these ranges increase. The soil on the equator may warm to above the melting point of water. At one of the poles, the carbon dioxide in the atmosphere freezes out of the air, giving the south pole a cap of dry ice. The average surface temperature is about −60 degrees Fahrenheit.

The atmosphere is an unbreathable mixture in which carbon dioxide predominates, making up about 95% of the total. Nitrogen (2.7%) and argon (1.6%) account for the rest, with traces of oxygen, carbon monoxide, water vapor, and other gases contributing less than 0.7%. Not only is the atmosphere unbreathable, it's thin. At the surface, the pressure is 0.7% of the earth's sea level pressure. Because the boiling point of a liquid decreases with decreasing air pressure, a jar of water placed on the Martian surface would boil, even at the subzero temperatures that prevail. What water Mars has retained is frozen into its polar caps, and perhaps is locked beneath the surface in a layer of permafrost.

The thin atmosphere is still thick enough to support weather, such as the great dust storms that can blanket the entire surface with a thick cloud of rusty dust. The dust actually *is* rusty: large quantities of iron oxides turn the deserts of Mars the familiar orange of rusting iron. Carried aloft by winds in excess of 200 miles per hour, tiny particles of Martian soil can stay airborne for months, turning the sky a salmon pink color. These global storms are seasonal, occurring for the same reasons that March winds are so blustery here on earth.

Until the first probes from Earth reached Mars in the mid-1960s, our maps of Mars were based entirely on telescopic observation and were heavily influenced by the long-standing belief in the observation of thin, dark lines stretching across the surface. They were called *canali* by the Italian astronomer Schiaparelli, who observed them at a series of perihelion oppositions in the late 1870s. This Italian word for "channels," or straight lines, was picked up by the wealthy American amateur Percival Lowell. He insisted, for several decades and with great persuasiveness and charm, that these markings were precisely canals: great engineering works of a dying civilization, seeking to channel its diminishing water supplies from its polar ice caps to its spreading deserts. Lowell was right about the ice caps and the deserts, but he was wrong about the canals (as several more sober astronomers of the time were quick to point out, only to be ignored by a public enthralled by Lowell's imaginative descriptions). To be visible at a distance of 35 million miles, such canals would have to be 30 to 50 miles wide. We know now that such a body of open water would not last long in the thin atmosphere, but even assuming more favorable conditions, it's hard to see why an advanced civilization would build so wastefully.

But Lowell, who throughout his long and otherwise productive career produced increasingly detailed and complex maps of an irrigation system that never existed, provided an important lesson for observers since: we see what we believe. If you want canals on Mars, you will find them there, as Lowell and hundreds of other observers demonstrated for decades. Moreover, the findings of the first instrumented probes to photograph the Martian surface confirmed a basic reflex of the human image-processing system. We have a passion for order. A random series of markings—tiles on a floor, impact craters strewn across a plain, or the spaces between words on a page of print—will arrange themselves in our mind's eye until they form a satisfyingly regular pattern *even if that pattern does not exist in the original image.* When the Mariner probes flew by and orbited Mars, they found no canals. In the places where two generations of observers had mapped them, they found no geological features remotely like straight lines—only random scatterings of impact craters, linked up unconsciously by astronomers who glimpsed them just beyond the limits of resolution.

The actual surface of Mars holds wonders enough. Globally, Mars exhibits two distinct kinds of terrain. The northern hemisphere is generally composed of smooth lowlands, apparently caused by geologically recent outflows of volcanic lava. The southern hemisphere comprises heavily cratered, ancient highlands. About a half-dozen major features dominate this basic division. Two of these are the polar caps. Like the great ice fields covering Earth's poles, the Martian caps expand and retreat with the seasons. Unlike Earth's, however, Mars's caps are thin sheets, composed not only of water ice, but of frozen carbon dioxide: dry ice. The southern cap is the colder of the two and is predominantly dry ice. It can melt away entirely when it faces the sun at perihelion, releasing great quantities of carbon dioxide into the atmosphere. At its maximum extent, the cap can reach as far as 55 degrees south latitude. The northern cap is mainly water ice and is usually the smaller of the two, bespeaking the small quantity of water vapor available in the Martian atmosphere. Because the melting point of water is higher than the normal temperature on Mars, this cap shows less seasonal change than the southern one.

Another large feature of the surface may be related to the polar caps. These are the volcanoes of the *Tharsis Ridge,* which rises some 10 kilometers (6 miles) above the smooth plains around it. Three large volcanoes stand 20 kilometers (12 miles) above the ridge. To their west stands the largest single volcano in the solar system, *Olympus Mons,* 26 kilometers (16 miles) tall, 600 kilometers (370 miles) across, surrounded by a cliff 8 kilometers (5 miles) high (Fig. 19.1). From its summit, plumes of cloud sometimes trail. Visible in earthly telescopes, these faint wisps of white gave rise to the volcano's pre-*Mariner* name, *Nix Olympica,* "the snows of Olympus." These volcanoes, and others like them scattered across Mars, are probably largely responsible for the water vapor and carbon dioxide in the atmosphere and frozen at the poles. These gases were probably vented in steam from volcanic eruptions billions of years ago.

East of the Tharsis Ridge for 4,000 kilometers (2,500 miles) runs the *Valles Marineris,* the Grand Canyon of the solar system, cutting almost 4 miles into the Martian crust in a gash as much as 500 kilometers (310 miles) from rim to rim. Its walls are scarred with signs of ancient landslides and the sinuous rills of watercourses formed by streams that evaporated eons ago. Two other major features are the basins of the southern hemisphere, *Argyre* and *Hellas,* both of which are great dust bowls; windblown dunes march across their surfaces. They are apparently major impact basins, similar to the lunar maria and the Caloris Basin on Mercury.

All of these features speak of a geological history more complicated than those of the other planets we have seen so far, and as yet poorly understood. The tantalizing signs of water erosion, seen in the many sinuous, braided channels that cross the surface, suggest that at some time in Mars's past, rivers flowed (Fig. 19.2). This suggests in turn an atmosphere thick enough to prevent water from boiling away, warm enough to allow it to melt. The enormous gash of Valles Marineris may also be a sign of large-scale movements in the planet's crust. Earth shows vigorous crustal movements; they are the force behind mountain building, volcanoes, and earthquakes. The motions of Mars are apparently less energetic and extensive, but any such sign of activity bespeaks a geology far more active than that of Mercury or the Moon.

Finally, there is the last great question of the solar system: is there life

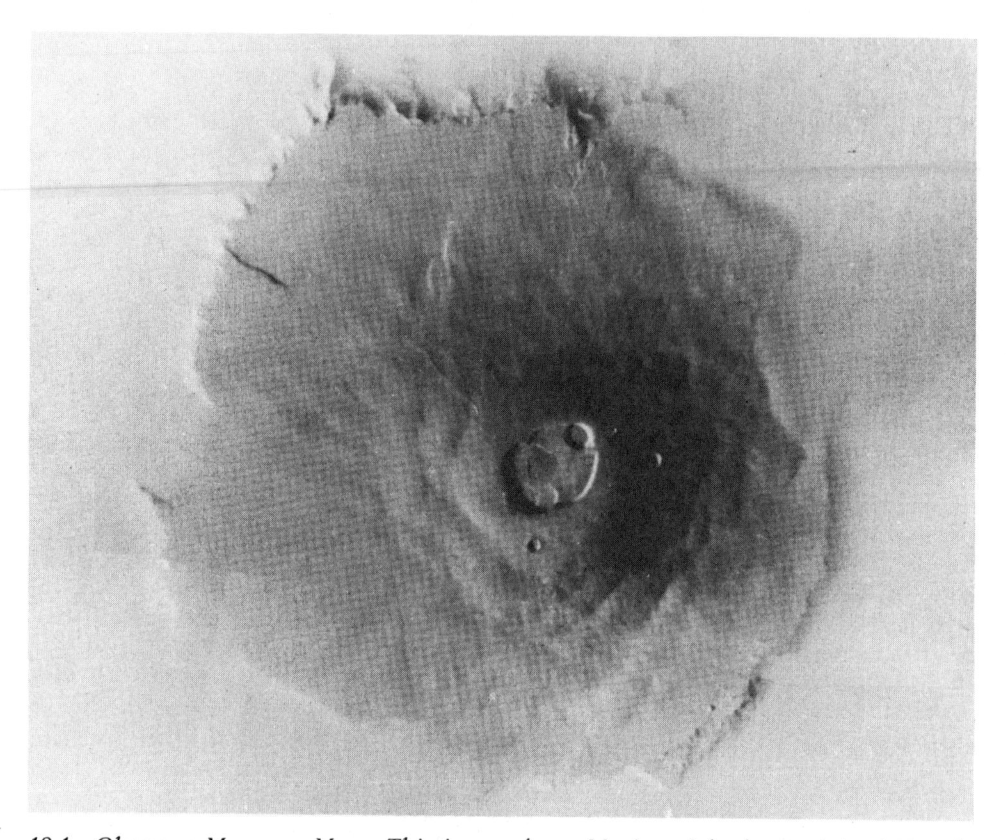

19.1 Olympus Mons on Mars. This image from Mariner 9 looks straight down the throat of this volcano, which dominates the Tharsis region of Mars. Six hundred kilometers (372 miles) across and 23 kilometers (14.2 miles) high, it is surrounded by a steep cliff. The summit caldera is 65 kilometers (40 miles) across. Cirrus clouds streaming from the peak of this volcano have been glimpsed by terrestrial observers for centuries; the mountain itself is not visible in amateur telescopes. (NASA)

beyond Earth? The results of the *Viking* lander experiments appear negative, but they sampled only an infinitesimal fraction of the total surface. The presence of water, in climatic conditions not much harsher than some in which earthly microbes have been found to thrive, suggests the possibility that, somewhere on the Martian desert, life may dwell. Should a future mission to Mars, such as a computerized roving vehicle, discover such small living things, evolved independently of Earth's teeming kingdom, surviving under such harsh conditions, the implications for the existence of life elsewhere in our universe will be immense. The answers may lie only a few tens of millions of miles away.

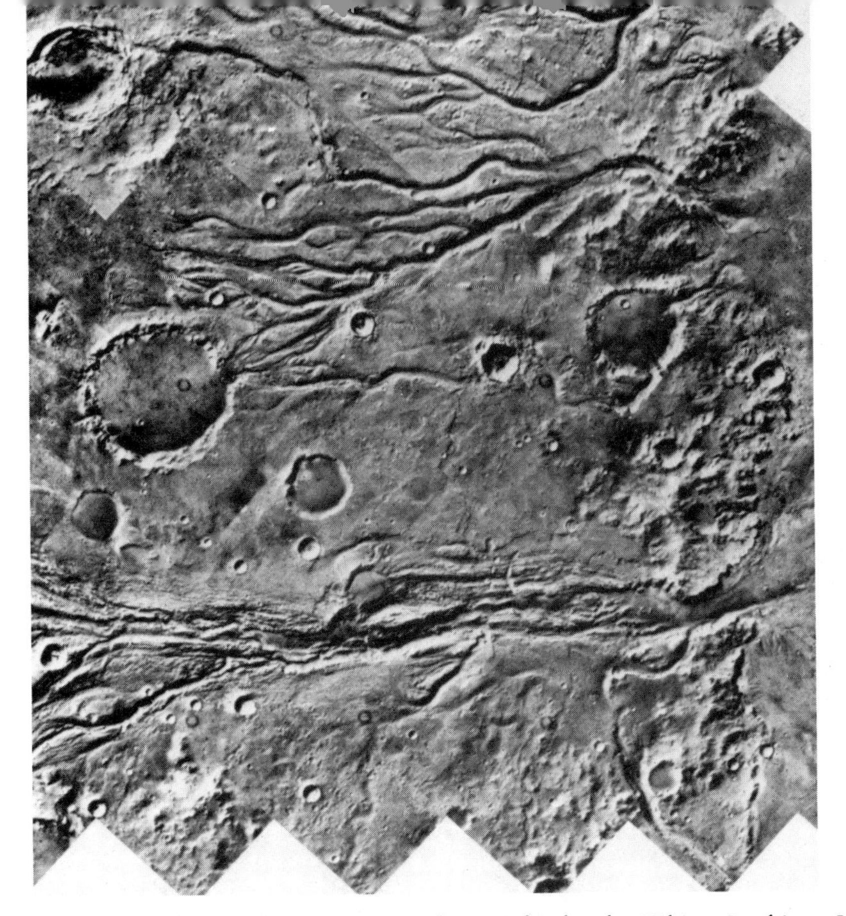

19.2 The surface of Mars, as seen from orbit by the Viking I orbiter. In addition to impact craters, channels suggestive of a massive flood cut across the terrain. The region shown is called Chryse Planitia. It is visible left of center on Martian map 1 (Fig. 19.3). Such detail is not visible in amateur telescopes; chains of large impact craters, glimpsed during moments of good seeing, probably gave rise to the belief in martian canals. (NASA)

OBSERVATION

Mars is unmistakable to the naked eye for its red color, unique among the planets and unmatched by any star along the ecliptic save Antares. It is also the fastest moving of the superior planets, marching visibly east across the sky from night to night, or swinging nimbly through its retrograde loop at opposition. It also shows the widest range in brightness, changing from magnitude 2.0 at conjunction to a brilliant −2.8 at its closest oppositions. To the telescopic observer, however, Mars can be extremely difficult, tantalizing, and at times astonishingly rewarding (Figs. 19.3, 19.4, 19.5).

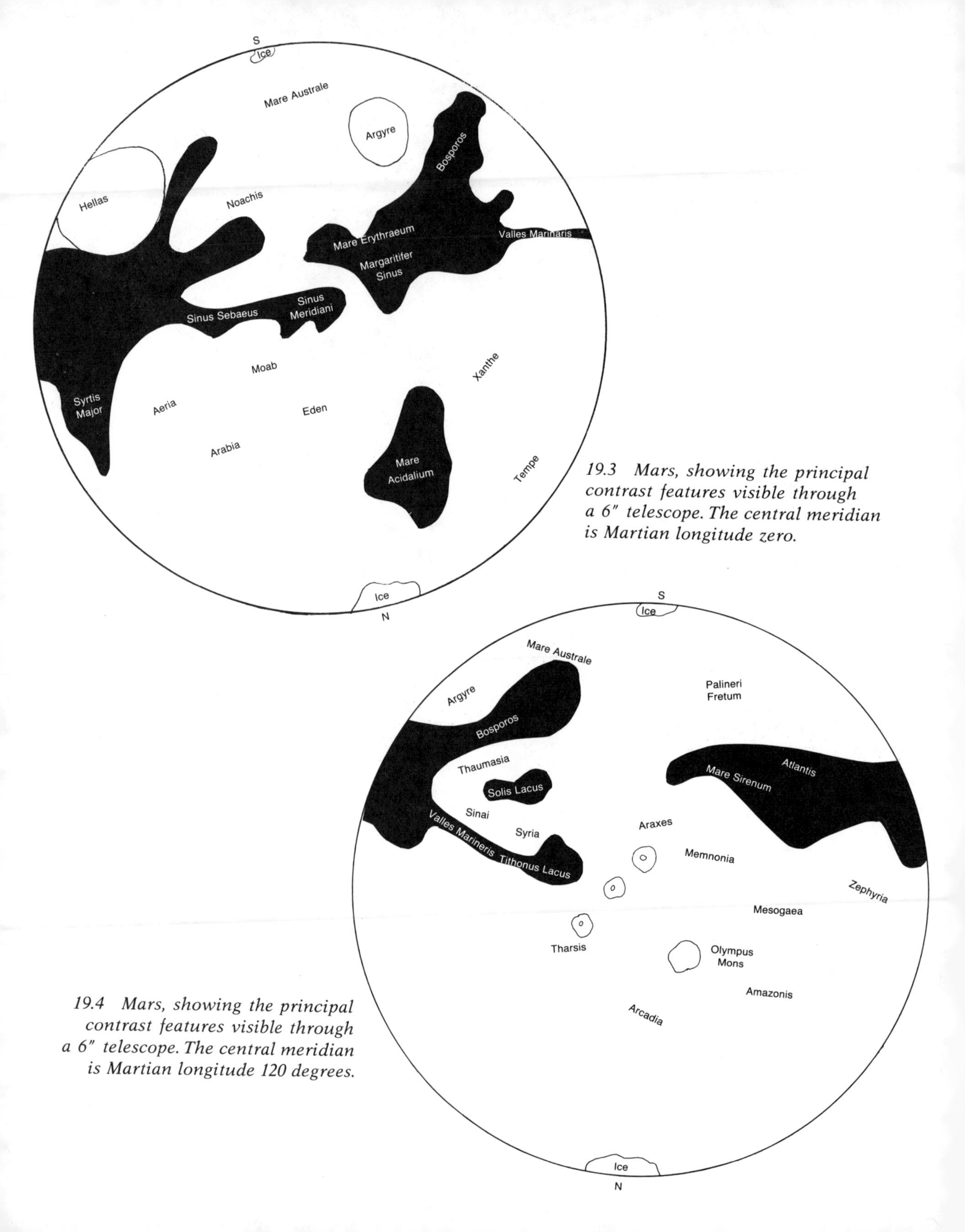

19.3 Mars, showing the principal contrast features visible through a 6″ telescope. The central meridian is Martian longitude zero.

19.4 Mars, showing the principal contrast features visible through a 6″ telescope. The central meridian is Martian longitude 120 degrees.

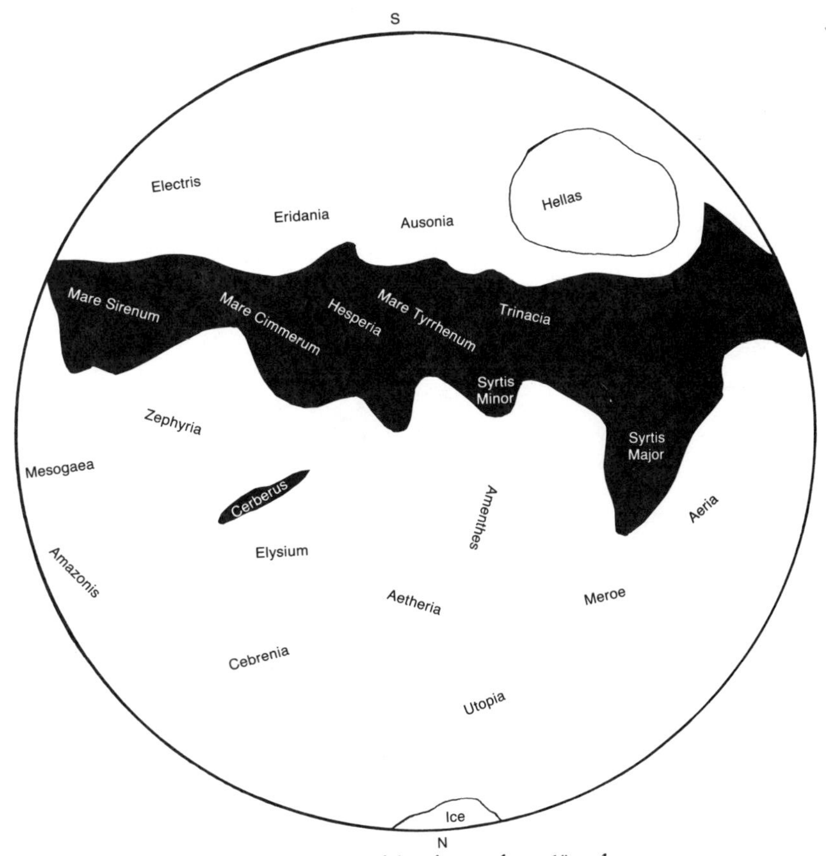

19.5 Mars, showing the principal contrast features visible through a 6" telescope. The central meridian is Martian longitude 240 degrees.

The difficulty of observing Mars stems from several factors. First, its orbit. Opposition occurs only in alternate years, leaving long months when the Martian observer must look elsewhere. But even during oppositions, conditions can be dismal. At aphelic oppositions, which occur three times for every two perihelic oppositions, the planet's disc reaches only 14 arc-seconds across. In a small telescope, such a tiny image yields very little detail. The rapid relative motion of Earth and Mars makes each opposition painfully brief, no more than 3 months before its disc shrinks too small to see well. Around conjunction, the apparent size of the disc is only a few seconds; it can be difficult to distinguish, then, from a star. Perihelic oppositions occur every 16 years, however, and conditions improve radically. The disc swells to 25 arc-seconds across, almost double its aphelic size. But even under ideal conditions, Mars mixes its blessings. The geometry of its orbit is such that it comes to perihelion opposition only in the month of August. It appears then in the southern arc of the ecliptic, low in the southern sky where atmospheric haze dims the view.

Observation of Mars is sensitive to haze and turbulence in the atmosphere because of the fineness of the detail on its surface and the high magnifications necessary to bring them out from the small disc. In a way, this gives the amateur with a small telescope an advantage. Typical atmospheric conditions limit the amount of detail to the limiting resolutions of a 6-inch telescope. Further resolution is useless, except in rare moments when the atmosphere settles, the image of the planet clarifies dramatically, and detail that was invisible a moment before leaps out at you. These moments of good seeing have kept planetary astronomers glued to the eyepieces of large telescopes through long, frustrating hours of waiting, and for decades limited the best records of the planet's surface to drawings: photographs don't work quickly enough to take advantage of these brief moments of revelation (Fig. 19.6). During such moments, use the highest magnification your eyepieces will allow. In a 6-inch, 250× may be useful at these times.

Generally, however, the fainter details tend to disappear at such high magnifications, so the optimal view, even at moments of good seeing, will probably appear at magnifications of 150× to 180×. For ordinary viewing, you will find that lower powers, down to 100×, show all the detail available. At these powers, getting the proper focus becomes more important, especially if your telescope is relatively short focus. The surface of the planet itself can be difficult to focus on; the image will frequently be fuzzy even if your focus is correct. Use whatever faint stars are visible in the field; when those are focused to points, Mars should be focused as well.

Because Mars is exceptionally bright, it responds well to filtering, which can bring out a number of details on the planet's surface. Subtle changes at the boundaries of the polar caps as they thaw and retreat, or freeze and advance, can be brought out by neutral density or polarizing filters, which cut the glare of sunlight off the ice fields. Clouds, high in the Martian atmosphere, tend to form wherever the terrain slopes sharply upward. The Tharsis ridge and the slopes of Olympus are fertile hunting grounds for these short-lived features, which a light blue filter can bring out. Dust storms can be global or local; local storms can be brightened against the background with a light yellow filter. The dark markings commonly called canals can vary in apparent color, but are actually a tawny brown; a light brown, blue, or green filter will increase their contrast against the prevailing russet surface.

One of the most startling differences between Mars and the other bodies we have observed so far is that Mars's features change from night to night, as the planet rotates. Because Mars takes 37 minutes and 23 seconds longer to rotate than Earth, the same Martian feature will cross the central meridian of the disc 37 minutes and 23 seconds later each night. The hemisphere visible at the same time each evening will gradually change, as new terrain becomes visible on the eastern limb and familiar scenery slips off the west. After two weeks, the scene has changed entirely, and the hemisphere that was hidden from you on the first night is now facing you, and vice versa. The rotation of Mars is fast enough to be noticeable over an evening's observing:

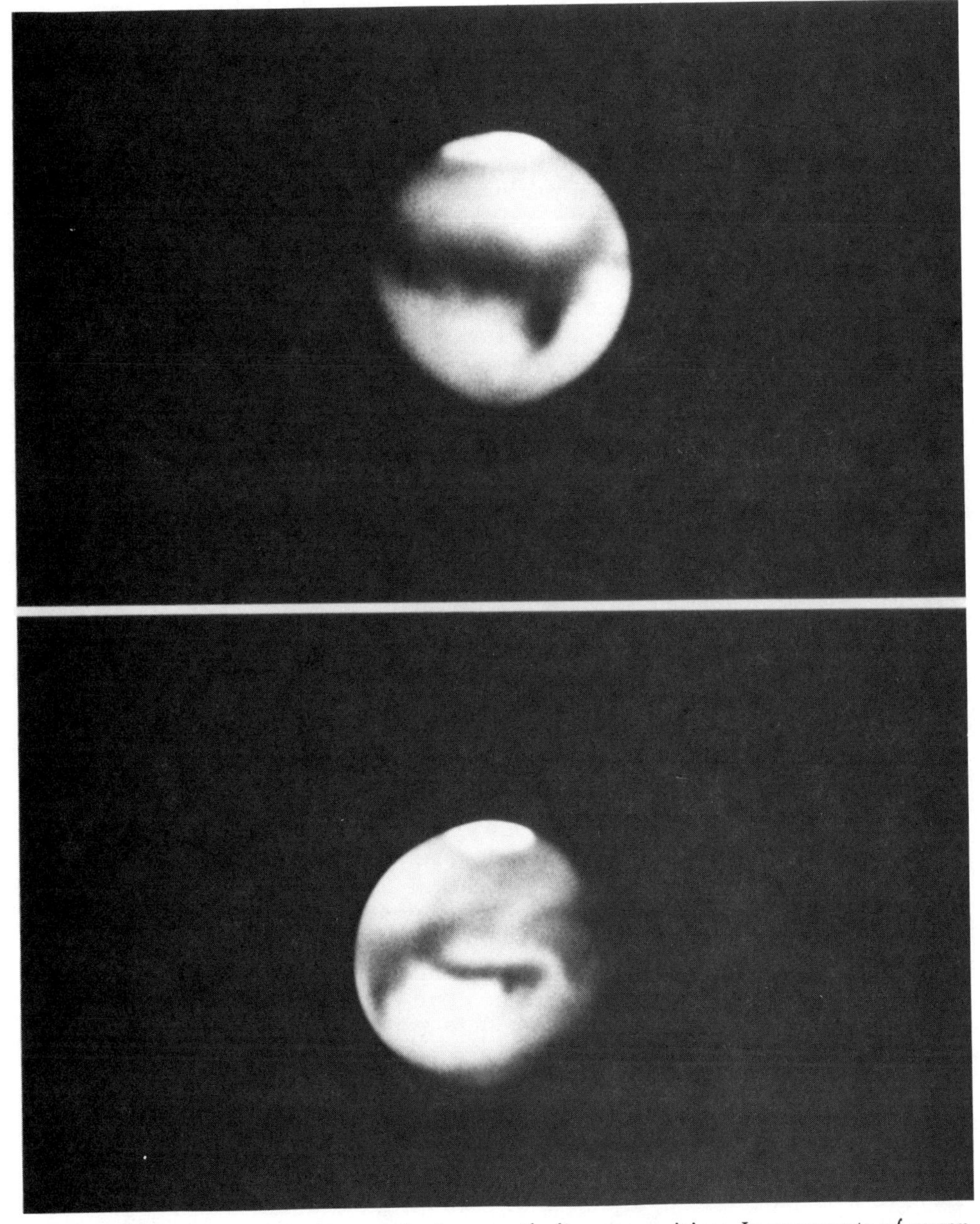

19.6 Mars: two views, taken at the last perihelion opposition. In moments of exceptional seeing, an amateur observer can glimpse many of the features shown here. Compare with maps on pp. 170–171, Figures 19.3–19.5. (Lick Observatory photograph)

a feature you noticed first at the center of the disc will be far toward the western limb a few hours later.

The maps of Mars accompanying this chapter can help you locate the large-scale features. The feature you will most likely identify first is the dark triangle of Syrtis Major; once you have that landmark, locating the others is mainly a matter of watching the planet turn. Remember, however, that the visibility of these features depends primarily on their contrast with their surroundings. Wind-borne dust can change the pattern of contrasts from week to week, as dark areas disappear under a blanket of dust, or wind scours away an obscuring layer elsewhere.

You can start to acquire necessary skills by looking for the most accessible surface features. The biggest of these isn't actually a surface feature, but a phase effect, such as that responsible for the changing phases of the inferior planets. Even though Mars is a superior planet, its orbit is close enough to Earth's to grant us a glimpse of part of its unlit face around quadrature. The phase can be as small as 85%, at which times the planet looks distinctly gibbous.

The polar caps are the most noticeable surface features. Their presentation toward us varies at different times of the Martian year. When Mars is in Sagittarius, Capricornus, and Aquarius, the southern pole tilts toward us, and the northern cap can be hidden beyond the planet's northern limb. When Mars is in Gemini, Cancer, and Leo, the north pole tilts our way. Seasonal temperature variations can dramatically alter the size and shape of either cap, the southern being the more volatile of the two. At times, it can even break up into distinct fragments. As each cap shrinks, it may be surrounded by a darker region, *Lowell's Band.* Once thought to be caused by melted water, this is probably a haze formed as frozen carbon dioxide sublimes into a dense, low-lying fog, sliding down the valleys of the Martian polar wastes.

The dark areas of Mars, the most prominent of which is *Syrtis Major,* were once thought to be oases of vegetation. Others likely to stand out are the *Sinus Sabaeus,* a long, narrow stain stretching west from Syrtis. These are areas of rough highland terrain, scoured clean of brighter dust particles by the winds that sweep the Martian surface.

Wind erosion also explains the *wave of darkening* that occurs around the Martian equinoxes, causing darker features to stand out more sharply against the lighter background. They were once thought to be a vernal greening, as water flowed from the melting polar caps toward the equator. Seasonal variations in wind velocities are the real cause. Watch for changes in the shapes of large dark areas; they may be accompanied by the appearance of yellow dust clouds. Often a thousand miles across, these clouds can move across the surface day by day. Around the Martian equinoxes, small dust clouds can grow to hemispheric or global size. Major dust storms can change the shape of the large basins of the southern hemisphere, *Hellas* and *Argyre,* which are ordinarily visible as roughly circular, light-colored regions. Other transient phenomena to watch out for are ordinary clouds, much like the high, thin "mare's tails" of our own skies. These can be most visible along the limb of the planet, but you may catch a glimpse of them along the Tharsis ridge, or downwind (south and east) of Olympus.

A survey of Mars isn't complete without mention of the two moons of Mars, Deimos and Phobos—from the Greek words for "terror" and "fear." They are far too small to be visible in amateur telescopes. Even though they are bright enough, at magnitudes 12.8 and 11.6, respectively, the glare of the planet drowns them out entirely. Until recently, we knew little about them, but the *Viking* orbiter managed to obtain pictures good enough to reveal them as tiny, rocky worlds, pitted with craters. They are probably asteroids, captured by Mars from the asteroid belt that fills the vast emptiness between its orbit and Jupiter's.

The next perihelic opposition is rapidly approaching. For the rest of the 1980s, conditions at opposition will be good. After the start of the next decade, however, conditions decline appreciably. Forthcoming oppositions, and conditions, will be:

Date	***Constellation***	***Angular Diameter (arc-seconds)***	***Magnitude***
7/10/86	Sagittarius	22	−2.4
9/28/88	Pisces	24	−2.6
11/27/90	Taurus	18	−2.0
1/7/93	Gemini	14	−0.9

Asteroids

20

PHYSICAL CHARACTERISTICS

Also called the *minor planets,* the asteroids are a collection of small, rocky bodies, ranging in size from 1,000-kilometer (620 miles) Ceres to untold millions of boulder-sized fragments. The majority of these worldlets orbit the sun in a region called the *asteroid belt,* a doughnut-shaped space lying along the ecliptic between the orbits of Mars and Jupiter, between 2.5 and 3.5 astronomical units from the sun. How did this belt get there? Astronomers have long suspected that Jupiter, with its enormous tidal influence, has played a role in the formation of the asteroid belt, by preventing planetisimals in that region from gathering into a single planet.

Our knowledge about the asteroids is limited because they are small, dark worlds, orbiting much farther from us than Mars. Only a handful of them are larger than 161 kilometers (100 miles) across. They are important because at least some of them appear to have endured for billions of years, unchanged by the geological forces of pressure, melting, impact, and erosion that have radically changed the chemical compositions of the planets. If we could obtain samples of those asteroids, we would have some of the original material from which the solar system formed, with many of its components intact. At present, the most detailed theories about their chemical makeup are inferred from analyses of meteorites found on earth, which are probably asteroidal fragments.

Meteorites occur in two major groups, the metallic and stony. Some members of the latter class, rich in carbon, water, and other light materials, appear to be the remnants of the original solar nebula. There are several clues toward identifying asteroids with these classes of meteorite. One of the most important is the observed *albedo* of the asteroids—their ability to reflect sunlight. A stony, or *chondritic,* meteorite has a very low albedo, about 3.5%; it reflects poorly, and so appears dark. Seventy-five percent of the known asteroids share this albedo, suggesting that they are made of the same dark material. The rest of the asteroids have higher albedos, roughly five times brighter than the chondrites. These are apparently rocky or metallic bodies, and may be the fragments of asteroids that were once large enough to melt and separate, like the planets, into metallic core and rocky outer layers. Such large bodies would have been ground to pieces by collisions in the crowded belt. Fragments from the core of a large asteroid would appear today as predominantly metallic; pieces from the crust would be stony.

Asteroids tend to cluster within the belt at certain orbital distances. Between these thickenings in the asteroid belt are noticeable gaps, called *Kirkwood gaps,* where few asteroids appear. These are regions where *orbital harmonics* occur between the orbits of the asteroids and Jupiter. At these orbital radii, an asteroid would revolve at a rate that would bring it into opposition with Jupiter at the same spot in its orbit every few revolutions. This constantly repeated tug from Jupiter's gravity, coming at the same part of the orbit, soon moves the asteroid out of that region of the belt. Eventually that region is swept clear, leaving the gaps we see today.

There are three groups of asteroids that orbit outside the belt. These are the Trojan, Apollo, and Amor groups. The Trojans share Jupiter's orbit, at the *Lagrangian points* of the giant planet's orbital path. The Lagrangian points are the two regions 60 degrees ahead or behind any planet in its orbit. There the gravitational pull of a planet and the sun balance, and a body can maintain a stable orbit indefinitely. The Trojans are asteroids that wandered into the Lagrangian points of Jupiter's orbit and stayed there. The Apollo and Amor asteroids, however, occupy unstable orbits and will eventually leave them. These orbits are unstable because they cross the orbit of a planet—Earth and Mars, respectively. Eventually, a close encounter between such an asteroid and a planet will (1) drop the asteroid into a lower-energy orbit, closer to the sun, or (2) whip it out farther away from the sun, or (3) send it crashing into the planet's surface.

Collisions between the earth and a sizable asteroid happen rarely, but they happen. We take a direct hit from an asteroid about 16 kilometers (10 miles) across, every hundred million years or so. Such a collision leaves an impact crater over 100 miles across; Hudson Bay may be one. It also launches a cloud of material—pulverized rock, water vapor, molten metal—into orbit around the earth. The cloud can linger for several years, blocking out so much sunlight that plants die—and most large animals that depend on plants for food. Many scientists now believe that such an event was responsible for the "great dying" that ended the Cretaceous era on earth—the catastrophe in which the dinosaurs disappeared. There are several dozen known asteroids in the Apollo class. Not all of them will necessarily collide with earth: near misses are far more likely. But sometime over the next few dozens of millions of years, our turn will come again.

OBSERVATION

Even with very large telescopes, observing asteroids is more a matter of hide-and-seek than anything else. Too small to show a disc, they are distinguishable from stars only by their motion across the sky from night to night. This motion can be too slow to notice over the course of one evening, but it can be fast enough to move an asteroid completely out of the field if you don't watch for it the next night. Observing an asteroid involves some careful detective work.

With the four brightest asteroids, all of which reach about magnitude 6 or 7 at opposition, your sleuthing is simplified. The *Observer's Handbook* and *Sky & Telescope* provide finder charts for the brighter asteroids. There are more known asteroids than you could find and follow in a lifetime, but the ones bright enough to track with modest equipment are far fewer. Asteroids are iden ified by a name, usually that of a mythological or historical character, and by a number. Numbers are assigned in order of discovery: "minor planet #1," for instance, denotes Ceres, the first asteroid to be discovered. Most of the minor planets visible in amateur equipment have low numbers. To find the dimmer asteroids, start by getting their coordinates from an ephemeris and plot that position on your atlas. (Because these coordinates are often given for the current epoch, there will be a slight discrepency with the coordinate system of your chart.) For example, on 9 November 1980, Vesta was 2 degrees north and 0.5 degrees east of Regulus. A passage close to a bright marker star such as this is an ideal time to try to find an asteroid. At other times, you will need to use dimmer stars as reference points, offsetting with your setting circles or hopping field widths until you reach the asteroid's position. Compare the pattern in the eyepiece with the pattern of stars on the chart: the "extra" star in the telescopic field is your asteroid.

There are scores of asteroids within reach of a 6-inch telescope. They can be bright and at times, pass close enough to earth for their motion to be easily visible against the stars. Most are dim and distant, moving only slowly over the weeks. Their orbits can be regular circles or highly inclined, eccentric ellipses. Once you find your first, you may find that the sensation of seeing one of the loneliest, smallest objects you can detect in space is addictive. Some amateurs do nothing but find and track asteroids, collecting them the same way birders build life lists. One aspect of asteroid watching has some immediate scientific interest as well. This is the observing and timing of asteroidal occultations of stars. A recent occultation has indicated the possibility of double asteroids, or asteroids with tiny moons. Occultations are the best way of detecting such pairs, and many are within reach of amateur equipment. See the section on occultations in chapter 15 for more on occultations, and watch the astronomy magazines for periodic bulletins on upcoming asteroidal occultations.

The five brightest asteroids are:

1. *Vesta (minor planet no. 4)* has a diameter of 378 kilometers, but is almost a full magnitude brighter than larger Pallas (below) at opposition. Its closer opposition distance (stemming from a shorter orbital period, 3.6 years), accounts for some of this brightness, but a high surface albedo must also be involved.
2. *Pallas (no. 2).* Pallas, 485 kilometers in diameter, reaches magnitude 6.3 at opposition, making it an easy binocular object at those times. Its orbital period is 4.6 years.
3. *Iris (no. 7)* reaches magnitude 6.7 at opposition, yet its diameter is a scant 125 kilometers. Its orbital period is 3.7 years.
4. *Juno (no. 3)* is the fourth largest known asteroid, at a 118-kilometer diameter, and achieves an opposition magnitude of 6.9. Its orbital period is 4.4 years.

   The last object on this list is not among the brightest asteroids, but it deserves mention for two reasons: it is the largest of the minor planets and was also the first discovered.
5. *Ceres (no. 1).* Diameter, 768 kilometers; opposition magnitude, 7.0; orbital period, 4.6 years, almost exactly that of Pallas.

Jupiter

21

PHYSICAL CHARACTERISTICS

More than ten times the earth's diameter, 318 times more massive, Jupiter is the largest planet in the solar system. Yet all that bulk spins on its axis once every 9 hours, 51 minutes. That combination of size and energy is typical of Jupiter, which is both an enormous and enormously active world.

Some 133,200 kilometers (82,600 miles) from pole to pole, Jupiter is 9,500 kilometers (5,890 miles) broader through its equator. Its speed of rotation is so high (almost 45,160 kilometers, or 28,000 miles, per hour at the equator) that it warps the planet from a sphere to a distinct ellipse. Its great mass is mainly gas and liquid, however, so slushy that it cannot even rotate as a solid. The equatorial regions complete a rotation five minutes faster than the rest of the planet. Jupiter's average density is only 1.3 times that of water—less than a fourth of the earth's. That average figure covers an enormous range of densities, however; from a thin upper atmosphere of light gases and ices to a core so tightly compressed that hydrogen becomes a metal.

Orbiting far from the sun's heat (5.2 astronomical units, or almost a billion kilometers), Jupiter's great mass helped it to hold onto all of its primordial atmosphere. Its atmosphere today is a mix of gases dominated by hydrogen (almost 90% of Jupiter's upper atmosphere) and helium (10%), with less than 1% composed of water, methane (swamp gas), and ammonia, and other gases—all molecules that are abundant among the stars. In fact, the composition of Jupiter's atmosphere varies by only 1% from that of the sun. Unlike the terrestrial planets, Jupiter and the other Jovian planets, because of their great mass, retained these volatile gases. Instead of boiling off into space, much of these gases froze, forming the ices of water, methane, and other compounds, crystals of which give the planet's banded atmosphere its distinctive colors.

The Jovian planets do not have solid surfaces. The atmosphere of these worlds simply thickens into slush, which thickens further, until, deep inside, the enormous pressure forces the hydrogen into a solid, metallic form. At the very center of Jupiter lies a "tiny" core of earthlike, rocky material—about three times the mass of the earth.

Along with a range of densities, Jupiter's sphere also exhibits an extreme range of temperatures. At its uppermost levels, Jupiter's atmosphere is less than 100 degrees Kelvin (−279 degrees Fahrenheit). As in any atmosphere, the temperature rises with increasing depth, so that, 100 kilometers (62 miles) below the cloudtops, the temperature rises above room temperature. Increasing pressure heats things further, so that, at its center, Jupiter shines with an infernal heat of 25,000 degrees Kelvin. The pressures there are 80 million times that of the air around you. Under these conditions, strange changes occur in the ordinarily light gases that make up the planet, which is why the bulk of Jupiter's interior is apparently *solid* (not frozen) hydrogen. It acts as a metal, capable of generating an enormous electromagnetic field as the planet spins, like the molten iron core of the earth.

THE JOVIAN ATMOSPHERE

The face of Jupiter visible in a telescope (Fig. 21.1) is nothing but cloud. These clouds are a mix of many gases, containing traces of complex molecules, many of which share a fascinating element: carbon. Carbon-containing molecules are the basis of life on earth. The abundance of carbon compounds on Jupiter, along with its wide range of temperatures, suggests that Jupiter may be able to evolve life. Carbon forms a variety of compounds in the Jovian atmosphere, among them acetylene, ethane, ethylene, and methane. These different chemicals, along with others such as phosphine, hydrogen-sulfur compounds (including the chemical responsible for a rotten egg's smell, hydrogen sulfide), and several molecules containing ammonia, churn up and down in the winds of the Jovian atmosphere. They alternately freeze in the upper clouds and break down in the deeper furnace. This cycle of burning and freezing drives a complex chemistry, in which chemicals crack and combine to give the Jovian clouds their many subtle colors, from the salmon pink of the Great Red Spot to the cream, white, and brown of its bands and belts.

The gases of Jupiter's atmosphere lie in a series of layers. Uppermost, in the planet's *exosphere,* is a haze of hydrogen and helium, perhaps containing a thin layer of meteoritic dust. This layer is transparent and does not block our view of the uppermost visible layer, where clouds of frozen ammonia crystals float at a subarctic 150 degrees Kelvin (−190 degrees Fahrenheit). Another clear layer of light gases lies above the next deck of clouds, which is

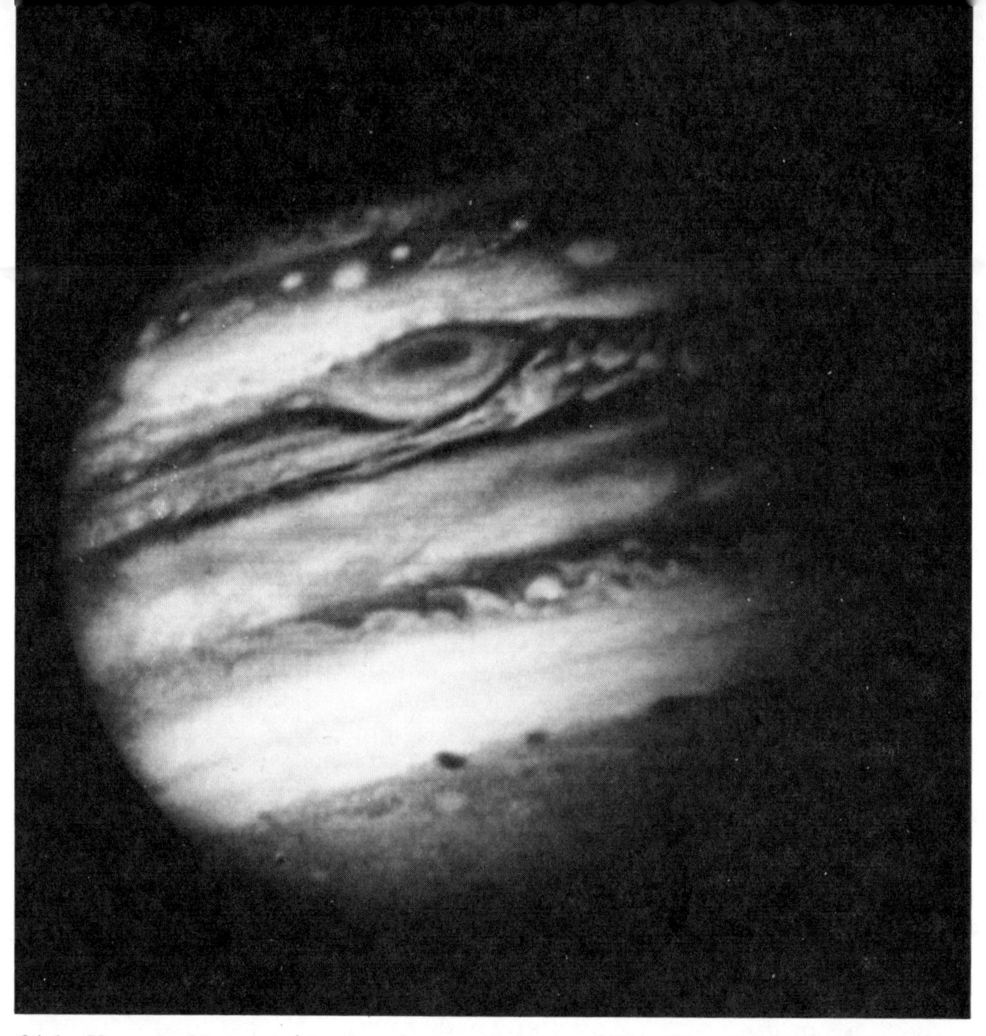

21.1 Voyager 1 image of Jupiter, from a distance of 54 million kilometers (32 million miles). The Great Red Spot, surrounded by turbulence, appears below center. The resolution of this image exceeds that of the finest earth-based photographs, although traces of the fine detail shown here can occasionally be glimpsed by the amateur observer during moments of good seeing (NASA/JPL).

a mixture of frozen ammonia and hydrogen sulfide. These give the tan color that dominates the planet's appearance in the telescope. The lowest visible layer, at temperatures around 273 degrees Kelvin (0 degrees Celsius, 32 degrees Fahrenheit), would look familiar: high cirrus clouds of frozen water, thickening into a fog of water droplets. Below this layer, the atmosphere clears as it thickens. Those clouds we see from earth are from 100 to 300 kilometers (62 to 190 miles) deep. They form only the thinnest skin of an atmosphere that continues for thousands of kilometers beyond our sight.

The innermost layer of Jupiter's atmosphere is difficult to define. If we were to probe deeply enough, we would find a region where the mixture of hydrogen and helium gradually condenses into a thick haze, which simply thickens into a sea of liquid hydrogen. On earth, we use this compound as rocket fuel. This sea is about 20,000 kilometers (12,400 miles) deep; its floor is metallic hydrogen, which occupies the bulk of the planet, down to the small, solid core, composed of quartzlike rock.

JOVIAN WEATHER

The Jovian atmosphere churns continually up and down, rising from the warmer lower levels to the icy upper layers, from which it sinks again. The planet's rotation whips these vertical air currents out into great bands rushing parallel to the planet's equator. Wind speeds, generally from the west, rise above 362 kilometers (225 miles) per hour. These are not hurricane winds; they're of tornado force, and they blow perpetually, around the entire planet. Moreover, the winds in one band of latitude can blow in the direction opposite to those in the adjoining region. At 20 degrees north Jovian latitude, for instance, winds blowing at a relative speed of 362 kilometers (250 miles) per hour pass each other in opposite directions.

When this happens on earth, at speeds of 64 kilometers (40 miles) per hour or so, the resulting storm can spawn dozens of tornadoes. On Jupiter, whirlwinds also form at these boundaries. The largest example of these is the Great Red Spot, a vortex larger than the earth. Churning incessantly at hurricane force, it has been observed in the Jovian atmosphere for centuries and shows no sign of blowing itself out. The Spot, 40,000 kilometers (25,000 miles) long, 14,000 (8,700 miles) across, is simply the biggest of hundreds of similar storms, some of which could swallow the entire earth.

THE MAGNETIC FIELD

The magnetic field generated in Jupiter's metallic core surrounds the planet, trapping charged particles in belts above its equator. Within Jupiter's belts, particles can reach temperatures comparable to that of the sun's corona. They are so thinly dispersed, however, that the *Pioneer* and *Voyager* probes flew through them relatively unscathed. The magnetic poles of Jupiter, like Earth's, are tilted from its rotational axis. As the planet rotates, this tilt causes its magnetosphere to gyrate wildly. At the magnetic poles, charged particles spiral inward from space until they cause the upper atmosphere to glow in giant aurorae. The brief, stormy night on Jupiter is also punctuated by brilliant flashes of lightning.

THE RINGS

A set of dark, narrow rings (Fig. 21.2), much smaller and darker than Saturn's but essentially similar, extends from 53,000 kilometers (33,000 miles)

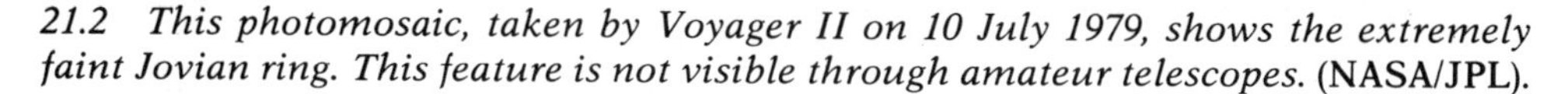

21.2 This photomosaic, taken by Voyager II on 10 July 1979, shows the extremely faint Jovian ring. This feature is not visible through amateur telescopes. (NASA/JPL).

to perhaps as far as 290,000 kilometers (180,000 miles) above the clouds. There are apparently three main rings lying at the outer edge of this region. The broadest of these is 6,000 kilometers (3,800 miles) across; the narrowest, one-tenth that size. At their thickest, however, these structures are less than 30 kilometers (19 miles) thick. These immensely flat, thin rings are composed of grains of dust about a micrometer (one-thousandth of a millimeter) wide. They were detected indirectly by the *Pioneer 11* probe, but the discovery was not confirmed until *Voyager 1* sent back an image of a delicate arc of light across the Jovian sky. So small and faint, these rings are visible from earth only in the largest telescopes, using the most sophisticated of techniques. Their importance to astronomers lies in their being one example of a phenomenon that at least two other Jovian planets (Saturn and Uranus) share.

THE GALILEAN MOONS

Named for their discoverer, Galileo Galilei, the Galilean moons are also named, in order of distance outward from Jupiter, Io, Europa, Ganymede, and Callisto. They are but the four largest members of a numerous family of satellites that orbit Jupiter. We now know of well over a dozen members, ranging in bulk from Ganymede, which is as large as the planet Mercury, to several that are only small asteroidlike bodies. The Galilean moons are all at the large end of this range; if they appeared on their own anywhere else in the solar system, we would call them planets.

Io, the innermost of the four, orbits Jupiter at a mean distance of 413,000 kilometers (256,000 miles), taking 1 day and 18.5 hours to complete one revolution. So close to Jupiter, Io's orbit lies inside the Jovian magnetosphere, and its surface is bombarded by charged particles. The impact of these particles is enough to etch away the material that coats the Ionian landscape, launching particles of Io into space. As this moon orbits Jupiter, it leaves a trail of sodium, sulfur, potassium, and other elements in a great doughnut-shaped path around the planet.

The spectacular images of Io returned by the *Voyager* probes revealed a bizarre world. Io's face is predominantly orange. Its color is thought to come from a thin surface coating of sulfur, which spews constantly from the interior of Io through the vents of active volcanoes. Although Mars has volcanic cones, Io is the only known world in the solar system (besides earth) with active volcanoes. The force responsible for this volcanic activity is apparently the immense gravitational pull of Jupiter. Tides within Io so stress its interior that much of it must be molten. This molten material generates pressure, as molten rock does within the earth, and bursts to the surface.

Europa is the next moon out from Jupiter, orbiting at a distance of 671,000 kilometers (416,000 miles), taking three days, 13.22 hours to complete one revolution. The radius of Europa, 1,563 kilometers (969 miles), makes it the smallest of the Galilean moons. Europa's surface, as revealed by the *Voyager*s, is a delicate, golden shell of ice, streaked by a fine network of cracks. These are fractures in the icy surface. The ice crust is less than 100 kilometers (62 miles) thick, overlying a rocky interior. One of the smoothest bodies in the solar system, Europa's surface has almost no significant hills or valleys and few visible impact scars. Apparently the ice is not strong enough to sustain mountains or valleys, rills or craters. It flows slowly, flattening any irregularities left by local impacts or global stresses.

Ganymede, orbiting Jupiter 1.07 million kilometers (660,000 miles) out, requires one week, 3¾ hours to complete a revolution. At 2,638 kilometers (1,636 miles) radius, it is the largest satellite in the solar system. Unlike Europa, its surface is dark and spotted with impact craters. Ganymede once had a liquid inner layer underlying its solid crust. This mantle was not molten rock, however, but liquid water. Gandymede was at one time one of the few bodies of the solar system to possess seas, but they surged underground, forever shut away from the sky. They have since frozen, but even so they occasionally manage to escape to the surface. Whenever a large meteorite strikes the icy crust, ice from the interior flows through the crater and the surrounding cracks, overspreading the surface with bright, fresh ice. These bright spots stand out against the predominantly dark surface, which is composed of ancient, brutally cratered terrain. Cutting across the pitted waste are vast arrays of parallel grooves, apparently the result of stretching forces within the crust.

Callisto (Fig. 21.3), outermost of the Galilean worlds, stands 1.88 million kilometers out from Jupiter. At that distance, it requires 16 days, 16½ hours to complete one orbit. Callisto's surface is the darkest of the four. Even with a diameter of 4,820 kilometers (2,988 miles), making it the second largest of Jupiter's moons, it appears from earth as the dimmest of the lot, reaching only magnitude 6.1 at brightest. Its structure is apparently similar to Ganymede's. Callisto, too, once had subsurface seas, but with a thicker crust of ice and rock, less fresh ice escapes to brighten its surface.

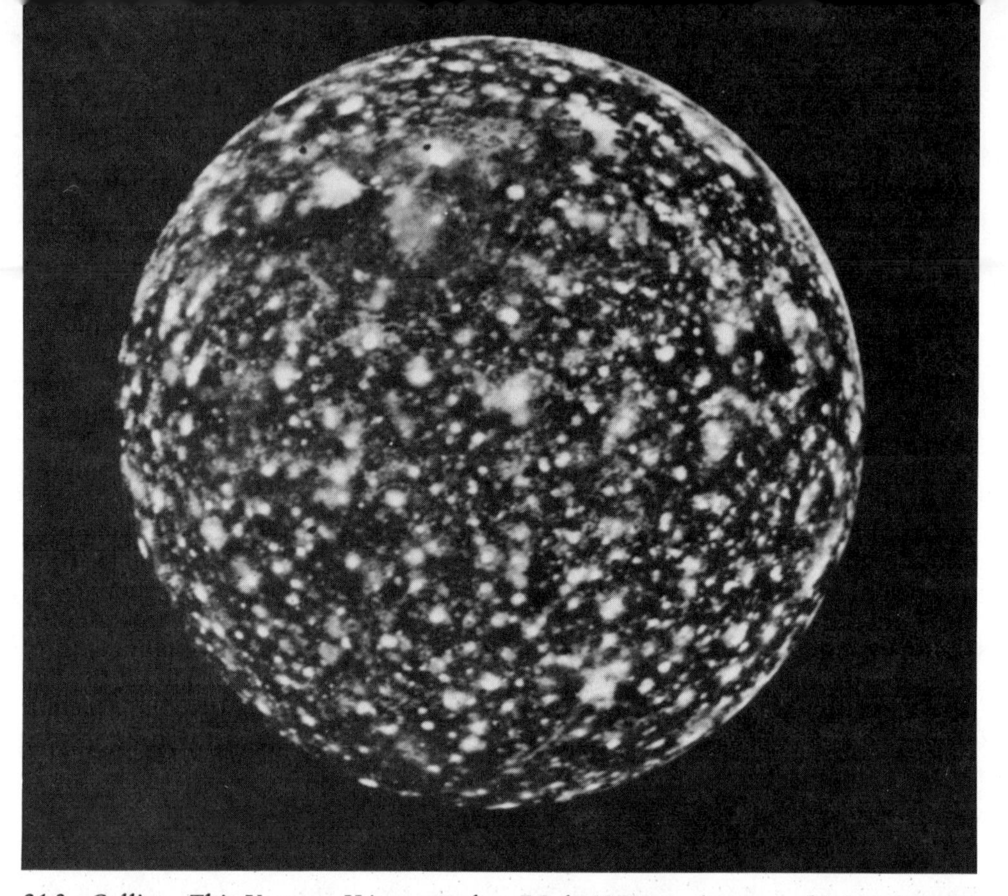

21.3 Callisto. This Voyager II image, taken 7 July 1979 at a distance of 1.1 million kilometers (675,000 miles), shows the brutal cratering of Callisto's dark surface. The surface has apparently not changed since Callisto's formation, over 4 billion years ago. Bright spots are fresh ice, which oozed out of Callisto's interior after impacts. Callisto appears in amateur telescopes as a featureless spot of light. (NASA/JPL)

OBSERVATION

To the naked eye, Jupiter is rivaled in brightness only by Venus and Mars. It is distinct from these two in its color, which is a faintly creamy white, and in its stately motion among the constellations. The orbit of Jupiter lies within 1.3 degrees of the ecliptic, so it frequently passes close to the several bright stars, such as Regulus and Antares, along the ecliptic's path.

Taking 11 years, 315 days to complete one orbit around the sun, Jupiter spends one year in each of the constellations of the zodiac. Its synodic period is 398.88 days. For the rest of the decade, it will appear in the constellations indicated in the following table:

Date	Constellation	Angular Diameter (arc-seconds)	Magnitude
8/4/85	Capricornus	49	−2.3
9/10/86	Aquarius	50	−2.4
10/18/87	Pisces	50	−2.5
11/23/88	Taurus	49	−2.4
12/27/89	Gemini	48	−2.3

Jupiter is so large and bright that it offers good viewing almost all year. The advantage to observing near opposition is that Jupiter will be visible in a dark sky all night long.

Its brightness at opposition ranges from −2.0 to −2.5, at which time it achieves an angular diameter of from 44.2 to 49.9 arc-seconds. These are average figures; because of Jupiter's equatorial bulge, the angular diameter varies, depending on where you measure it, by almost 3 seconds. At conjunction, the planet dims only to magnitude −1.2, and its apparent size decreases to around 30 arc-seconds.

Jupiter is large enough to appear as a visible disc through binoculars. If you look closely on either side of the planet, you will see as many as four tiny points of light, the Galilean moons. Over successive nights, you can watch their positions change as they pursue their orbits around Jupiter.

Through a 6-inch telescope at powers of 30× or less, the first detail you are likely to notice is a brownish band crossing the surface of the planet parallel to its equator and slightly north of center (Fig. 21.4). With higher magnifications, more of these dark bands, or *belts*, appear—as many as seven of them under good conditions, as well as the dark polar regions. These belts, and the intervening, lighter-colored *zones*, mark the wind currents that blow

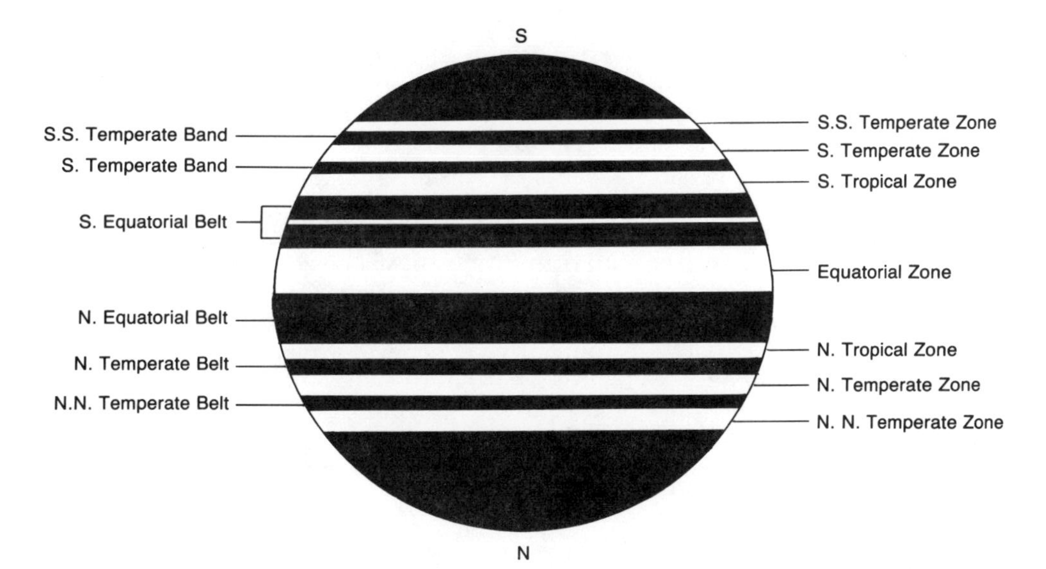

21.4 Bands and zones in Jupiter's atmosphere.

around Jupiter. The winds tend to blow toward Jupiter's west in the belts, and eastward in the zones, with the highest velocities in the equatorial and north tropical zones.

The number and clarity of these bands varies with the quality of your seeing. On some nights, only the north equatorial belt will stand out clearly. Focusing the disc can be especially difficult at such times. Even with good seeing and sharp focus, the planet's limb will still look fuzzy. Jupiter's atmosphere absorbs light, and when we look toward the limb we are looking through a greater thickness of atmosphere, causing the limb to blur and darken. The starlike points of the Galilean moons are the best guides to focus on.

Your first views of the planet will probably look fuzzy no matter what; as with Mars, the best views come only during moments of good seeing. A 6-millimeter eyepiece, giving 127× in a 6-inch, *f*/5, gives good results most nights; you will very occasionally find a night steady enough to make a 4-millimeter, at 190×, useful in a 6-inch.

After the belts and zones, the most noticeable marking is likely to be the Great Red Spot, which lies on the border between the south tropical zone and the south equatorial belt. The Spot is not as easy to see as you would think, despite its size, primarily because its color is not as distinct as it is often pictured (especially in the computer-enhanced pictures returned by the *Voyagers*). Another reason, frequently forgotten, is that the odds are even that the Spot will be on the far side of Jupiter any time you go looking for it.

The color of the Spot varies from red to salmon to gray. At times it has disappeared entirely. The spot is only one of several such disturbances you may glimpse during periods of exceptionally good seeing (Fig. 21.5). These take the form of dark- or light-colored ovals, similar to the Spot, although usually less than half the size. They appear most often on the borders of the south and north tropical zones, around latitudes 20 degrees north and south. Also occasionally visible are *festoons,* elongated whorls on the borders of the zones and belts. They form as the colliding winds set up eddies across the disc of the planet. These occur most commonly trailing west of the Great Red Spot in the equatorial belt.

One of the most pleasing aspects of Jupiter's visible face is its changefulness. Any of the belts or spots is liable to darken or lighten over the course of an apparition. An entire belt can split in two lengthwise, or fade so much as to disappear into the surrounding zones. The south equatorial belt is especially volatile, sometimes fading entirely away. Occasionally, it darkens, churning out dozens of eddies along its edges. You won't notice these changes until you become familiar with the general appearance of the disc. Filters can be helpful here. Light yellow, light brown, pink, and light blue filters can all accentuate various features, especially the Red Spot. Because the background of the planet is so bright, you might want to accentuate the darkness of the spot by using a blue filter. Red can be useful for locating the lighter oval disturbances.

The Galilean satellites orbit above Jupiter's equator. Jupiter's equator is tilted slightly more than 3 degrees from its orbit, and Jupiter itself can be as much as 1.3 degrees from the ecliptic. These angles affect the visibility of eclipses and transits of the satellites. Depending on the angle between Jupiter's equator and Earth, the phenomena of the satellites include:

1. Occultations, in which the satellite disappears behind the disc itself
2. Eclipses, in which the satellite, as it goes around the back of Jupiter, remains visible north or south of the disc, but passes into the shadow of the planet
3. Nonevents, in the case of Callisto, which is far enough from Jupiter to miss the shadow entirely at times

21.5 The changefulness of the Jovian disc is plainly visible in these three images, taken by Voyager spacecraft several months apart. Note the many different light and dark ovals and the changed shape of the white festoon north of the Great Red Spot. Such changes are also visible in the less dramatic detail in amateur telescopes, especially during moments of good seeing. The two larger shots were returned from Voyager II on 9 May 1979; the smaller image was taken by Voyager I on 24 January 1979. Distance for all three was around 25 million miles. (NASA/JPL)

Before opposition, when Jupiter rises after sunset, we look at Jupiter with the sun over our left (east) shoulder. The shadow of the planet lies to its west, and eclipses therefore take place on the western, or preceding side of the disc. After opposition, the shadow lies east of the planet, and eclipses occur on the following side of the disc. At opposition, Jupiter's shadow lies directly behind the planet.

The satellites can also *transit,* moving across the face of the planet (Fig. 21.6). Transits are fascinating to watch and can be of two varieties. In *disc transits,* the moon itself passes between us and Jupiter. These occur more often, especially for the outer satellites, when Jupiter is near the nodes of its orbit, lying on the ecliptic. In *shadow transits,* the shadow of the moon passes across the Jovian cloudtops, appearing as a distinct dark shadow, in an event analogous to a solar eclipse here on earth. In either case, the satellite moves across the disc from east to west. Before opposition, the shadow transits ahead of the satellite itself; after opposition, the satellite precedes its shadow.

The astronomy magazines and annual ephemerises publish charts identifying the positions of the moons around the planet for each night of the current apparition. They will also note the times and locations of occultations, transits, and eclipses. Even without such advance notice, these events are common enough that you will likely chance across one soon after you start observing Jupiter. Whenever you find Jupiter accompanied by fewer than four moons, look carefully at the planet's disc. If you see a dark spot crossing it, a shadow transit is taking place. Look carefully ahead or behind the shadow, and you may see the satellite itself. If nothing appears on the disc, an eclipse or occultation may be in progress. Look to the east of the planet (after opposition) or to the west (before opposition), and wait. Shortly, you may observe one of two things. Before opposition, the moon may emerge suddenly from the western limb of the planet, ending an occultation. After opposition, at some distance from the eastern limb (the distance is least just after opposition, increasing until quadrature), a point of light will brighten, ending an eclipse. The satellite will take several seconds to reach full brightness as it emerges through the hazy borders of Jupiter's shadow, as light filters through the high cloudtops of Jupiter's limb to flood the satellite's surface with light (Fig. 21.7). The disappearance of a satellite, especially if into the shadow of Jupiter, is equally dramatic. Before opposition, any time you find a satellite lying just west of the planet, watch to see if it is heading east, behind the disc of Jupiter. If it is, follow it until it disappears; you will see the same rapid fading, like a spark dying on a hearth.

21.6 Jupiter, in blue light, showing the Great Red Spot, its satellite Ganymede, and the satellite's shadow in transit. (Hale Observatories)

21.7 Jupiter and Io, from Voyager II at a distance of 24 million kilometers. The shadow of Io is plainly visible on the planet's disc, a phenomenon also visible from earth. (NASA/JPL)

Saturn

22

PHYSICAL CHARACTERISTICS

Second largest of the Jovian planets, Saturn (Fig. 22.1) is similar to Jupiter in form, composition, and structure. Its differences are primarily matters of degree. Its gorgeous rings, for instance, are the largest of the three known ring systems of the solar system. Its winds are far more violent, with speeds approaching 1,774 kilometers (1,100 miles) per hour. Orbiting the sun at a distance of 9.54 astronomical units, its atmosphere is colder. The temperature of the uppermost cloud deck is 80 degrees Kelvin (−315 degrees Fahrenheit). In the lower layers, formed of ammonia, ammonium hydrosulfide, and water-ice clouds, essentially similar to those on Jupiter, the temperature rises, reaching about 20,000 degrees Kelvin at the core. Saturn's lower temperature accounts for the slightly different mix of gases in its upper atmosphere. With 94% hydrogen, Saturn has retained an even higher percentage of the lightest element than Jupiter. Helium, with trace elements including methane and ammonia, makes up the rest.

The mass of Saturn, 95 times that of Earth, is less than a third of Jupiter's. Its radius is 60,000 kilometers (37,200 miles). Saturn's density, however, is less than that of water; if you could find an ocean large enough to launch it in, it would float. Like Jupiter's, the planet's atmosphere bulges outward under the centrifugal effect of its spin.

Saturn's most noticeable feature is its rings, which orbit the planet above its equator, starting about 13,000 kilometers (8,100 miles) above the cloudtops and extending to an altitude of 420,000 kilometers (260,000 miles).

22.1 Saturn. Voyager II recorded this image while yet 56 million kilometers (35 million miles) from the planet, yet detail visible in the image, such as the fine texture of the rings, already surpasses that visible in earthbound telescopes. Note the appearance of the planet's visible disc: it is much blander than Jupiter's, a difference also apparent in amateur telescopes. (NASA/JPL)

22.2 In this nearly edge on view, Voyager II demonstrates the thinness of the Saturnian rings. The small black spot near the bottom is the shadow of Rhea. (NASA/JPL)

Astonishingly, this entire structure is no more than 2 kilometers thick, forming a knife-edge so thin it disappears entirely when seen edge on from the earth (Fig. 22.2).

Although the rings appear to be solid sheets of material circling the planet, we now know they must actually be composed of small orbiting particles. Gravitational stress will tear apart any body that approaches within a certain distance (called *Roche's Limit*) of a planet. The rings of Saturn lie within the Roche limit of Saturn so they cannot be solid. Spectroscopic studies have identified the material as ordinary ice, ranging in size from small, cold-drink-sized bits, to bergs the size of automobiles.

The ring particles tend to gather at different orbital distances, much as the asteroids clump at different distances from the sun. Gaps similar to the Kirkwood gaps divide the rings into three major subdivisions, and into many hundreds of minor divisions within each. The innermost visible subdivision, called the *C*, or "crepe" ring, is 18,000 kilometers (11,200 miles) wide. It appears as a ghostly gray band of light that merges, some 31,800 kilometers (19,700 miles) above Saturn, into the brighter *B* ring. *B*, the larger of the two main rings of Saturn, extends out to 55,800 kilometers (34,600 miles). There it ends abruptly in the *Cassini Division,* an apparent gap, 4,800 kilometers (3,000 miles) across, between *B* and the outer, *A* ring. Outside this division, the *A* ring stretches almost 16,000 kilometers (9,900 miles). It contains, about ¾ of the way out, a hairline crack similar to Cassini's Division, called *Encke's Division*. The *Pioneer* and *Voyager* probes found several other regions, inside and outside these three main regions, and revealed that the entire structure is composed of hundreds of fine gradations of density, so that the entire ring system looks very much like a phonograph record.

The *Voyager* encounters with Saturn revealed an astonishing amount of hitherto unsuspected and as yet uninterpreted information about the rings. There are mysterious dark spokes radiating out across the rings, possibly caused by static electrical charges that build up among the ring particles. There are rings that orbit off-center, and rings that seem to intertwine as they are shepherded in their orbits by tiny moons. Theorists will be busy for years to come trying to puzzle out the exact causes of the ring phenomena recorded by the *Voyager*s.

New information on the moons of Saturn was another revelation of the *Voyager* missions. The *Voyager*s discovered a system of small worlds no less complex and wonderful than Jupiter's. Of these, five are easy targets with a 6-inch telescope, and two more are within reach, so Saturn actually offers more opportunities for moonwatching than Jupiter does. The following table lists the brighter satellites and their characteristics.

Satellite Name	***Orbital Radius[1] (kilometers)***	***Period (earth days)***	***Radius (kilometers)***	***Magnitude[2]***	***Elongation[3] (arc-seconds)***
Mimas	185,000	.942	195	12.1	30″
Enceladus	238,000	1.370	250	11.7	38″
Tethys	295,000	1.888	525	10.6	48″
Dione	377,000	2.737	560	10.7	1′00″
Rhea	527,000	4.518	765	10.0	1′25″
Titan	1,222,000	15.945	2,560	8.3	3′17″
Iapetus	3,560,000	79.331	720	9–12[4]	9′35″

[1] Above the surface of Saturn.
[2] The average value, at opposition.
[3] Average angular separation from Saturn at greatest elongation, at opposition.
[4] Varies between minimum at eastern, maximum at western elongations.

The moons of Saturn are, like Ganymede and Callisto, composed primarily of water-ice around cores of rock (Fig. 22.3). Their surfaces are generally less active than those of the Galilean satellites and more scarred. Mimas, in particular, bears an enormous pit from a collision with an asteroid, which almost shattered it into flinders.

The most intriguing of the moons is Titan, not only for its size but for its atmosphere. Composed of nitrogen and traces of methane, the atmosphere of Titan surrounds it so thickly that the *Voyager* cameras were unable to show us the surface of this world. This atmosphere is literally thick—

22.3 This montage combines a number of images returned by Voyager I from the Saturnian system, showing the planet and some of its moons: (right) *Tethys;* (second from right) *Mimas, showing its enormous impact crater;* (below rings) *Enceladus;* (above rings) *Titan, its surface obscured by atmospheric haze;* (far left), *Rhea, showing bright markings of fresh ice upon its surface;* (lower left) *Dione, with many impact craters and bright markings. These satellites are visible in amateur telescopes as featureless points of light.* (NASA/JPL)

enough to make Titan seem the largest satellite in the solar system. Actually, stripped of its gaseous shroud, the solid body of Titan is slightly smaller than Ganymede. From space, Titan presents an impenetrable reddish veil, similar to the silver clouds that swathe Venus. Were it not so cold, Titan might be a potential breeding ground of life; there is an atmospheric pressure at the surface somewhat more than earth's, and an abundance of carbon-containing molecules. Oxygen and liquid water are missing, however, from this deep-freeze world colder than 100 degrees Kelvin (−279 degrees Fahrenheit).

Iapetus has long intrigued observers with its variable brightness, which increases by three magnitudes from eastern to western elongation. The *Voy-*

ager missions confirmed an early theory, that one side of Iapetus is significantly darker than the other. Orbiting Saturn in spin-orbital lock, Iapetus always presents its darker face to us when it appears east of the planet. The pattern of bright and dark markings is apparently caused by fresh ice overlying a darker surface.

OBSERVATION

With its slow sidereal period of 29 years, 167 days, Saturn lingers several years in each sign of the zodiac. In the 1980s, it swings from western Virgo into Libra in mid-decade, and enters Scorpius as the decade draws to a close. Its orbit can stray as much as 2.5 degrees from the ecliptic. Saturn's light is dimmer than Jupiter's and distinctly yellower to the naked eye.

Saturn's brightness depends on a feature peculiar to it among all the bright planets. Because its equator tilts fully 29 degrees from its orbit, the rings go through a cycle over the long Saturnian year, during which we see them edge on, then tilted broadly toward us, then edge on again. Seen edge on, the rings disappear entirely, and Saturn appears to the naked eye at its dimmest, with a magnitude as low as 1.5 at conjunction. With the rings opened up at their broadest, sunlight reflects toward us from their full surface, brightening Saturn to as much as magnitude −0.2 at opposition. At such times, which occur every 14.7 years, we see the rings to their greatest advantage (Fig. 22.4).

The rings appear edge on when Saturn is in central Virgo and Pisces—around 12 and 0 hours of right ascension, respectively. During these oppositions, glare from the rings is least, and your chances of seeing the dim, innermost satellites are best. The rings appear broadest when Saturn passes through Sagittarius and Gemini—at the latter time, when Saturn attains its highest declination and rises well above horizon haze at transit, opportunities for seeing detail in the rings are greatest. Saturn has just passed through Virgo, and there was an edge on presentation of the rings in 1980. The rings are currently opening and will be at their broadest in 1987. After that year, they will narrow again to another edge on presentation in 1994/5.

Observing Saturn is much like observing Jupiter, with the difference that the "surface" detail on Saturn is much subtler. The contrast between the zones and bands is mute, and the disc of the planet most often seems unmarked. The lightest-tinted filters can help to bring out detail, especially on the borders of zones and bands where circular markings and festoons appear, similar to those on Jupiter. Saturn's family of satellites offers more to the amateur observer than Jupiter's: there are more within reach of a 6-inch telescope; they stray farther from the main body at greatest elongation; and because of the extreme tilt of their orbital planes, the more distant moons pass much farther north or south of the planet than Jupiter's moons ever go. This last difference makes eclipses, transits, and the like generally rarer.

Ring phenomena are by far the most interesting challenge Saturn has to offer. Can you see the crepe ring? Many observers have seen it, after some

22.4 Rings of Saturn, seen from Earth. Top left: *the ball of the planet, the rings edge on.* Top right: *edge on rings are visible in this overexposed image.* Bottom: *the range of ring presentation, from* (left) *edge on to* (right) *wide open.* (Lick Observatory photograph)

searching, in an 8-inch telescope; it has been seen in a 6-inch. But one of the greatest observers of all time, William Herschel, studied Saturn carefully for years with a 49-inch reflector and never noticed it. High powers can aid your search by increasing contrast, but powers too high will dim the image too much. You may find it helpful to move the body of the planet off-field to reduce glare from the bright disc. Encke's division is always difficult, requiring an 8-inch.

Uranus, Neptune, Pluto

23

PHYSICAL CHARACTERISTICS

Opportunities for amateur observation dwindle rapidly as the solar system spreads through a gap of 1.5 billion kilometers after Saturn. There, the gas giants Uranus and Neptune, and the tiny, enigmatic double planet, Pluto, mark the vast emptiness before the solar system finally dissipates, one or two light years from the sun. Uranus, the closest and brightest of these cold, dim worlds, is visible occasionally to the naked eye, as a sixth-magnitude star that crawls agonizingly slowly among the constellations. Neptune hovers around magnitude 7.7, visible in a 6-inch telescope as a disc—but barely. Pluto, at fourteenth magnitude, is beyond the reach of all but large amateur instruments.

The total scientific knowledge about these worlds is also small. We know that Uranus and Neptune are both Jovian worlds. They are much smaller than Jupiter and Saturn, having radii of about 25,000 kilometers (15,500 miles), and only one-twentieth the mass of Jupiter. The exact composition of their atmospheres is unknown, but their greater densities suggest that, besides hydrogen and helium, heavier gases such as nitrogen, methane, and ammonia may be abundant. Brilliant aurorae have been recently observed on Uranus, suggesting the presence of a strong electromagnetic field. The upper atmospheres of these worlds are cooler than those of the inner gas giants.

In even the largest telescopes their surfaces appear only as vaguely banded, greenish blue discs, with so little detail that the determination of their rotational periods—even the angle at which their poles stand—has been difficult. The latest figures indicate a 24-hour rotation period for Uranus, and around 18 hours for Neptune. The angle of the poles of Uranus is the oddest thing we know about the planet. Its north pole is tipped 98 degrees from the north pole of its orbit.*

This inclination makes the Uranian seasons extreme. For about one-quarter of its 84-year orbit, its north pole points toward the sun and its southern hemisphere rotates in uninterrupted night. Dawn comes for the south pole with the Uranian equinox, as the planet spins with each of its poles along the terminator, the sun rising and setting over the entire planet. But soon, as Uranus slides around its orbit, its north pole slips into darkness, and the south pole has a "midnight sun" of 42 years' duration. In 1985, the north pole faces us straight on, the disc rotating counterclockwise. In A.D. 2007, the equator will be turned toward us, with the north pole on the eastern edge of the disc and the south pole on the west.

Uranus has five satellites (Fig. 23.1), called in order outward from the planet, Miranda, Ariel, Umbriel, Titania, and Oberon. Their orbits are evenly spaced, from 130,000 to 586,000 kilometers (80,600 to 363,000 miles) above the planet. They are small bodies, averaging around 900 kilometers (560 miles) across. Because of the weird position of the Uranian poles, the satellites never appear to shuttle east-west around the planet. When the poles point toward the sun, we look "down" on the moons from the pole of their orbits and see them performing full circles, neither eclipsing or transiting the Uranian sphere. At the Uranian equinoxes, they move north and south, performing the usual eclipses and transits.

Neptune (Fig. 23.2) has an inclination almost exactly equal to Earth's 24 degrees. Its two moons, Triton and Nereid, are mismatched and have unruly orbits. Triton, the larger of the two, orbits in a retrograde direction 355,000 kilometers (220,000 miles) from Neptune. Its diameter, approximately 4,000 kilometers (2,500 miles), puts it in the planet-sized class of satellites. A marked contrast is 900-kilometer-wide Nereid, which orbits Neptune in the most elongated orbit of any satellite, ranging from 15.7 million to 2.3 million kilometers from Neptune.

These orbital quirks, and the equally strange orbit of Pluto, have for many years given rise to speculation that Pluto may once have been a satellite of Neptune that somehow managed to escape and set up on its own. Recent thinking is that this is unlikely. The forces required to wrench a sat-

*This means that the pole on the south side of the ecliptic is called the north pole. The reason for this apparent inconsistency is that the north pole is by convention (except in the case of Venus) the pole around which a planet rotates in a counterclockwise direction. That is, if you were above the north pole of Uranus, the planet would rotate counterclockwise below you, just as it would on any of the other planets (except Venus).

23.1 Uranus and its satellites. (Lick Observatory photograph)

ellite from its orbit—whatever one might imagine capable of doing the deed—would be unlikely to permit the renegade to settle down into a stable orbit of its own.

The orbit of Pluto is eccentric, carrying it from a perihelion distance of 4.5 billion kilometers to an aphelion distance of 7.5 billion kilometers. For several decades at a stretch Pluto can swoop inside the orbit of Neptune, temporarily becoming the eighth most distant planet. Such is the case right now, making Neptune the farthest planet from the sun. "Swoop" may be too lively a term, however, for a planet with a sidereal period of 248 years. Pluto is currently headed toward its perihelion, which it will reach in 1989. The orbit is also inclined more from the ecliptic than any other planet's, carrying Pluto at times as far as 17.2 degrees north or south of the ecliptic and often entirely out of the zodiac.

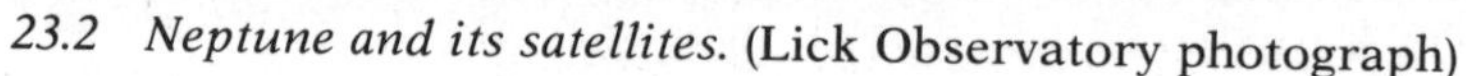

23.2 Neptune and its satellites. (Lick Observatory photograph)

Our knowledge of Pluto (Fig. 23.3) is even scantier than what we know about Uranus or Neptune. It is small, but just how small we still can't say for sure; it has a radius of between 1,200 and 1,900 kilometers (750 and 1,200 miles), and a mass about 0.2% of the earth's. Were it closer to the sun, in orbit around a larger planet, it could easily pass itself off as an unexceptional moon, except for one hitch: it has a moon of its own, called Charon. With a diameter about one-third of Pluto's, it apparently occupies an orbit only 17,000 kilometers (10,500 miles) from Pluto itself—a mere stone's throw, as these things go. Charon takes about 6.5 days to orbit Pluto, which is apparently the period of Pluto's own rotation. This is the only case we know of a planet's spin being locked to the orbit of its satellite: the same face of Pluto always turns toward its moon. This situation is hardly surprising when we consider that Charon is the largest satellite, relative to its planet, in the solar system.

Pluto differs from the rest of the outer planets in a number of ways besides its eccentric, inclined orbit. It is not a gaseous world, and may be more similar to the moons of Saturn that anything else. It seems to be a world of ice, perhaps possessing a thin atmosphere of neon and methane. Much of this methane, however, would have actually frozen out of the atmosphere; lakes of frozen methane, or great drifts of natural-gas snow, cover the Plutonian landscape. The freezing point of methane is 60 degrees Kelvin (−351 degrees Fahrenheit). Pluto is so far from our inner regions of the solar system that from its surface the sun looks little brighter than a star. Under the dim bulk of its monstrous moon, its icy surface revolves slowly in perpetual night. It may share the outer solar system with other worlds, similar to it-

23.3 Pluto. Two exposures, showing motion over 24 hours. The gleam of light at upper right is a galaxy. (Lick Observatory photograph)

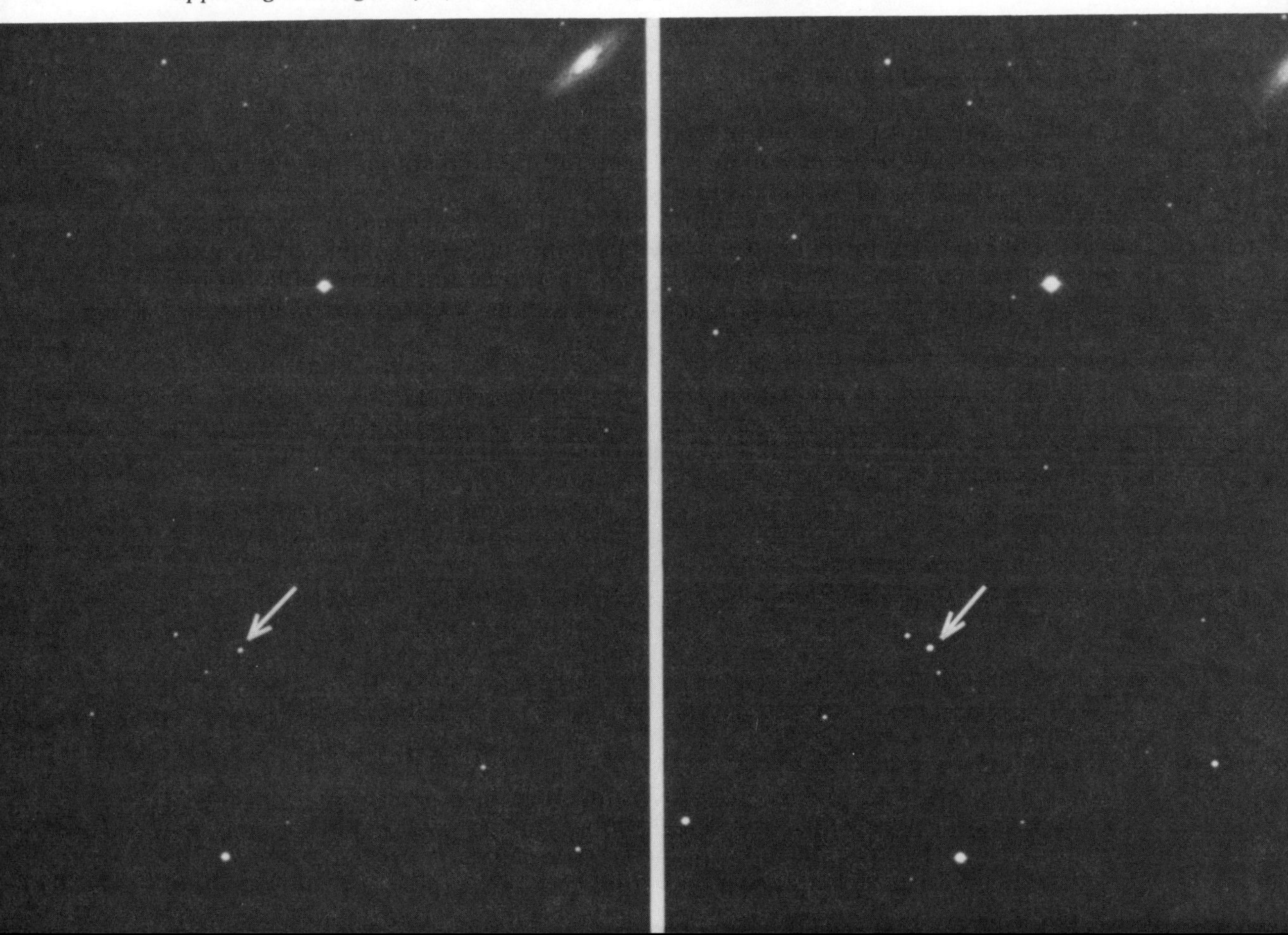

self, isolated and difficult to detect, so large is space in these outer regions and the illumination of the sun so dim.

As the *ultima Thule* of the known solar system, the coldest storehouse of the materials from which the solar system formed, even Pluto has much to tell us. The space telescope, scheduled for launch from a space shuttle in 1986, should be able to reveal more to us about this world, but until funds become available to send instrumented probes to the far reaches of the sun's domain, Pluto's mysteries will remain. The situation is brighter—somewhat—in the case of Uranus and Neptune. The United States currently has one mission underway, which should encounter both these worlds in the next decade. This is none other than the doughty *Voyager*, which has already returned so much information about Jupiter and Saturn. Never intended to survive past Saturn, its camera platform crippled but still usable, the *Voyager* continues on a course that will—if its systems continue to function—return to us the first detailed images of worlds that have been ciphers to us since the beginning of time.

OBSERVATION

Opportunities for observation of the outer planets are poor but not entirely nonexistent. It is possible to track Uranus with binoculars as it moves slowly among the stars, remaining years in each zodiacal constellation. Currently in Scorpio, it will pass into Sagittarius in the late 1980s, where it will remain until near the end of this century. Even at opposition, its disc is so small (4 arc-seconds) that you will need detailed information to find it. The astronomy magazines and the *Observer's Handbook* publish finder charts for each of the outer planets. The January edition of *Sky & Telescope* usually contains detailed information for such objects.

In a large amateur telescope, the surface occasionally shows faint belts—although it is interesting that some observers have reported belts crossing the planet from east to west, as with Jupiter and Saturn, when such bands could only appear on north-south lines or as circles around the poles. Observers with telescopes of moderate size can occasionally see Titania and Oberon, the third and fourth moons out from Uranus. Titania, the brighter of the pair, can be as bright as magnitude 13, and reaches as much as 33 arc-seconds from the planet at greatest elongation. The main problem with glimpsing these moons, even at that distance, is the glare from Uranus itself. Even at only magnitude 6, the planet's disc is still more than 100 times brighter than its satellites. Hiding the planet off-field is your best bet here. Oberon, at magnitude 14, is beyond the reach of a 6-inch telescope, even on the best nights. If you have the equipment to try for it, however, at greatest elongation it appears as much as 44 arc-seconds from the disc of Uranus.

Neptune, of course, is an even tougher nut. Never brighter than magnitude 7.6, nor larger than 2.5 arc-seconds, finding it is challenge enough for most people. With help from one of the annually published finder charts, look for it as a faintly bluish, starlike object. It will appear slightly out of focus even when the surrounding stars are sharp points. It is helpful to make sure your optics are well collimated before embarking on a hunt for Neptune, and to wait for a night of steady air. If you think you've found it, switch to the highest magnification you have; if the suspect expands to a detectable disc, you can congratulate yourself on your skills as a planet-hunter. Moving only 2 degrees around the celestial sphere each year, Neptune's motion among the stars is visible only to a painstaking observer willing to chart the surrounding starfield every night as if hunting an asteroid.

If you find Neptune, you might want to look for Triton. A few observers have reported sighting it in apertures as small as 6 inches, but I'm not one of them. Many observers have missed it with a 12-inch, and a 14-inch doesn't make it exactly easy. The problem, of course, is the dimness of Triton—it can attain magnitude 13—compounded by its closeness to Neptune, from which it never strays by more than 17 seconds of arc.

At magnitude 14, Pluto is a rare sighting for any amateur. I understand it looks reddish. It should, theoretically, be visible in a telescope of 8.5-inch aperture, but, because it shows no discernible disc in an amateur instrument, picking such a vanishingly dim speck from the thousands of stars of similar brightness is a heroic task. If you want to devote the time, remember that it took an observer at a well-equipped research observatory months of full-time searching to discover it, and he was using a camera to simplify the task. Use the techniques described for finding asteroids, choosing guide stars from the finder chart published annually in the *Observer's Handbook* and the astronomy magazines. Pluto's slow motion past a star will give it away. Even though it appears as nothing but a spot of light, the sight of Pluto can be worth the effort, if you like to know you have seen the dimmest planet. But Pluto is not the farthest object in the solar system, nor is Neptune. That distinction goes to even more mysterious objects, from the edges of interstellar space—the comets.

Comets

24

PHYSICAL CHARACTERISTICS

On 16 October 1982, astronomers on Mount Palomar detected the first signs of one of the major astronomical events of your lifetime: on February 9, 1986, Comet Halley will reach perihelion again. The discovery photograph obtained at Mount Palomar showed only a single speck of light. But when it last appeared, in 1910, Comet Halley was a spectacular sight, with a tail of glowing gases long enough to reach from earth to the sun.

Since then, this mountainous chunk of ice 5 kilometers (3 miles) across, has coasted on an elongated orbit almost as far as the orbit of Pluto. On the edges of the solar system, in 1948, it turned, and has since been falling toward the sun, gathering speed. At perihelion passage, it will pass inside the orbit of Venus. It has repeated this voyage every 76 years since at least 240 B.C. It was not until the eighteenth century, however, that the English astronomer Edmund Halley noticed the recurrence of great comets at these intervals, and deduced that each of these apparitions was a single comet. Halley predicted the comet's next appearance, and although he died before the comet could return, his prediction proved true, and so the comet bears his name.

In the middle of the 1980s, Comet Halley (Fig. 24.1) returns once more to the warm regions of the solar system. As it nears the sun, heat and the charged particles of the solar wind will evaporate the icy surface of the comet's nucleus. Molecules of light elements and organic chemicals will stream out into a glowing tail that may grow as much as 150 million kilometers long. At its peak, Halley's Comet can be a spectacular sight, visible by day,

24.1 Comet Halley, 6 and 7 June 1910. (Lick Observatory photograph)

and hanging above the horizon at twilight, the pearly tail flaring over 45 degrees.

But you don't have to wait a lifetime to see a comet. In the past 10 years several comets, such as Comet West (1976), have risen to naked-eye brightness. Comet Halley is simply one of the better known of a numerous class of solar-system objects. In its general behavior it is typical of all comets, which also have in common its basic structure and composition.

Not all comets, of course, are as obvious as Halley's. Most are strictly telescopic phenomena, glimpsed as faint, tailless gleams in the dark sky, where they drift nightly among the stars. Some come close enough to the sun to display the bright tail that makes Halley's so spectacular, but even of this group, the majority are faint, binocular objects at best. Only a few ever become obvious to the naked eye.

About a dozen comets appear each year. Some are *periodic comets,* traveling on orbits that bring them back into the sun's vicinity time and again until eventually their icy substance evaporates entirely away. Within this group, there are the *short-period comets,* whose orbits are small and not much more elliptical than those of the planets. A typical short-period comet has a period from 3 to 10 years long. Some may remain entirely outside earth's orbit, shuttling back and forth between us and the asteroid belt. Because their materials are exhausted by frequent exposure to the sun, such objects rarely become impressive as naked-eye objects (Fig. 24.2).

Long-period comets, such as Comet Halley, are the more spectacular of the periodic comets, especially those whose orbits make them *sun-grazing comets.* The orbits of this last class have perihelia so close to the sun that some occasionally fall into the solar atmosphere and vanish. Moving on orbits from 50 to 500,000 years or more long, the long-period comets spend most of their time in the outer reaches of the solar system, beyond the orbit of Saturn. Because they are exposed less often to the sun, they have retained more of their original material. When it evaporates during their brief perihelion passages, this material can grow into a visible *tail.*

At its maximum length, which occurs usually as the comet passes through perihelion, such a tail can be tens of millions of kilometers long. The tail of a comet has two parts, composed of two different kinds of mate-

24.2 Comet Ikeya-Seki. A member of the sun-grazing class of comets, this was one of the brightest comets of the past several decades. (Official U.S. Navy photograph)

rial. The longer, brighter part, is called the *plasma tail.* It glows, for much the same reason as an aurora.

A *plasma* tail is composed of electrons and other charged particles of gases. A particle usually becomes charged when something (either another particle traveling at high speed or sunlight of the right wavelength) knocks off one or more of its electrons. An atom without its ordinary number of electrons is called an *ion.* Because each electron carries a negative electrical charge, their loss leaves an atom with a net positive electrical charge. Objects carrying electrical charges behave somewhat as do the poles of small magnets: a positively charged particle will attract negative charges, and vice versa, just as the N-poles of two bar magnets will repel, while a N-pole of one magnet and a S-pole of another will attract. This behavior becomes apparent in a comet's plasma tail.

When a comet draws close enough to the sun for its ices to melt, they also feel the full force of the solar wind. Charged particles in the solar wind travel at high velocities; they are frequently highly energized as well. When these particles strike the evaporating gases of the comet's head, the collision strips electrons from the gases, ionizing them. Under the continued bombardment of the solar wind, the ionized gases absorb energy (by a process described in chapter 15), causing the gases to glow.

The magnetic attraction of the charged particles of the solar wind draws these glowing gases out across millions of miles of space, creating a tail that always points away from the sun. Before perihelion passage, when the comet is approaching the sun, the tail streams out behind it. After perihelion, as the comet retreats, the tail *precedes* it, pointing incongruously ahead of the fast-moving comet, like a pennant waving from the mast of a sailboat. Because the solar wind is not a steady stream but has gusts and eddies much like a terrestrial breeze, the plasma tail of a comet can sometimes be whipped around, changing its direction and shape in a matter of minutes.

The other part of the comet's tail, the *dust tail,* is composed of extremely small, fine particles of solid matter, essentially ultrafine grains of sand. It shows less variation, and is from one-tenth to one-hundredth the size of the plasma tail. These small particles are driven away from the sun by "light pressure": the unimaginably small impact of light waves or photons. The dust particles in the tail shine by reflected sunlight. Without an electrical charge, the dust tail is not susceptible to the minute variations in the solar

The source of both tails is the comet's head, which has two parts: the *coma* and the *nucleus* (Fig. 24.3). The coma is a cloud of evaporating gases that surrounds the solid ice nucleus. The nucleus is probably very similar to an iceberg, around several kilometers in diameter. Most of what we know about its composition comes from observation of the gases in the coma. The coma usually appears when the nucleus comes within 450 million kilometers of the sun. It is composed of gases melted from the icy nucleus; they are hot, but not electrically charged. These gases evaporate outward from the center at speeds up to 500 meters per second, in what would be a roaring

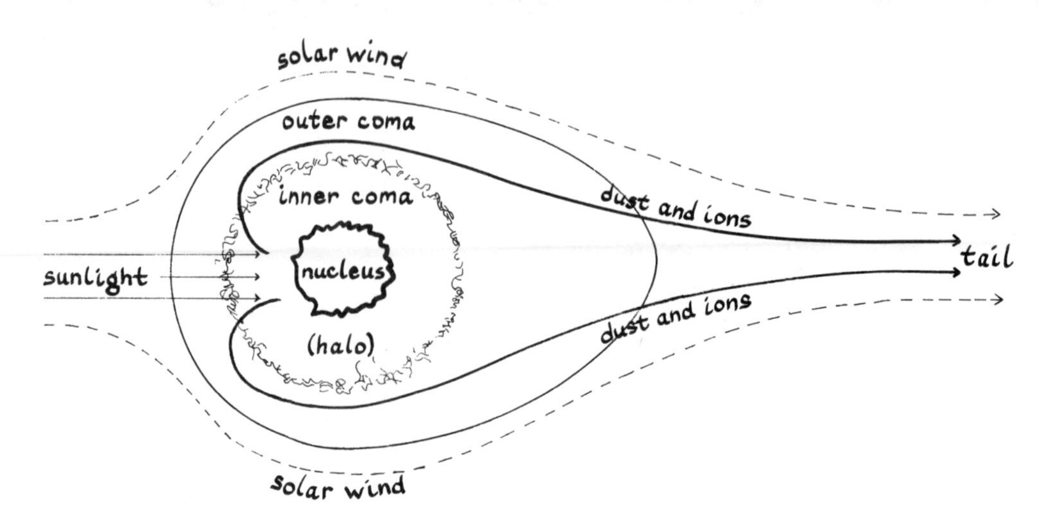

24.3 Structure of a comet. Scale is distorted here to show the inner regions of the coma. The diameter of the nucleus is typically 10 kilometers; the outer coma can reach tens of millions.

wind except that these gases are so thin. If you could bring a sample into a laboratory, it would be difficult to distinguish from a good vacuum. The gases of the coma are different from those of the plasma tail. Neutrally charged gases, predominantly hydrogen, oxygen, carbon, and nitrogen constitute the coma, as well as more complex *radicals* (incomplete molecules) such as hydroxyl (OH), and molecules such as carbon monoxide (CO). This mixture of lightweight gases and carbon compounds may sound familiar; it is akin to the mixture of gases found in the atmospheres of the Jovian planets. Akin, but not alike: exotic compounds like cyanogen, and metals such as sodium, iron, cobalt, nickel, and copper also appear.

In the early history of the solar system, comets probably formed near the orbits of Uranus and Neptune. Repeated close encounters with these larger bodies accelerated them out to their present position, the *Oort cloud.* This is a region 10,000 to 150,000 astronomical units out from the sun, extending halfway to the nearest stars. Here, at the utter edge of the solar system, the materials in comets have never felt the heat of the sun. Never forming into large, solid bodies, neither have they been altered by the heat and pressure of planet-forming processes. Comets, many astronomers think, may be the only entirely unaltered remnants of the original solar nebula. Held in deep-freeze these past 4½ billion years, they occasionally descend from their dark fastness just long enough to blaze brightly in the sun, tantalizing us with glimpses of their composition before they recede back into the night, or fall into short-period orbits and evaporate their secrets away.

The peculiar orbits of the comets are owing to their distant origins. Orbiting so far from the sun, most of the many millions of these small, icy bodies would stay where they were. Occasionally, however, something disturbs them. The sun sometimes brushes close enough to another star for its weak gravitational hold over the comets to slip. Disturbed from their circular orbits, they fall into unstable paths. Many of them eventually settle back into circular orbits, nudged into line by near misses with other comets. But some start falling, and their new perihelia bring them into the solar system, falling for millennia out of the cold and darkness toward the bright star at its center.

For most comets, it's a round trip. After a brief, blazing passage near the sun, they return, on a path that is open-ended, taking the shape of either a hyperbola or a parabola. In either case, the comets return to the Oort cloud, where near encounters with other cometary nuclei recapture them. Some actually punch right on through the cloud and go into interstellar space. But for some comets, the orbital geometry is just right, and after their perihelion passage they pass close to Jupiter or one of the other planets, which swings them back again toward the sun. These comets are captured and take up an elongated orbit among the planets. These are the periodic comets. Any comet on such a path is doomed. Eventually, its ices will all evaporate, and only the solid rocks and gravel once embedded in its core will remain, perhaps to fall to earth as meteors in some future era.

OBSERVATION

Because comets can enter the inner solar system practically anywhere, their paths can take them almost anywhere on the celestial sphere. The brighter ones, however, appear near the sun. Once a comet has been discovered (or, in the case of a returning periodic comet, "recovered"), the astronomy magazines often publish predictions of its positions during its passage. The easiest way to find such a comet is to mark those predictions on your star charts and note the time when the comet is due to pass near a bright star that you can use as your guide to the position.

Before going out to look, however, you need to check on several things. First, is the comet above the horizon after dark? You may find that it sets before the sun, in which case you will need to get up before dawn, and watch for it to rise before the sun. It may be too close to the sun to be visible at all, in which case you will have to wait until it passes perihelion. After perihelion, the comet will appear on the far side of the sun, passing from evening to morning skies, or vice versa (and usually from north to south of the sun, and also vice versa). If the comet is up, will the moon interfere? A dim comet can be very hard to distinguish if the sky background is lit up by the moon.

Finally, how bright is the comet? The finder chart or listing of coordinates will usually include magnitude estimates, which should give you a general guide to what kind of instrument you need, or whether the comet is visible in your telescope at all. Some of the fainter, short-period comets can be below the limiting magnitude of a moderate telescope. Others can be bright enough to show up well in binoculars. Such magnitude estimates can be misleading, however, for several reasons. First, they usually refer to the *integrated magnitude,* a measure frequently used for the brightness of large objects. The integrated magnitude of such an *extended object* is the brightness it would have if it were compressed into a starlike point. Since the total brightness of extended objects is spread out across the sky, they always look dimmer than their integrated magnitude would indicate. Another source of confusion in such estimates for comets is that comets have been known to change brightness by several magnitudes in a single night.

When you know the location of a comet, start looking for it as soon as stars of comparable magnitude become visible. You may find yourself racing against time, as the comet starts to set before the sky background is fully dark. Using guide stars, star-hop to the comet's predicted position. Use your lowest-power, widest-field eyepiece. A light-polution-rejection (LPR) filter may help to make the comet visible, especially if it is showing a tail.

Most comets are faint and appear simply as a dim gray fuzz, brightening toward the center. Even the brighter comets before they have grown a tail have the same appearance. Such comets are difficult to recognize, especially if the sky is not fully dark; anything in the air, such as thin wisps of cloud, can throw you off. A good set of charts will help you verify whether you have found the comet and not a star cluster or nebula. The only way to be certain you have found a comet is to watch it over time, either for several hours (in the case of short-period comets in a dark sky), or on successive nights. If the object you've glimpsed in the sunset glow reappears the next night shifted slightly farther along the comet's predicted path, you've found it.

Every 2 or 3 years, a comet comes close enough to the sun to grow a noticeable tail, becoming bright enough to be visible to the naked eye. In those cases, scanning the coma and tail through a variety of powers can reveal a wealth of fine detail as the charged gases of the plasma tail mingle with the solar wind. Use filters if necessary to cut skyglow (again, an LPR filter, which is designed to pass the kind of wavelengths emitted by ionized gases, is ideal for this). The shape, length, and thickness of the tail can change as you watch (Figs. 24.4a and 24.4b). The coma, too, can change size and shape. Occasionally, even more spectacular effects appear. Comets have been known to fragment during close perihelion passages, splitting into two or more pieces, with separate or mingling tails.

Bright periodic comets are rare, returning only at scattered intervals. But new comets are constantly appearing and often put on spectacular displays. It takes, on average, 300 hours of searching to find a comet (an average figure that conceals some incredibly lucky finds after only a few dozen hours, and some decades-long intervals of bad luck). The average observer need only persist, and soon will find his or her name in lights. Moreover, amateur equipment is all that's necessary. The most prolific discoverers of comets have used instruments smaller than 6 inches.

The ideal instrument for comet hunting may be the 6-inch, *f*/5 reflector described in Appendix 1, mounted on the Dobsonian mount in Appendix 2. Such an instrument answers most of the optical and mechanical requirements of comet hunting. Hunting for any object requires the widest field possible. The typical comet is about magnitude 9 and 2-minute angular diameter at discovery, so optical requirements are wide field combined with

24.4a and 24.4b Comet West, 1975. Two views, showing the change in the structure of the tail over 48 hours. (Lick Observatory photograph)

light-grasp, plus some capacity to magnify. Of course, these requirements are contradictory, so a compromise instrument like the short-focus 6-inch is ideal. Most comets are discovered near the horizon, so the mount must allow for horizontal scanning. The Dobsonian is better at this than anything else.

Horizon scanning at dusk or dawn is the most productive method of comet hunting. If you have a choice, your chances of discovering a comet are at least three times better if you search before sunrise. Your competition is less, for one thing. In the evening, start with the strip of sky just above the western horizon, and about two-thirds as wide as the field of your eyepiece. Scan *slowly* along this strip from north to south until you have covered an arc about 45 degrees on either side of the ecliptic. When you have completed one sweep, move the tube up from the horizon by a degree or so and sweep back toward the north. The strip of sky in the second sweep should overlap the first. This overlap will bring most of the area you cover through the bright center of your field. Continue this process, moving back and forth across the sky, until you have reached an altitude of 20 degrees.

For morning searches, start at least an hour before twilight begins, sweeping up again from the horizon. Try to time your series of passes so that you finish about 10 minutes before twilight. Reserve those last minutes for close inspection of the horizon. Here is where any comet just past perihelion is most likely to show up. Any object appearing in this region is likely to have been hidden by the sun for some time, out of reach of amateurs and professionals alike.

Should you sight a suspected comet, your first step should be to consult a good atlas. The odds are good that the first several "comets" you find will turn out to be dim galaxies or nebulae. If the chart shows nothing in the location of your find, there is still the possibility that you have found a comet that has already been recovered. The circulars distributed by *Comet News* (see below) will help you to eliminate this possibility. If you still think you have a comet, watch the suspect as long as you can. If it shows any motion against the background stars, you can be reasonably certain that you have a comet. If it shows no motion over a few hours, it could still be a comet, one that is yet so far from the sun that its orbital speed is low.

Your next step is to determine the object's celestial coordinates. Mark the position of the comet on your charts relative to nearby guide stars, and read the coordinates from there. Record the universal time at which you took the position. If the object shows motion, record two such positions and the universal time for each. While taking the position, get an estimate of the object's integrated magnitude. Defocus the eyepiece until nearby stars are blurred to the size of the *focused* image of the comet, then compare the brightness of these blurs to that image, focusing and defocusing to make the comparison. (Chapter 17 gives more information on estimating brightness by the use of comparison stars.)

Once you are positive you've found an unreported comet (the opinion of a more experienced amateur is advised), send all of this information to Dr. Brian Marsden, at the IAU Central Bureau for Astronomical Telegrams, 60 Garden Street, Cambridge, Massachusetts 02138. The Bureau prefers to be notified by telegram, but you can leave a 30-second message on their telephone-answering machine; the number is 617/864-5758. In either case, your message should specify:

1. What you've found (for example, a suspected comet)
2. Its position (if derived from charts, specify the epoch of the chart's coordinate system)
3. The universal time at which you found the comet in this position
4. The estimated integrated magnitude
5. Your own name, address, and telephone number

Typographical errors are common in telegrams, especially in strings of numbers. There are two ways to protect your message from garbling in transit. The easier one is slightly expensive: simply spell out all the numbers, "one" instead of "1", and so on. The cheaper method involves a slightly complicated code, which I can't imagine being able to manage in any state of excitement. If you're interested, however, the code, and other details of Dr. Marsden's work, are explained in an article in *Sky & Telescope* (August 1980, page 92).

For timely reports of comet sightings, your best bet is to subscribe to *Comet News Service* (listed under "magazines" in the bibliography). This newsletter covers newly discovered comets and returns of periodic comets, giving the casual comet fancier a chance to follow up on the discoveries of the more determined observer.

HALLEY'S COMET: THE 1985/86 PASSAGE:

At present, Comet Halley is far out in space, in the direction of southern Gemini. Approaching at over 33,000 kilometers (20,500 miles) per hour, it was still beyond the orbit of Saturn when astronomers on Mount Palomar detected it in October 1982. Since 1911, it has receded to its aphelion distance of over 5 billion kilometers. Late in 1984, it swung westward through northern Orion almost to Aldebaran, then looped back east and north again toward Gemini. That motion signaled the beginning of the current passage.

In 1985, Comet Halley crosses the orbit of Jupiter, the asteroid belt, and then the orbit of Mars. It moves across the sky as well, following the ecliptic several degress to its north. It reaches magnitude 12 in September and shortly afterward starts retrograding in Taurus, passing just north of the Hyades the second week of November 1985. By that time, its magnitude is about 10 and the plasma tail starts to form. Passing just south of the Pleiades on the night of 16 November, it moves through Aries over Thanksgiving, coming to opposition in that constellation before the end of the month. It dives through Pisces during December, and rises to naked-eye magnitude in the evening sky. As it crosses the equator hard by the Water Jar of Aquarius around the turn of the year, it begins to sink into the sunset. At this point, the tail, visible in binoculars, is 2 to 3 degrees long. From the time it passes the orbit of Jupiter, its nucleus starts to disappear behind a growing coma, spreading out into a tail as it passes the orbit of Mars.

Comet Halley passes earth twice—once inbound, its tail streaming behind it, and once outbound, its tail preceding it in its slow nightly passage across the sky. Its closest approach on the inbound leg will occur on 27 November 1985, 92 million kilometers from earth. Throughout December, its tail will appear above it as it hangs over the southwestern horizon after sunset. After the New Year, it will move closer to the sun each night, hanging lower in the southwest at sunset. Still in Aquarius, its magnitude will rise to around 5, but it will become increasingly difficult to make out against the solar glare. After passing Jupiter at the end of the second week in January, Comet Halley should disappear. On 9 February it will reach perihelion on the far side of the sun. At that time, its tail will grow to its maximum length, perhaps 80 million kilometers.

During this period, Comet Halley's apparent position will change little from night to night, because it is over 1.5 astronomical units away on the far side of the sun. The tail will seem to shrink as well, even as it grows to its greatest actual length. From our point of view, the tail will appear head-on, pointing directly away from us. This angle and the comet's great distance will combine to make the perihelion passage a less spectacular sight than at some previous apparitions.

After perihelion, the comet will reappear in the last week of February. It will rise tail-first at dawn, 10 degrees long and around magnitude 3. Standing far south of the sun, Comet Halley will become increasingly difficult for northern hemisphere observers to make out as it passes below the teapot of Sagittarius at the end of March, below the tail of Scorpius, and finally into Centaurus in April. All this time, it will be approaching ever nearer the earth, its tail stretching out west of it as the comet appears to pick up speed. The end of March will see the comet grow to 25 degrees in length, its brightness holding steady around magnitude 2. In April, it will move fully one-quarter of the way around the sky, reaching its closest approach to earth, 62 million kilometers, on 11 April 1986. The comet will appear largest and brightest at this time, and well up in dark skies—but only from the southern hemisphere. In the north, it will still be well below the southern horizon.

Comet Halley's return to the north will not come until the beginning of May 1986, when it enters eastern Hydra. By that time, it will have faded by several magnitudes, the tail shrinking to perhaps 10 degrees. The comet will continue to dim rapidly, dropping below naked-eye visibility within a few weeks. For the next month or so, it will continue to be within grasp of amateur instruments, until conjunction with the sun at summer's end. And that will be all for Halley's comet until around the year A.D. 2061.

The Sun

25

The sun is essentially a controlled nuclear reactor, generating electromagnetic energy over the entire spectrum by a process called *thermonuclear fusion.* Before discussing that process, however, we should look at a few of the means by which the sun's energy moves from its central reactor out into space. The processes of energy transfer include *conduction, convection,* and *radiation.*

You are already familiar with the first of these. Conduction involves the mechanical transfer of energy between particles. On this level, energy is motion: the energy any particle has makes it either vibrate or travel. In the conduction of energy, for instance, the vibration of one particle gets passed to those next to it. The buzz of energy travels through the material. Conduction is the dominant process in solids, especially metals. You are most likely to notice it when you burn yourself on the handle of a spoon that was left too long in a pot of soup.

Convection occurs in liquids and gases, where particles are freer to travel. There, energetic particles actually travel from one place to another, carrying their energy with them. Again, you can see this process in the kitchen, as soup near the bottom of a pot heats and rises to the surface in convection currents. In a gravitational field, heated matter rises because hotter particles generally collide with each other more violently and thereby occupy more space than quieter, cooler ones; a similar volume of heated material contains fewer particles, weighs less, and therefore literally floats above the denser, cooler material.

The third important method of energy transfer, radiation, is harder to see. We are most familiar with infrared radiation, and you are most likely to notice it when warming yourself at a fire. The side facing the fire gets much hotter than your other side because the fire has become hot enough to emit electromagnetic radiation. Radiation does not require a solid, fluid, or gaseous medium; it works most efficiently in a vacuum.

Heated matter emits electromagnetic radiation. In most cases, the amount emitted is very small. The frequency of the radiation depends on the temperature of the matter, in a relationship commonly called the blackbody law. A "blackbody" is a theoretical beast; it is anything that radiates energy perfectly. It is perfect because the kind of electromagnetic energy it radiates depends purely on its temperature. Every object that radiates electromagnetic energy radiates most strongly at a peak wavelength, and radiates weakly at other wavelengths. This gives each object a characteristic *energy curve.* A fireplace operates at a temperature of around 450 degrees Kelvin (350 degrees Fahrenheit), and therefore radiates most strongly in the infrared, and quite weakly in the X-ray regions of the spectrum. A woodstove behaves much like a blackbody; so does a star (Fig. 25.1).

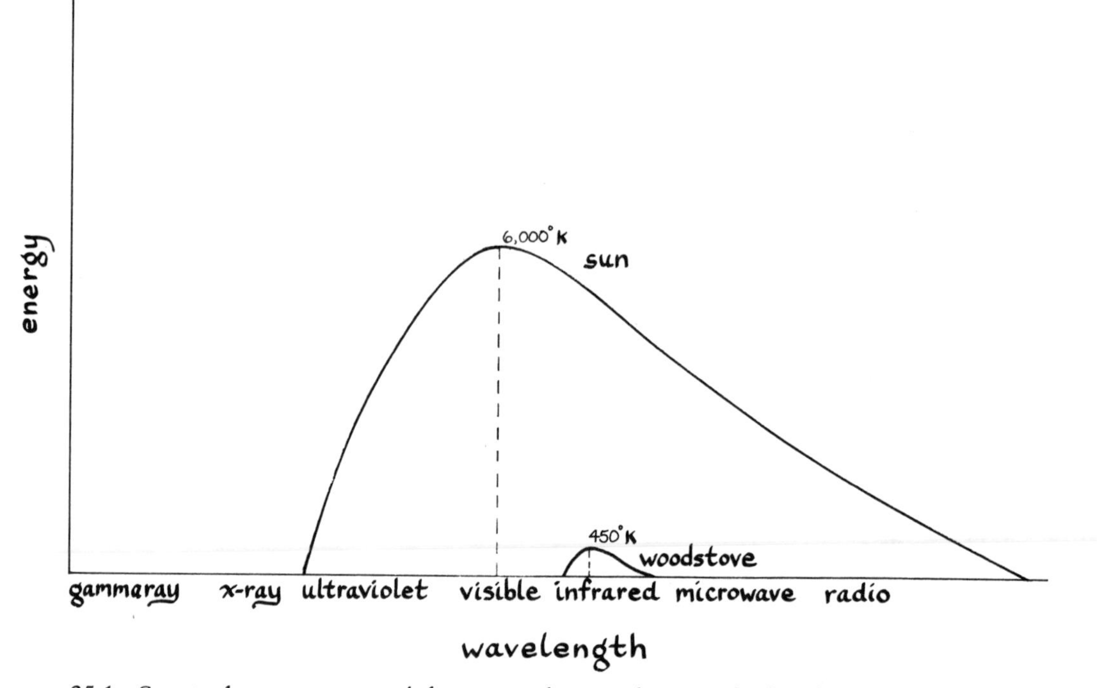

25.1 Spectral energy curve of the sun and a woodstove. The height of the curve (not to scale) indicates total energy; the location of the peak indicates the surface temperature.

Consider, then, the sun. All of its heat reaches us by radiation, over a distance of 150 million kilometers. We take it for granted, not even stopping to consider how it performs a few simple tricks: lifting millions of tons of water each day from the oceans to the sky; raising weeds and corn, wheat, and sequoias out of earth; warming our planet from the absolute chill of the void; burning the top layer off of your skin—all at a distance of 93 million miles. If you set out to drive from the earth to the sun at turnpike speed, your children's children's children would not complete the trip. And yet over this great distance, the sun radiates heat and light enough to warm not only our planet, but to warm 2.2 billion earths (which is the number of planets our size that would fit onto a spherical shell 150 million kilometers from the sun). Of all that energy (382 trillion, trillion watts, to be exact), all but the billionth part of it radiates eternally out into space, where it joins the combined flood of light and heat from the billions of other stars in our home galaxy.

The sun is a star, and an average one at that. Its importance to astronomers lies in its unique position. It's the only star we have at hand to study. To an amateur, the sun is a ceaselessly changing display, the only star that comes out by day. We know that it is made primarily of hydrogen, the simplest, lightest, most abundant element in the universe, comprising nearly 94% of the sun's total number of atoms. Six percent of the remaining atoms are helium (the second-lightest, simplest, and most abundant element). Heavier, more complex elements make up less than 0.1%; these are oxygen, carbon, nitrogen, silicon, magnesium, neon, iron, and sulfur, in order of decreasing abundance.

But it's a bit absurd to talk of the sun atom by atom. There are 1.99×10^{30} kilograms of matter in its bulk, 300,000 times the mass of earth, all contained in a sphere 1,392,000 kilometers across. Its total volume is 1.3 million times that of the earth. Despite its enormous mass, the entire sun has an overall density of only 1.4 grams per cubic centimeter—only slightly greater than that of water. Like the Jovian planets, the sun does not rotate as a solid body. It takes 25 days to complete a rotation at the equator, but at the poles its motion is slower, completing one turn in 33 days.

The visible surface of the sun has a temperature of about 6,000 degrees Kelvin. This surface, called the *photosphere,* is simply the lowest layer of the sun's outer atmosphere. It is the level from which the energy generated in the interior radiates off into space. Beneath the photosphere, in levels invisible to us, is an outer shell, the *convection zone,* overlying a much thicker region, the *radiation zone.* In the convection zone, heated material rises to the photosphere, gives off some of its heat, and sinks to the bottom of the convection zone, where it heats up and rises again. In the radiation zone, energy must punch through the tightly packed sea of hydrogen and helium nuclei, taking hundreds of thousands of years to make the journey from the core to the surface. At the sun's *core,* a region perhaps 140,000 kilometers across, the pressures and temperatures rise high enough to support the thermonuclear processes that generate the sun's energy (Fig. 25.2).

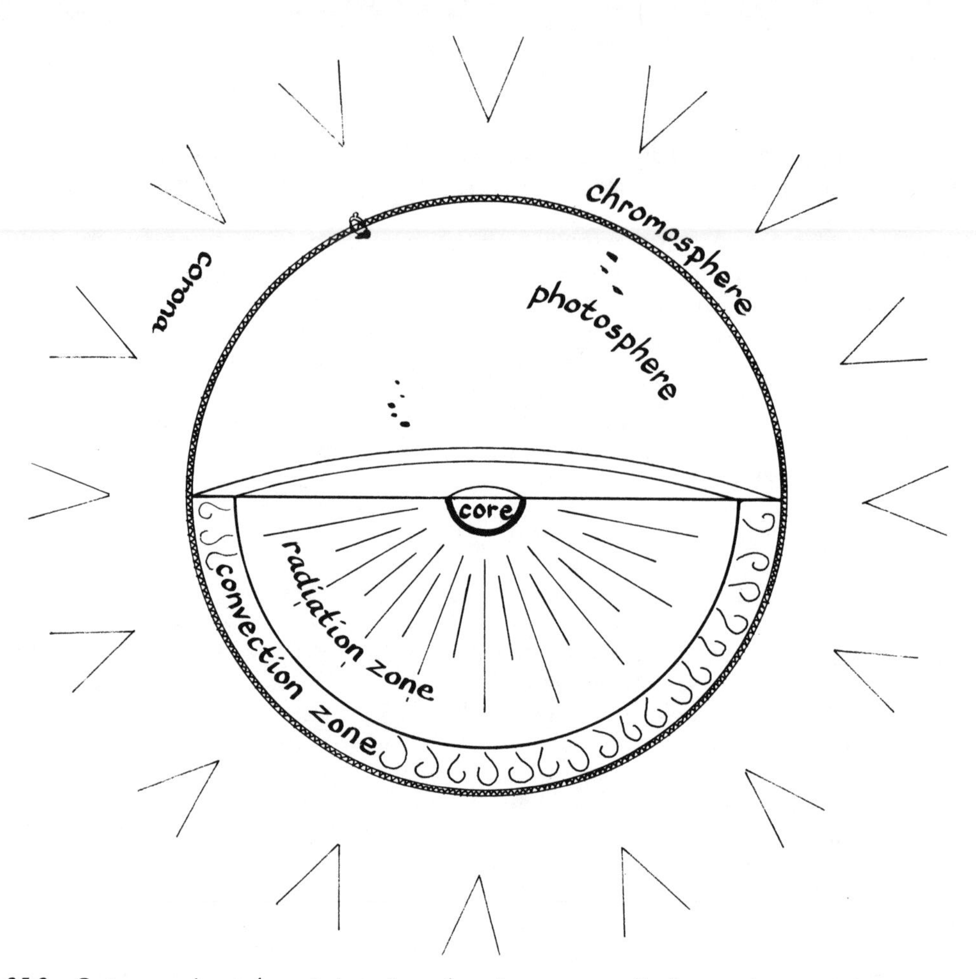

25.2 Cutaway view of sun's interior, showing core, radiation and convection zones, photosphere, chromosphere, and corona. Chromosphere is actually thinner than shown.

At the sun's core, the temperature is 15 million degrees Kelvin. The hydrogen there is squeezed to a density 160 times that of water. Under these conditions of heat and pressure, a chain of events occurs at the subatomic level. The process, called *thermonuclear fusion,* is the source of the sun's energy. As the name suggests, the process involves high temperatures, and occurs in the nuclei of atoms. In its simplest form, fusion takes four hydrogen nuclei—four protons—and fuses them together to make one nucleus of helium. The nucleus of helium contains not four protons, but two protons and two neutrons. If we were to measure the mass of the helium nucleus, we would find that it has less mass than the four hydrogen nuclei. What happened to the missing mass? It was transformed into energy.

The exact details of the process as performed inside most stars are as follows. At temperatures from a few million up to about 16 million degrees Kelvin, fusion proceeds through *proton-proton* reactions. The high pressure and temperature of the solar interior accelerate protons to speeds so great as to overcome the strong repulsive force that normally keeps them apart. In the first stage of the process, two protons crash together, changing one of them into a neutron. The result is a proton-neutron pair, called a *deuteron.*

When a deuteron forms in this way, the breaking of the repulsive force between two protons produces a *positron,* the antimatter double of an electron, and a *neutrino,* which is a massless, chargeless, ghost of a particle. The positron soon meets an electron and the two annihilate each other, producing two very short-wave photons of light, *gamma rays.* The neutrino speeds away, unhindered by the dense material around it. Both the gamma ray and the neutrino carry energy away with them.

Soon the deuteron collides with another proton, forming a lightweight nucleus of helium—two protons and a neutron. Another gamma ray escapes in the process. Finally, this nucleus collides with another such nucleus, releasing two protons and forming a full-weight helium nucleus composed of two protons and two neutrons.

One gram of hydrogen transformed in this way produces only 0.9929 grams of helium. That missing seven-thousandths of a gram makes all the difference, as it radiates off in a shower of gamma rays and neutrinos. It has been transformed into energy according to Einstein's formula $E = mc^2$. The "m" in that equation is the very small 0.007 grams. But the "c" is the enormous speed of light in a vacuum, making "E", the yield from the reaction, 6.4×10^{18} ergs per gram of hydrogen. An erg is an extremely small unit of energy; a small caterpillar crawling an inch expends about an erg. But imagine 6,400,000,000,000,000,000 caterpillars crawling, and you get some idea of the vast energy liberated by the thermonuclear "burning" of one gram of hydrogen. And the sun has 1,800,000,000,000,000,000,000,000,000,000,000 grams of hydrogen.

This is the fundamental form of thermonuclear fusion. At temperatures higher than 16 million degrees, a more complex reaction starts, involving nuclei of carbon, nitrogen, and oxygen.

The radiation generated in the sun's core changes considerably as it struggles outward through the surrounding layers. As the energy moves outward, the gamma rays are changed into less energetic forms of radiant energy, but the total energy remains the same. The peak in the spectral energy curve slides down the scale until, by the time the rays emerge at the photosphere, they peak in the yellow part of the visible region of the electromagnetic spectrum. On the blackbody energy curve, this indicates a temperature for the photosphere of about 6,000 degrees Kelvin.

The visual appearance of the photosphere is complex. On the largest scale is the *limb darkening.* The solar disc looks dimmer at its edges, because we are looking through a greater thickness of the sun's upper layers. The most noticeable features of the photosphere are the *sunspots,* dark areas that grow from specks no more than 1,500 kilometers (930 miles) across to cover regions as much as 50,000 kilometers (31,000 miles) wide—four times the width of the earth. They tend to congregate in groups stretched horizontally across the solar disc. At times, a group can become large enough to be visible to the naked eye.

Sunspots are not actually dark—they only look so compared to the brilliance of their surroundings. Spots represent distortions in the sun's powerful magnetic field. Because the sun rotates unevenly, the lines of force in the magnetic field can become twisted, and eventually kinked. When this occurs, the lines can loop above the surface, as if a giant horseshoe magnet were being held just above the photosphere. These magnetic loops restrain the customary seething motions of the solar plasma, preventing heat from rising to the surface. The result is a sunspot, a region where the solar surface has managed to cool slightly. The magnetic loop responsible for one sunspot usually causes another at the loop's other end.

Other facets of sunspot behavior are less well understood. The biggest mystery is the *sunspot cycle,* a variation in the number of sunspots. Every 11 years, at *solar maximum,* dozens of large sunspots march across the solar surface; at *solar minimum,* the photosphere can appear blank for months. The most recent peak in the solar cycle, which seems to have occurred in 1980, was the second most active on record; the cycle coasts to minimum around mid-1985.

At the beginning of each cycle, the first spots appear around 30 degrees north and south solar latitudes. Within 3 or 4 years, more spots appear, anywhere from 10 to 50 degrees from the sun's equator. As the cycle passes its peak, the number of sunspots in the higher latitudes declines sharply. In the closing years of each cycle, there are fewer sunspots, and almost all of them appear in a band 15 degrees on either side of the equator.

Every 11 years, as the sunspots disappear from the solar disc, the poles of the solar magnetic field reverse, the north pole taking on a south magnetic charge, and vice versa. This change influences the magnetic behavior of the sunspots of the next cycle. During each cycle, the eastern spot or group of spots in a magnetically linked pair will always have the same magnetic polarity as all the other eastern spots in that hemisphere: north in one hemisphere, south in the other. During the next cycle, the relationship is reversed. The western spot in any group is always of the opposite magnetic polarity from the eastern spot. (Because the spots move with the sun's rotation, the eastern spot is usually called the leader; the western spot follows, or trails.)

Even though sunspots represent relative calms on the seething surface of the sun, they are the focal points for intense activity in the regions around them. The entire solar surface, or course, bears the marks of the convection currents, which look exactly like the mottled, cellular markings on the surface of a slowly boiling, thick soup. On the sun (Fig. 25.3), these marks are called *granulation,* and appear as bright areas with dark borders, around 1,000 kilometers (620 miles) wide. The darker border is cooler matter, sinking into the convection layer. The bright center of each granulation cell consists of hot gases upwelling from the solar interior. Around sunspots, even brighter areas, called *plages,* appear. They usually occur over a region where a sunspot group is about to emerge, and sometimes linger after the spot has subsided. While the sunspot is present, they can surround its entire area, giving an even brighter border to the dark group. Dark *filaments* can cut across such plages (Fig. 25.4). They are actually great, arcing loops of solar plasma, also called *prominences.* They rise thousands of kilometers above the photosphere, and stretch for hundreds of thousands of kilometers across its surface.

Plages and prominences are not ordinarily visible to terrestrial observers. They can be seen, however, in a telescope equipped with a special filter, which passes only a specific wavelength of red light (called hydrogen-alpha, because it is the strongest wavelength produced by hydrogen atoms under extremes of energy—for more about this effect, see the next chapter). When a large plage occurs near the solar limb, where the darker surroundings provide contrast, it can be visible in ordinary white light; plages seen in this way are sometimes called *faculae.* Sometimes, a particularly large and violent outburst around a sunspot can release a *solar flare,* a gust in the solar wind in which vast quantities of the solar plasma, heated to temperatures of 5,000,000 degrees Kelvin, are ejected at speeds fast enough to reach the earth in a few hours. In the most violent cases, a flare becomes bright enough to be visible as a bright spot in ordinary light; they are more commonly seen in H-alpha filters. Typically, such a storm on the sun lasts several hours, and its effects are felt here on earth 1 to 2 days later, as charged particles spiral in along the earth's magnetic lines, causing aurorae in high latitudes, and disrupting radio and telephone transmission worldwide.

Such a flare rises through several outer layers of the solar atmosphere, the innermost of which is called the *chromosphere* for its bright pinkish color when it becomes visible around the limb of the moon during a total solar eclipse. About 2,000 kilometers (1,200 miles) thick, the chromosphere is the region where plages and prominences occur (Fig. 25.5). It is hotter than the photosphere, its temperature rising with increasing altitude, until, at its upper limits, it reaches 100,000 degrees Kelvin. The mechanism behind this rising temperature is apparently energy transferred by shock waves from the photosphere. Violent activity on the solar surface causes the entire chromosphere to throb like a drum, and the energy involved accelerates the particles in the chromosphere to higher levels of activity and higher temperatures. Heated so much, the gas in the chromosphere glows red, in a pattern visible at the limb, during eclipses or in hydrogen-alpha light, as a low, spiky hedge of red light called *spicules.*

Above the chromosphere, the solar *corona* expands out into space. The thinnest, uppermost region of the solar atmosphere, it is distinguishable from the vacuum of space by the extraordinarily high energies of the few particles present there. Its density is perhaps one ten-billionth that of earth's atmosphere, but it compensates for what it lacks in air pressure by

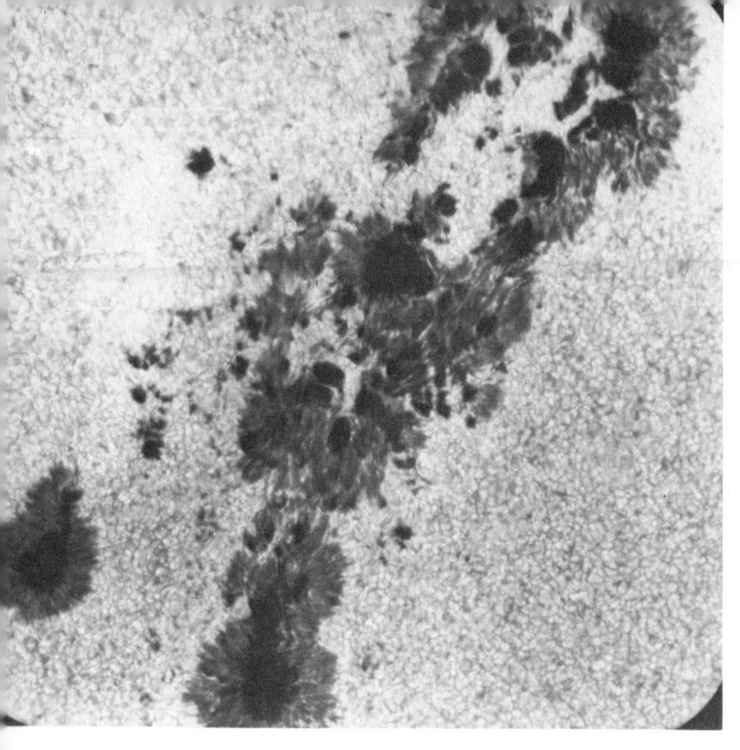

25.3 A complex sunspot group, close up. The several umbrae are connected by merged penumbrae. The fine cellular structure covering the surrounding surface is granulation—convection cells welling to the top of the convection layer. (Sacramento Peak Observatory)

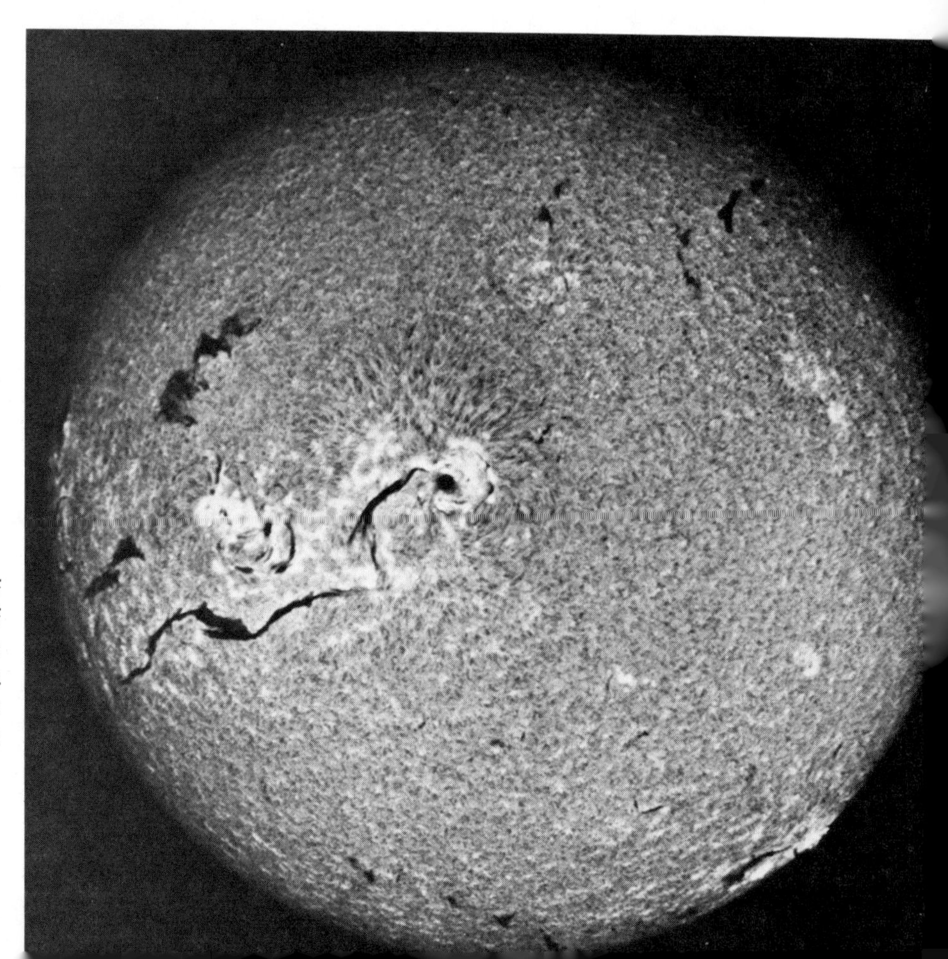

25.4 The full disc of the sun, in the light of hydrogen-alpha. The dark lines are filaments. (Sacramento Peak Observatory)

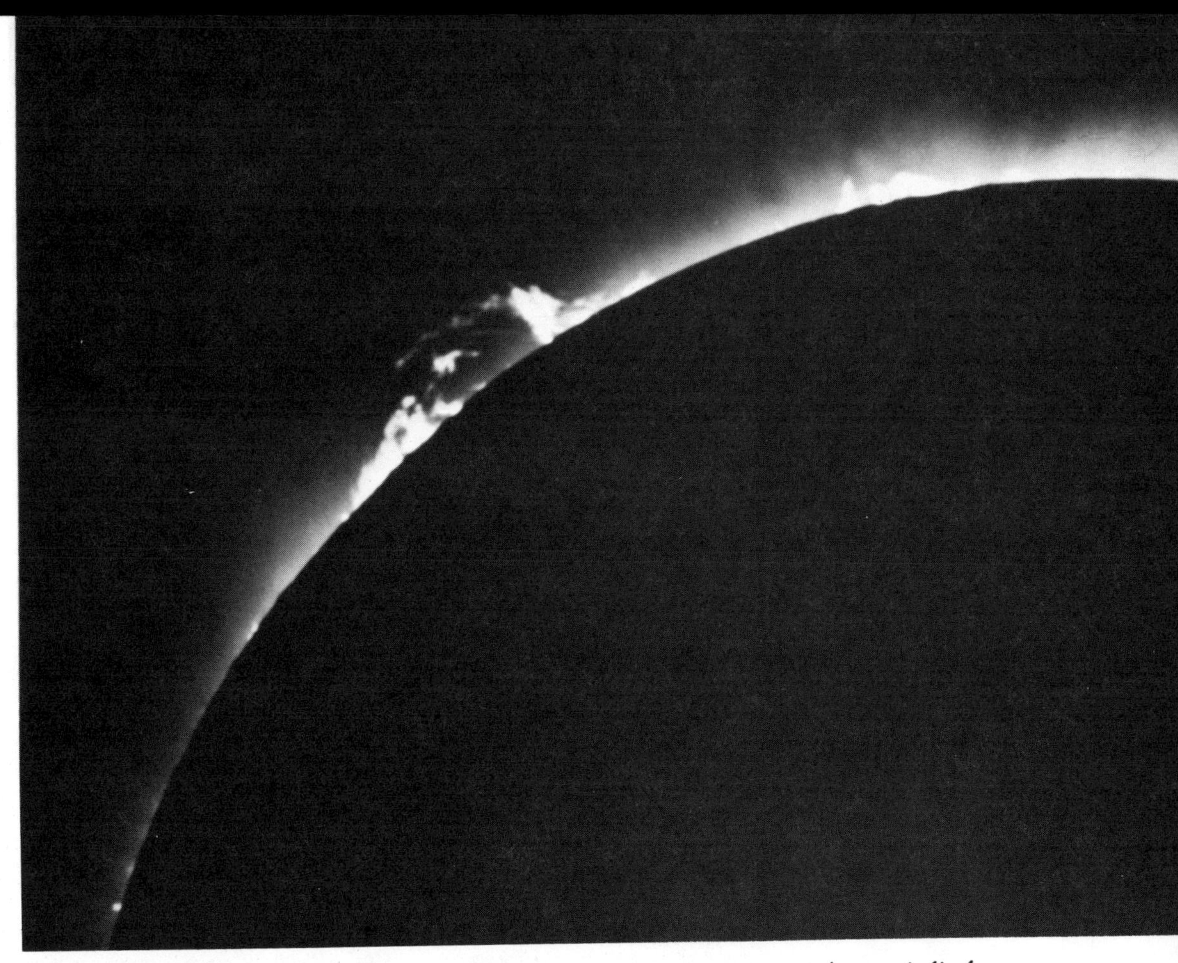

25.5 Prominences visible at limb of eclipsed sun. Note unevenness of moon's limb. (Lick Observatory photograph)

reaching temperatures as high as 2,000,000 degrees Kelvin. The reason for this enormous increase in energy, far from the sources at the sun's core, is now a mystery. Once thought to be owing to the same shock waves that heat the chromosphere, it is now known that such a mechanism could not achieve the necessary energy levels. The problem remains a mystery. Some involvement between the corona and the magnetic fields of the sun may give an explanation, however. This relationship is visible in the appearance of the corona during eclipses. At solar maximum, the corona is complex and spiky, with great rays extending out to several times the solar radius. At solar minimum, the corona is contracted and featureless. Observations made aboard *Skylab* showed that the corona heats up above active regions in the photosphere, and also has "holes"—magnetically disturbed regions, perhaps comparable to sunspots.

OBSERVATION

Amateur astronomers face few hazards in their hobby. The only real danger you are likely to encounter is while observing the sun. That's a rather large hydrogen bomb you're about to play with. Treat it with respect.

The cardinal rule in solar observation is: DO NOT DO ANYTHING THAT COULD CAUSE UNFILTERED SOLAR LIGHT TO REACH YOUR EYES. Through an ordinary magnifying glass, sunlight can burn wood, so imagine what an astronomical objective will do to your eyes. Take care whenever you are working with a telescope in daylight.

The safest method is to reduce the incoming light *before* it reaches the objective, which is best accomplished by the solar filters mentioned in chapter 11. If you cannot afford one of these, it is still possible to do some enjoyable observing with a simple projector. When using such a device, remember that unfiltered solar radiation is moving through your telescope. It will heat up your eyepiece, and can be dangerous if anyone should remove the projection screen. *Never* leave a telescope set up for projection unattended. It would be far too typical human behavior for a bystander to remove the screen and peer into the eyepiece for a look.

You can make a solar projector out of a cardboard oatmeal tube (Fig. 25.6). Cut one end to fit snugly against your telescope, and cut away the other end, leaving one side of the tube longer to act as a sunshade. Inside the tube, mount a frame made of two thicknesses of shirt cardboard; cut a 3″ to 4″ circle out of this frame. Over this cutout, stretch a sheet of drafting film (sold at art and office supply shops). Focus a low-to-medium-power eyepiece on some distant object. Set the projection tube over the focuser (you may want to cut a hole in it, so you can reach the focusing knobs). Adjust the focus so the solar limb appears sharp against the frosted screen; refine your focus by watching any sunspots that appear. To enlarge the image, move the projection screen farther from the eyepiece; the image will require slight adjustment in focus as it grows.

Such a system will not provide the ultimate in detail or resolution, but you're getting much better than you've paid for. Because the conditions of solar observing don't often favor good seeing, you may want to put a cardboard mask in front of the aperture of your telescope, leaving just a 2.5-inch hole to admit sunlight to the objective. You won't notice the loss in resolution, and you certainly don't need the extra light.

If you have a choice, the short-focus telescope described in Appendix 1 is not the ideal telescope for solar observation, although it will perform more than adequately. A long-focus telescope, such as an ordinary refractor, or a Schmidt-Cassegrain, will give the best views of small-scale solar phenomena (the smaller granulations in the photosphere subtend an angle of about 1 arc-second). Unfortunately, seeing conditions will rarely permit you to work at the limit of resolution of your telescope; the heated air simply churns too much. A 2-inch refractor probably offers the best performance-per-cost for viewing the sun.

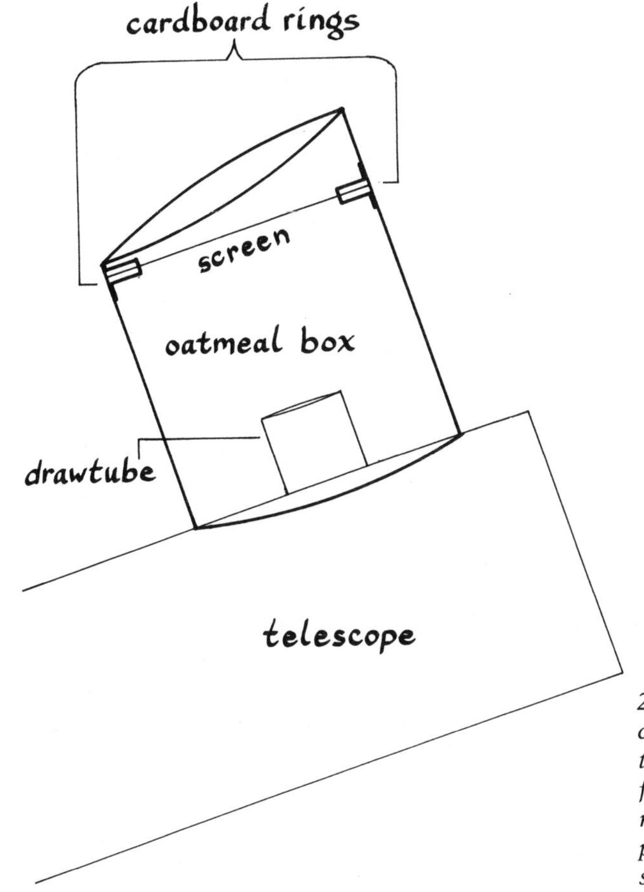

25.6 Solar projection screen. A large, cardboard oatmeal cannister, cut away to fit over a drawtube, holds a screen of frosted Mylar drafting film between two rings cut out of cardboard. The eyepiece projects the image of the sun onto the screen.

The first section of this chapter mentioned the use of a hydrogen-alpha filter. An H-alpha filter rejects all wavelengths except those emitted strongly by energetic hydrogen atoms. Many of the more interesting features in the photosphere and chromosphere, such as plages and prominences, appear best in such a filter. These filters are expensive, but necessary for serious solar observing. They cost several hundred dollars. You will find them advertised, under brand names such as "Daystar," in the astronomy magazines. Some amateurs find the sun an irresistible attraction, not only for its convenient schedule, but also because nowhere else in the sky can you see so many different processes in action at once, from the stately progress of the 11-year sunspot cycle, to the hourly changes of a prominence. The amount of detail, and the visible changes of the surface seen in the light of H-alpha radiation, are for many observers enough to justify the high cost of such a filter.

Until you make such a leap, you will find that most of your observing concentrates on sunspots. You may find, after observing sunspots for a time, that you want to contribute to a program of data recording, such as that of the American Association of Variable Star Observer's program of solar observation. Chapter 28 gives more information on this organization, which you can contact for details.

To begin with the basics, the first task is to find the sun. This is slightly less straightforward than finding the moon or a bright planet in the finder—because even a small finder will concentrate enough light on your eye to burn it severely. *Keep your finder capped while observing the sun.* Better yet, remove it entirely. To find the sun, move the telescope on its mount until the shadow of the tube shrinks to a circle. With a low-power eyepiece, you should find the sun easily.

Take an overall look at the disc before moving to higher power. Notice the general distribution of spots around the disc, and try to get a general idea where the sun's equator and poles are. The equator should be the midline of the broad band where the sunspots lie. Unless you are near minimum of the sunspot cycle, you should be able to see several large spots (up to a minute across), and a scattering of smaller ones, ranging down to tiny dark pores. They will usually group together. The sun's rotation, of course, is prograde, so sunspots first appear on the eastern side of the disc. They will show up at first extremely foreshortened, looking like dark, narrow ellipses. They will be difficult to see just after crossing the limb, because the darkening of the solar disc near its edges reduces the contrast between spots and the photosphere. The sun's synodic period is slightly over 27 days at the equator, so any one spot will take somewhat less than 2 weeks to cross the solar disc.

When you have located a sunspot group, start homing in on it with higher powers. The sun's disc averages about 32 arc-minutes across. Powers around 60× should give a full-disc view, with enough magnification to show some detail in the spots. The first thing you notice at this magnification is that a spot is not uniformly dark. It has a darker center, called the *umbra,* surrounded by a lighter gray aureole, called the *penumbra.* These regions apparently correspond to areas of greater or lesser intensity of the magnetic field around the spot. The interior of a sunspot can be simple, with one umbra surrounded by a circular penumbra, or it can be complicated, with multiple umbrae, and the penumbra bulging out into the surrounding photosphere.

At times, spots will congregate into groups that grow to cover several minutes of arc, over 100,000 kilometers (62,000 miles) across. Examine these with the highest powers you can, and you will find a wealth of detail. The region around the penumbra, for instance, can be mottled with gray regions that look like bits of the penumbra that have floated away into the surrounding photosphere. Look to the east or west of a large, well-defined spot or group for its companion, at the same solar latitude, traversing the disc anywhere up to a full arc-minute ahead or behind. Such a pair mark the points where a giant loop of the twisted magnetic field exits and enters the photosphere. Such pairs are also frequently scenes of prominences, and sometimes of flares.

Any spot you see will show evidence of the sun's rotation, appearing closer to the western edge of the limb on successive days, taking about two weeks to traverse the entire disc. You can see the different speeds at which different latitudes of the disc rotate by tracking two spots, each at different latitudes, across the sun. The spot at the lower latitude should gain a full day's march over the other in a single traverse of the disc. A healthy spot will persist for more than a single rotation, and often for as many as three or four. Because the spots are not anchored to any solid surface, they wander, falling behind or inching ahead of other spots at the same latitude.

One of the most impressive effects visible through an ordinary telescope is the *Wilson Effect.* Whenever you see a good-sized, simple and evenly circular spot crossing the middle of the solar disc, track it over the next week as it edges toward the western limb. As it nears the limb, its apparent shape will foreshorten from a circle to an ellipse. Watch as the umbra seems to go off-center within the penumbra, displaced toward the center of the solar disc so that the spot looks concave. Sometimes, the umbra displaces in the opposite direction, making the spot look convex. It is no illusion; spots actually do dimple into the photosphere, or bulge out in continent-sized plateaus. The center of a spot can be 800 kilometers (500 miles) or more above or below the level of the photosphere.

There is always the chance that you might see a flare. Rarely bright enough to be visible to the observer working in white light, lasting only a few minutes, a flare is for most of us a chance-in-a-lifetime opportunity, coming only after years of patient watching. They are more common near the peak of a sunspot cycle, so you can increase your chances by paying close attention to the sun in the year or so after the beginning of the next decade. Give close attention to any sudden brightening of the region around a sunspot. If its brightness continues to increase, outshining the photosphere, and seeming to illuminate the umbrae of nearby spots, you probably have a flare. A record of such an event, either a sketch or notes as to its time, location, duration, and general appearance, can be helpful to solar astronomers.

The Lives of the Stars 26

Looking up at the night sky, we can see that not all stars are like the sun; some must be brighter, some older, some younger, and many are different colors. More than anything else, the variety of stars reflects different stages in a life history most stars share. How long can we expect them to shine? Where do they come from? And what happens to them when they die? The understanding of these issues has been to a great extent the history of modern astronomy. The picture developed in this chapter is the result of an extended piece of detective work, an investigation remarkable for the great distances it has had to span.

STELLAR DISTANCES

Long before the invention of the telescope, astronomers reasoned that if the earth orbited around the sun, then the stars ought to show an apparent change in their positions over the course of the year, much like the retrograde motion of planets. Such displacements are common sights on earth and are generally called *parallax*. You can see parallax on a small scale by holding one finger about a foot in front of your face. Look at it first with your left eye closed, and then with your right eye closed. The finger seems to leap back and forth against the background.

When it involves a star, this apparent displacement is called the *stellar parallax* (Fig. 26.1). It was not until the nineteenth century that an astronomer detected parallax in a star. The star was 61 Cygni, a fifth-magnitude star in Cygnus, which was found in 1838 to show a parallax of less than a third of

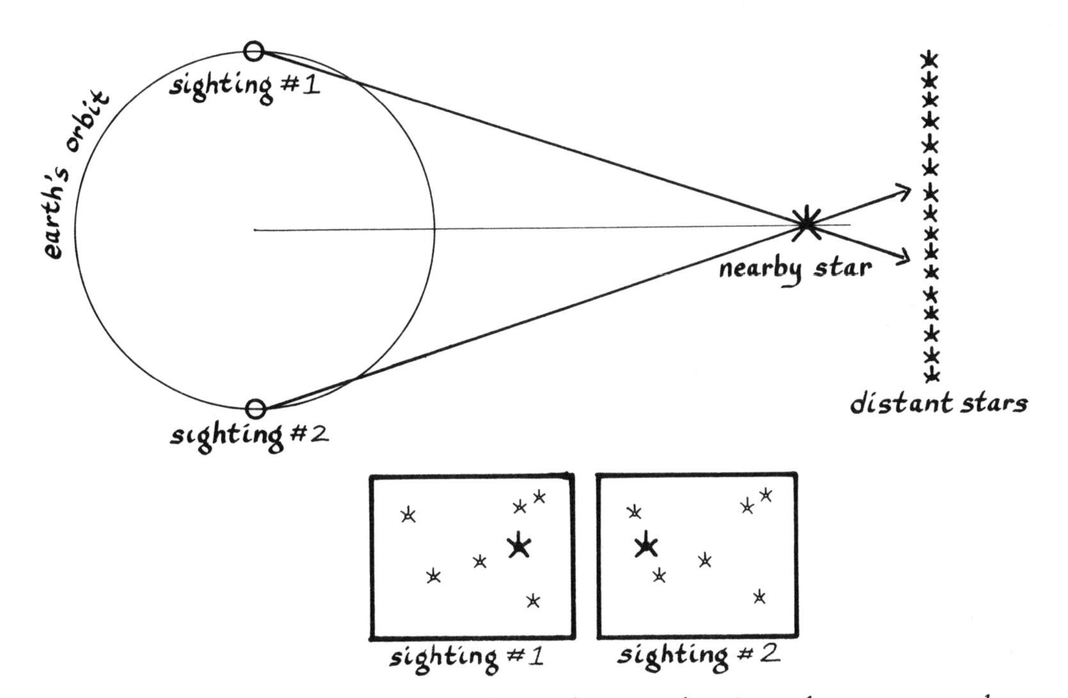

26.1 Stellar parallax. Two sightings of a nearby star, taken 6 months apart, cause the star to appear to shift against the background of more distant stars (inset). *Angular measurement of the shift gives the distance to the star by simple trigonometry.*

a second. The discovery was significant because it gave astronomers their first measurement of the distances between the stars. In principle, it is a relatively simple process, especially when augmented by photography. One simply photographs a star at six-month intervals and measures the apparent displacement of the star against the fainter, background stars, which are presumably too far away to show a noticeable displacement. That displacement, measured in arc-seconds along the celestial sphere, gives the parallax angle. This angle, along with our knowledge of the size of the earth's orbit, is enough to define the size of a long, skinny triangle, with earth's orbit as its base, and the distant star at its point. A simple trigonometric calculation gives the distance of the star.

The distances between the stars are so great as to require new units of measurement, the *parsec* and the *light year.* A parsec is the distance at which a star shows a parallax of exactly one arc-second; it is equal to 3.26 light years. A light year is the distance light travels in one year; it equals 9.5 trillion kilometers. The parallax found for 61 Cygni is exactly 0.292 arc-seconds. Trigonometry places its distance at 11.2 light years. The technique has been refined to measure the distance of stars as far as 300 light years away. Beyond that distance, the parallax angle is too small to measure reliably. Sixty-one Cygni is by no means the nearest star; Alpha Centauri, the third-brightest star in the sky, is 4.3 light years away. Sirius, the brightest star in the sky at magnitude −1.47, is 8.6 light years distant. There are some visible stars that show no parallax and must be more than 300 light years away; Deneb, in Cygnus, is one such star.

ABSOLUTE MAGNITUDES

To be twice as far away as Alpha Centauri yet shine more brightly, Sirius must be by several times the brighter of the two stars. Once we know its distance, the inverse-square law of light tell us the exact relationship between that distance and the true brightness, or *luminosity,* of any star. If we assign the sun a standard luminosity equal to 1.0, Alpha Centauri has a luminosity of 1.3 suns, and Sirius is 23 suns. Luminosity, unlike visual magnitude, expresses the actual energy output of a star, independent of the dimming effect of distance.

Another measure of luminosity is *absolute magnitude.* A star's absolute magnitude is the visual magnitude it would have if seen at a distance of ten parsecs. The absolute magnitude of the sun is only 4.8—a great drop from its visual magnitude of −26.7, but a more accurate picture of the sun's ranking in the big picture of the galaxy. On this scale, Sirius loses a bit of its glory; its absolute magnitude is only 1.4. Vega, brightest star in the summer sky, shines at an absolute magnitude of 0.5—one half-magnitude dimmer than its visual magnitude. This difference tells us that Vega is somewhat closer than ten parsecs—26 light years, to be exact.

The difference between a star's apparent and visual magnitude (in the case of Vega, 0.5) is also called its *distance modulus,* and can be another method of determining the distance to a star. So long as you have a way of determining its absolute magnitude, a simple calculation gives the distance to the star.

On the scale of absolute magnitudes, the true luminaries of our galaxy emerge. The brightest stars in the sky, we find, are not Sirius, Canopus, Arcturus, or the other stars that dominate the scale of visual magnitudes. They are Rigel, the blue-white diamond in the knee of Orion, and Deneb, the dimmest of the three stars of the Summer Triangle. Each shines with a light 50,000 times brighter than the sun's, giving them an absolute magnitude around −7.0. At a distance of ten parsecs, they would shine more than two magnitudes brighter than Venus at its maximum. To shine so powerfully and yet appear no brighter than they do, these two stars must lie far, far away. The distance modulus of Deneb, for instance, a full 8.36, gives a distance of 1,600 light years from earth; Rigel, 900 light years.

STELLAR CLASSES

How, if parallax can only give a star's distance out to 300 light years, did we derive the distances of Deneb and Rigel? If we had a means of determining a star's luminosity, then a reversal of the distance modulus gives the distance. There is a way to estimate a star's luminosity, using information contained in its light. This method involves the *spectral analysis* of starlight.

Like the sun, the stars have their spectral energy curves. Their black-body radiation varies according to their surface temperatures. The curves for hot stars peak in the bluer regions of the spectrum; cooler stars peak in the red. The part of the spectrum in which a star's energy curve peaks determines to a large extent the color it appears to the eye. Its color, therefore, is a surprisingly accurate gauge of its temperature. Of course, this relationship isn't limited to the visible range of the electromagnetic spectrum. Many stars are so cool as to have energy spectra peaking in the infrared. Extremely hot stars radiate strongly in the ultraviolet range.

Astronomers have grouped the stars into *spectral classes.* There are seven basic classes, with letter names: O, B, A, F, G, K, and M, in order from hottest and bluest (class O stars) to coolest and reddest (class M stars). A mnemonic for the order of these names is: "Oh, Be A Fine Girl [or "Guy"], Kiss Me!" A new class, C (for Carbon) stars, has been added to this series, after M. The full phrase may now go something like: "Oh, Be A Fine Girl/Guy, Kiss Me, Cutie!" For those who consider such a phrase irredeemably sexist, the best replacement I have been able to compose is: "Ollie Broached A Full, Gurgling Keg, Matey: Cheers!"

Because stars toward the blue end of this range tend to be young, and red stars tend to be old, the bluer classes are often called "early," and the red classes "late." Each of the classes is also divided up into ten subcategories, such as A7, B0, G3. Within the temperature range of that class, the lower the number, the hotter the star. A G2 star (the class to which the sun belongs) is therefore slightly hotter than a G3, but a G9 star is hotter than a K0. Within any class, the lower-numbered stars are said to be "earlier": a G0 or G1 star would be called an "early G star."

The following table lists some characteristics of the various classes.

Class	***Color***	***Surface Temperature (degrees Kelvin)***	***Examples***
O	intense blue	38,000	delta, lambda, iota, and zeta Orionis
B	blue	30,000	Rigel, Bellatrix (gamma Orionis), Regulus
A	bluish white	10,000	Sirius, Vega, Altair, Fomalhaut, Deneb, Castor
F	whitish yellow	7,000	Capella, Procyon, Polaris
G	yellow	5,800	sun, Alpha Centauri, Algenib
K	yellow orange	5,000	Arcturus, Aldebaran, Pollux
M	orange red	3,900	Antares, Betelgeuse, Ras-Algethi

Another way of expressing a star's spectral class is the *color index,* which is the difference between a star's magnitude at different wavelengths. In the most common color index, the *B-V index,* an O-type star has a value of about −0.3; an M-type star is about +1.4.

Starlight conveys more information than spectral class and surface temperature. Using the *spectroscope,* astronomers have been able to determine the chemical composition of stars and galaxies over distances as great as billions of light years. In its simplest form, a spectroscope is simply a prism placed in the light path of a telescope. Before encountering the prism, the light passes through a narrow slit. The prism then spreads the image of the slit out into the familiar visual spectrum. The spectrum can then be recorded on film, giving a *spectrograph.*

Any solid, liquid, or highly pressurized gas at high temperatures will produce a *continuous spectrum.* The sun, an incandescent lightbulb, and an electric heater all give off continuous spectra. But the spectrum of light emitted by a thin heated gas (such as a neon light) consists of a series of widely separated bright lines, each one corresponding to a particular wavelength of light. Such a spectrum is called an *emission spectrum.* Every chemical element has its unique pattern of bright spectroscopic lines, as distinctive as a fingerprint. The same pattern appears when a continuous spectrum passes through a cooler gas, except that, instead of bright lines, the pattern appears as *dark* lines against a continuous spectrum. Such a spectrum is called an *absorption spectrum,* because the wavelengths characteristic of the cool gas have been absorbed.

These lines are caused by the emission or absorption of photons by electrons in the gas. The atom of each element has a unique number of electrons (although the number can vary as the atoms are ionized). These electrons possess energy, but the amount they can possess at any moment is limited. Each electron around an atom can exist only at a specific energy level. If it changes its energy level by absorbing or emitting energy, it can do so only in certain amounts. When you buy soft drinks at the grocery store, you can purchase them only in 8, 12, 16, or 32 ounce bottles; no middle sizes are possible. So it is with electrons; when they encounter energy, they can absorb it only if it is in a "standard size." Likewise, when they emit radiation, they emit it only in those standard amounts. Imagine a piano keyboard with only a dozen or so keys. On such a piano, you could not play a smooth scale: you could only jump from one tone to the next. If an electron is to change its energy level, it faces the same limited choices. It can jump between distinct levels, but it cannot occupy an energy level in between.

When an electron moves from a high energy level to a lower, it emits a photon, and the frequency of that photon is determined by the change in the electron's level. The exact pattern of "standard" energy levels available to an electron varies from element to element. Each element, as it absorbs or emits photons, does so in a distinct pattern, as unique as a fingerprint. If the gas is absorbing energy, it produces a characteristic pattern of dark lines, in an absorption spectrum; if it is emitting energy, it produces an emission spectrum of bright spectral lines, again in a characteristic pattern. Detailed analysis of this pattern can reveal not only the element involved, but its temperature, pressure, and other characteristics. Spectroscopy gives astronomers a means of analyzing the chemical composition of stars (which produce primarily absorption spectra, as their light passes through relatively cool layers in their atmospheres), as well as the many hot and cold clouds of gas scattered throughout the universe.

From the pattern of dark lines visible in a star's spectrum, astronomers learn the following:

1. What elements are present. The pattern of spectral lines fingerprints each element involved (although teasing out the spectral signature of each element from a mix of gases can be arduous work).
2. The abundances of those elements. The intensity of absorption at each wavelength is a reliable guide to how many electrons, at what energy levels, are involved in the absorption. This effect, however, is interrelated with the next, causing some uncertainty in determining abundances.
3. The temperatures of the elements. This is one of the most important facts a spectroscope reveals. Each element, when heated violently enough, starts to lose its electrons entirely, in a process similar to a rocket blasting off at escape velocity. This is how atoms become ionized. Different elements ionize at different temperatures. The extent of ionization—whether the atom has lost one, two, three, or more of its electrons—depends on the energy it absorbs. By reading the pattern of absorption lines, a spectroscopist can tell what electrons are still present around heated nuclei. Once he or she knows the state of an atom's ionization, calculations can determine the temperature of the cloud of gas.
4. The pressure of the gas. The broadening of the spectral lines can give a good indication of the gas pressure. This, in turn, can indicate the surface gravity of the star involved, from which we can determine the luminosity class of the star.

Spectroscopic analysis of the chemical composition of the stars adds greatly to the usefulness of the system of spectral classes. We now know that a star's spectral class (and therefore its temperature) can be identified by the distinctive pattern of absorption lines characteristic of its surface temperature. O-type stars, for instance, have spectrograms dominated by the lines of ionized helium, oxygen, nitrogen and silicon, as well as un-ionized helium. A-type stars show the darkest hydrogen lines, as well as ionized calcium, magnesium, iron, and titanium. G-type stars, like our sun, show very dark lines of ionized calcium, and un-ionized lines of many metals. In the coolest stars, the temperatures are so low that many atoms keep their electrons, and even some molecules survive. M-type stars show the absorption lines of titanium oxide. The spectrograph is an accurate long-distance thermometer, not only within a lettered spectral class but in each of the ten numbered subcategories of that class. We can estimate a star's surface temperature to within a hundred degrees, over distances of thousands of light years.

THE DOPPLER EFFECT

Other information in a spectrum can be even more revealing. Much of our knowledge about the motions of the stars comes from applying the *Doppler effect* to their spectral signatures. You are already familiar with the Doppler effect if you have ever listened to the siren of a passing ambulance. As the sound approaches you, it seems to wail at a fairly constant pitch, but as it passes and speeds away, the frequency of the siren seems to drop. This effect applies to light waves as well. Any object emitting waves at a constant frequency will seem to emit them at a higher frequency when it is moving toward you, and a lower frequency when it is moving away (Fig. 26.2).

The mechanics of the Doppler effect are simple. Imagine a car moving toward you, blowing its horn at 500 hertz. It will put out a wave of sound 500 times each second. The sound waves move outward from the horn at a constant speed. But the car is moving, and tends to catch up with the waves spreading out ahead of it. At the same time, it outruns the waves spreading out behind it. Each new wave will therefore appear slightly closer to the waves in front, and slightly farther from the waves behind the horn. The sound waves ahead of the car have been compressed by the car's forward motion, and therefore they reach your ear at a higher frequency. The opposite is true as the car recedes: the waves reaching you are more widely spaced, and therefore sound at a lower frequency.

The same is true of light waves. When they are emitted from a source approaching us, the light shifts toward a higher frequency. Such a spectrum is said to show a *blue shift.* A receding light shows a spectrum *red shifted* toward longer wavelengths. Spectroscopists detect such shifts by comparing the positions of the lines in a star's spectrum with their positions in the spectrum in a laboratory sample. The amount of shift to the blue or red ends from the line's normal position gives an accurate gauge of the source's speed toward or away from the observer. Speed toward or away from us is usually called *radial velocity.* Unless the speed approaches that of light, the Doppler shift cannot measure speed in any other direction, such as across our line of sight. An object moving away and to the right will show a red shift, but the velocity indicated by the shift will not equal the actual motion through space of the source.

The Doppler effect has been of epochal importance in extragalactic astronomy, especially in formulating theories of the origin, size, shape, and ul-

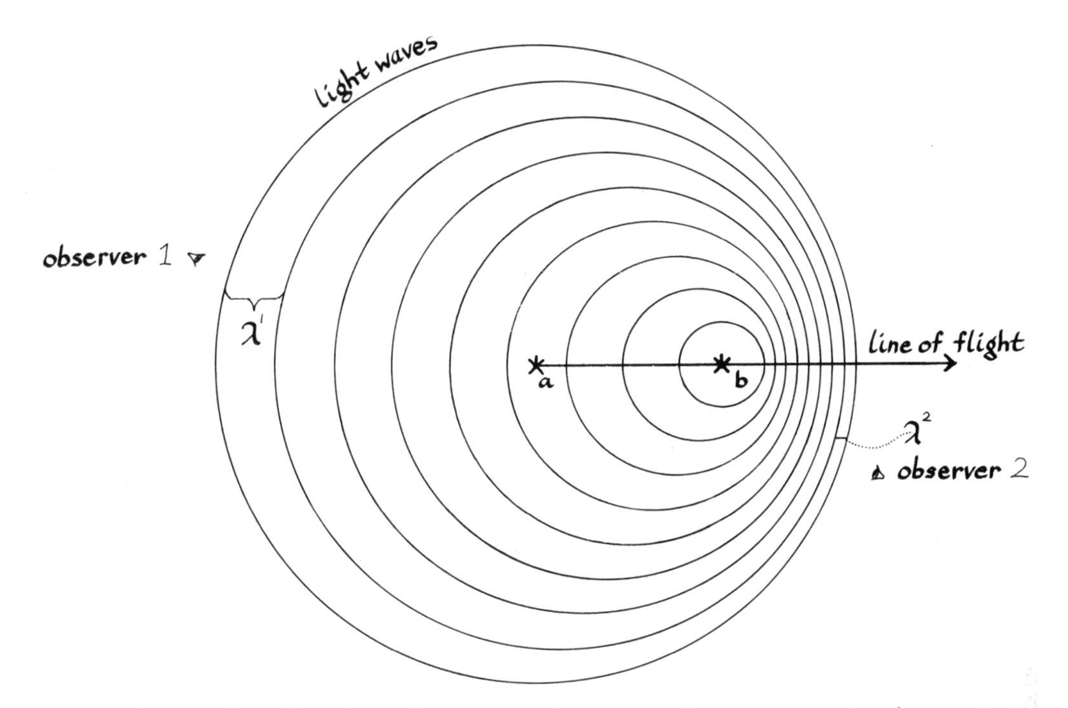

26.2 The Doppler effect. A moving source of waves, such as a star moving from point a *to* b, *catches up with wavefronts emitted in its direction of travel, and moves away from those emitted to its rear. Observer 1 sees light of increased wavelength, red-shifted light. Observer 2 sees light of reduced wavelength, blue-shifted light.*

timate fate of the entire universe. For stellar astronomers, the Doppler shift has been useful in determining the motions of stars, including their rotational periods. The lines in the spectrum of a rotating object include light from the side of the object rotating toward the observer (and therefore blue-shifted), and light from the the side of the object rotating away (and red-shifted). Such lines are wider than those from a stationary object. Details in stellar spectra can also reveal large-scale motions up and down within the atmosphere of a star. Strong magnetic fields, such as those around sunspots, cause spectral lines to split into two or more lines, forming closely spaced sets of lines in the position ordinarily occupied by one. This is known as the *Zeeman effect,* and is visible on some A-type stars as well as around sunspots on our local star.

LUMINOSITY CLASSES

The last significant piece in the stellar puzzle revealed by the spectroscope is the *luminosity class* of a star. Stars of the same temperature can be different sizes, and a large star of the same temperature will be brighter than a smaller one, simply because it has more surface area from which to shed light. The classes of stars include the very largest, brightest *supergiant stars,* called class I; two classes of *giant stars,* II and III; the *subgiant stars,* IV; and normal *dwarf* stars like the sun, in class V. The full formal classification of the sun is therefore G2 V—a normal-sized yellow star with a surface temperature of about 6,000 degrees Kelvin. The spectrogram of a star can also reveal, by subtle clues given by the pattern of absorption lines, the luminosity class to which a star belongs.

The spectroscope can tell us the true luminosity and hence the distance of a star even when it is far beyond the limit of the parallax method. Other techniques, both mathematical and observational, allow calculations of other factors significant in gauging the true nature of a distant star. For instance, it is possible to determine the radius of a star by several methods, giving additional clues to its luminosity class. Such information can be useful in determining whether a class G star, for instance, is a giant or a supergiant. Since a supergiant star has more surface area from which to radiate light, it will therefore have a higher total brightness than a giant star of the same class. Knowing a star's spectral class, then, and its radius, gives us a way of knowing its absolute magnitude, from which we can know its distance modulus. With these and other methods, astronomers have been able to learn an astonishing amount about thousands of stars. The vital statistics of size, temperature, luminosity, and chemical composition collected in this fashion have led to our present knowledge of *stellar evolution*—the many ways by which stars mature, grow old, and die.

STELLAR EVOLUTION

If we make a graph with the spectral classes running across the horizontal axis, and with the true luminosity on the vertical axis, and start putting the stars in their proper places on such a chart, we find that a pattern emerges. Such a chart, called the *Hertzsprung-Russell Diagram* (Fig. 26.3), has been one of the most important tools in our understanding of stellar behavior. The pattern tell us one important fact about the stars: there is a strong relationship between the star's temperature (or spectral class) and its brightness. Most O-type stars shine brightly; the typical members of the class have absolute magnitudes around −7. Class M stars tend to be not only cool, but small and dim as well. As we will see, this simple pattern has some important implications for stars in the upper left-hand corner of the chart.

The strong diagonal line on the H-R diagram is called the *main sequence.* Most stars in a random sample will tend to show up on that part of the diagram. The sun, as you can see, appears comfortably placed in the approximate middle of this middle-of-the-road group. Different regions of the chart correspond to the different luminosity classes. Across the top of the chart runs the sprinkling of stars called the *supergiant branch.* These are the stars of luminosity class I. A normal O-type star is naturally a supergiant; the main sequence runs up into the supergiant range in the O class. But what of the F, G, K, and M supergiants? How have they managed to shine so much more brightly than the normal members of their class?

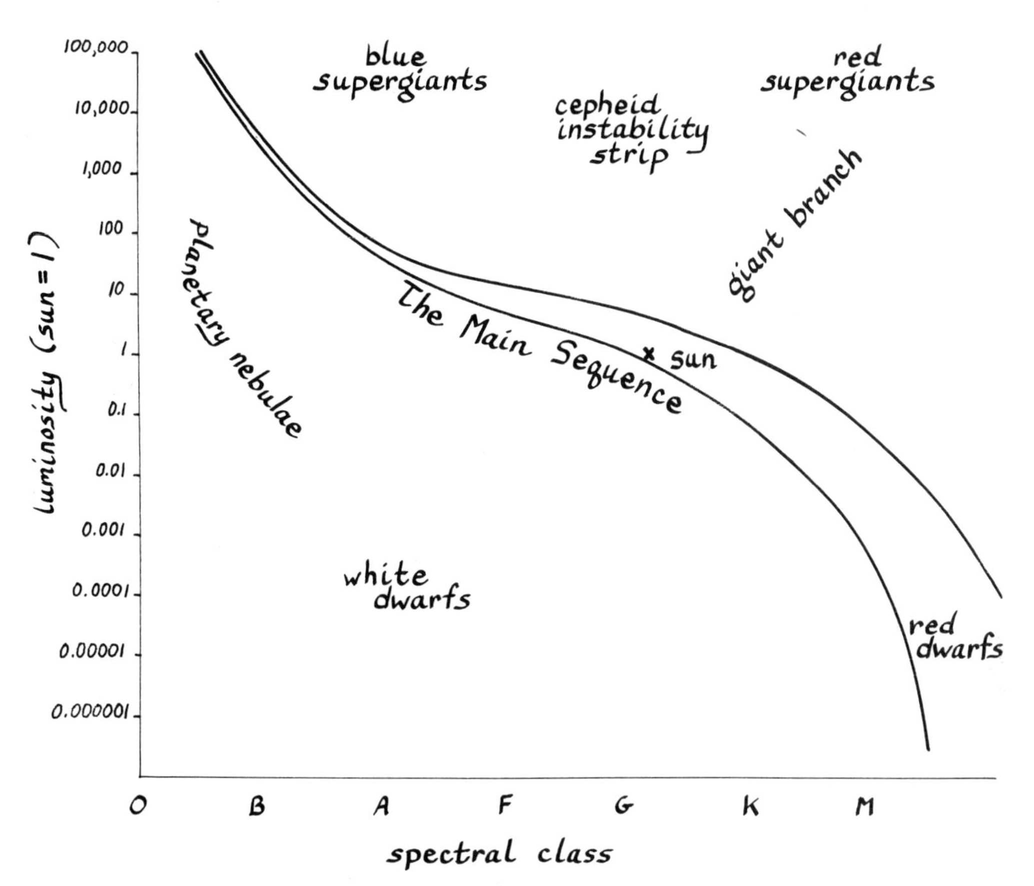

26.3 Hertzsprung-Russell diagram. When stars are plotted with luminosity on the vertical axis and spectral class on the horizontal axis, they tend to congregate in specific areas of the chart. Most normal stars, including the sun, lie on the main sequence. Blue and red supergiants appear at the top of the diagram. Red giants, and unstable stars called Cepheid variables, appear below them. The hot, shrunken central stars of planetary nebulae appear on the diagram's left edge, trailing down to the slightly cooler, tiny white dwarfs. The most common class of star, the dim, red dwarfs, congregate in the lower right-hand corner.

The spectral class of a star determines the temperature, and therefore the amount of energy emitted *per unit of surface area.* For a cool star to have a high total luminosity, it must therefore have a much larger surface area. An M-type star in the supergiant branch can shine 10 billion times more brightly than an M main-sequence star, and must therefore have 10 billion times more surface area. Something remarkable must be happening to pump a star up to that size. Below the supergiant branch runs the *giant branch,* representing luminosity classes II and III, and below them the subgiant (class IV) branch parallels the main sequence. Stars on the main sequence below the subgiant class are called *dwarf stars.* Despite their names, these are the healthiest individuals of the lot. The class includes not only our sun (imagine the size of a supergiant, if something the size of the sun is a "dwarf"!) but the main-sequence stars of class M, called red dwarfs. Red dwarfs, as the thickening of the main sequence in that region indicates, are the most numerous stars in the galaxy. Seventeen of the 25 stars nearest the sun are red dwarfs.

Below the main sequence appear other classes of dwarfs. The white dwarfs, stars of spectral classes A and B, somehow shine at a ten-thousandth, or less, the luminosity of ordinary stars of their class. "Blue subdwarfs," even hotter stars, shine more brightly than white dwarfs, but are vastly underluminous for their class. What produces these groupings of stars? Evidently, processes are at work that influence the size and temperature of the stars, causing them to migrate toward particular areas of the Hertzsprung-Russell diagram. As we shall see, most of these processes are those involved in the lives of stars. As a single star passes from infancy to old age, it takes on many roles, passing in stages through many of the regions of the diagram, before exiting, stage left and down, to darkness.

When the sun first formed in its proto-stellar nebula, it occupied a position on the far right edge of the H-R diagram, below the red giant region. Over the first million years of its life, it slowly contracted, and its luminosity dropped. For the next 10 million years or so, it shone erratically before achieving a stable balance between the expanding energy at its core and the gravitational pressure bearing in.

This balancing act is crucial in the life of a star, at its end as well as its beginning. When the gas in the center of a star reaches the temperature needed for nuclear fusion, the outward pressure of the gas trying to escape is so great that it halts the gravitational contraction. Yet this same expansion, by taking pressure off the stellar interior, reduces the star's inner temperature. As the temperature drops, the star contracts, and the temperature swings up again. During the early years of a star's life, this pressure balance is not yet stable, and the star's diameter and luminosity vary unevenly. Such variable stars, called *T-Tauri* stars for a typical member of the class, are often seen in association with hot, young O and B stars, usually in gas clouds. After the initial 10 million years or so, the forces finally do stabilize. Such a star contracts once more and settles into middle age, where it will remain for about 10 billion years. Its arrival on the main sequence is called its *main-sequence zero-age*. The hottest stars reach this point within 30,000 years of their initial condensation; the coolest stars can take one billion years.

For a star like the sun, arrival on the main sequence marks the start of the longest period of its life. For the next 10 billion years it will stay on the main sequence. The amount of time any star spends there is proportional to its mass divided by its luminosity. In other words, stars at the upper end of the main sequence spend far less time there than stars at the lower end. The more mass a star has, the greater the pressure on its interior. With more pressure on its interior, more of its total volume will reach the temperatures necessary to sustain fusion. Massive stars undergo fusion at a higher rate, reaching higher temperatures—and consuming their hydrogen more rapidly. When their hydrogen starts to run low, radical changes in the core catapult the aging star off the main sequence.

There are apparently upper and lower limits to a star's mass. The upper limit is more clearly defined, and known as the *Eddington Limit.* Stars more than 36 times the sun's mass generate so much radiation that they cannot contract into a single star. If a proto-star's contraction leaves it with too little mass, however, it won't generate temperatures high enough to sustain fusion. Such stars are known as *brown dwarfs.* A very small brown dwarf, some ten times the size of Jupiter, has been observed orbiting a nearby star. At such extremely small sizes, such objects may be more accurately termed large Jovian planets. True stars, producing energy by thermonuclear processes at their cores, shine with luminosities from 0.001 to 50,000 times the luminosity of the sun.

The mass with which a star forms is the single most important factor determining the course of its life. Too little mass, and the star never turns on; too much, and it blows itself away before its life is fairly begun. Between these two extremes, relatively small differences in mass can make an enormous difference in the longevity of a star, its behavior while it lives, and the way it dies.

As a solar-mass star reaches the end of its middle age, changes in the composition of its core signal the onset of old age. After billions of years spent converting hydrogen to helium, the supplies of hydrogen run low; the balance of elements in the core shifts in favor of the helium. The star has reached the point at which it turns off the main sequence, heading up and toward the right. It is beginning to become a red giant. Over the past 10 billion years, the star's radius has perhaps doubled. It has consumed only a tenth of its total mass of hydrogen, but in the region of the core where temperatures are high enough to support fusion, no fuel remains. Except within a thin outer layer of the core, fusion stops entirely.

The heat generated by that layer of the core heats the star's exterior still more. The outer layers continue to expand, but the core, too cold to stand up against the force of gravity, collapses. This collapse repeats, under even more extreme conditions of heat and pressure, the original contraction of the protostar, with similar results. The core temperature rises, and thermonuclear fusion begins once more. But now, with no hydrogen left to consume, the reaction shifts to the fusion of helium nuclei into carbon. Helium fusion requires much higher temperatures—around 100,000,000 degrees Kelvin. Gram per gram, helium fusion also produces less energy than hydrogen fusion. Internally, the star is on a course of diminishing returns. Externally, however, it is at the climax of its career. Swollen to 50 times its present size, 500 times its present brightness, in the far distant future our own sun will become this red giant, filling 25 arc-degrees in the sky of an earth whose oceans have evaporated, its great basins baking in the vacuum of space.

The sun's state will be more dire than earth's. Between the end of hydrogen fusion and the onset of helium fusion, the stellar core contracts, becoming so compressed that it no longer behaves like a gas. It reaches a density so great that a quart of it will weigh 2 million pounds. It becomes so dense, in fact, that it functions as an insulating blanket around the core. Temperatures at the core skyrocket in a few hours to the hundreds of millions of degrees necessary to start helium fusion. In a single *helium flash,* the core kicks up into the new form of fusion. The power of that flash unlocks the core, permitting it to expand enough to act like a gas again.

It regains stability, but at a price. The helium flash kicks the sun off the red giant branch of the H-R diagram, left onto the horizontal branch, well above the main sequence. For about 10,000 years, the sun's luminosity drops, and its bloated girth shrinks. For a time it quietly consumes the helium ash deposits at its core. But once it has passed through the furnace of the red giant stage, a star can never settle down into entire calm. Stars on the horizontal branch drift through a period in which they pulsate regularly in a seesaw battle between gravity and pressure. These are the variable stars of the RR Lyrae class. In a star the size of the sun, the horizontal branch marks the last stage in its active life. It may slide up and down along the branch several times, as further stages of contraction and expansion pump the last ergs of energy out of the helium available at the core. The core spreads and shrinks, but it is a losing struggle. Eventually, the sun swoops up and right from the horizontal branch into the *asymptotic branch,* becoming a red giant again. This time, it grows so huge that it cannot hold onto its outer layers. They evaporate into space, filling the solar system with a cloud of hydrogen mixed with light elements, metals, and molecules.

In the last stages of its life, the sun exhausts so much of its mass into space that, after a time, nothing remains but an exposed, diminished core, burning hydrogen and helium unsteadily inside an expanding envelope of what was once its outer layers. It can no longer return to the horizontal branch, but continues its upward swoop, looping around across the upper regions of the H-R diagram toward the *planetary nebula region.* Here, an abrupt outburst in the core sends a shock wave through the surrounding cloud of gas. The core has ejected in a single convulsion a large portion of its outer layers. This fast-moving, brightly glowing shell sweeps into the slower-moving cloud of the solar wind, piling it up like a snowplow in an expanding bubble of thick, highly energized gas. This bubble expands to as much as a light year in diameter. It will appear to the astronomers of that distant era as a beautiful planetary nebula, shining delicately for a few millennia before fading into space.

All that remains is the naked core. Blazing at a surface temperature of 100,000 degrees Kelvin, the sun fades rapidly, taking perhaps 500 years to drift down and to the right, well below the main sequence, into the realm of the *white dwarfs.* Here, the sun will die. As the reactions in the core weaken, gravity finally gains the upper hand in its 10-billion-year struggle. The outer layers of the core press in, until the whole has shrunken to a sphere the size of the earth. It cools slowly. The heat produced by gravity's squeeze is so high that the surface glows white hot, its spectral class late O to early F. But a white dwarf's tiny size prevents it from shining brightly. For all their heat, even the closest white dwarfs (such as the companions to Sirius and Procyon) shine with apparent magnitudes of only 9 and 10.

The star's history from this point is a catalog of the bizarre. The gravitational pressure is so great that at the surface crystals of diamond may grow from the carbon accumulated there. Like a spinning figure skater who draws in his arms to spin faster, a white dwarf rotates faster as it collapses, until the entire star completes a rotation in an hour or less. The nuclei and electrons in what were once its gases squeeze so closely together that they cannot move. The star, for so long a globe of gas, becomes solid. The electrons become *degenerate matter,* incapable of transporting energy by convection or radiation. Like a giant metal ball, the star can transport energy from its cooling interior only by conduction, an extremely efficient process for degenerate electrons. So great is the gravity at the surface of this whirling, crystalline sphere, that all irregularities slump. Its matter settles into its tightest shape. Miniature mountain ranges collapse, whisker's-width flaws shift, causing the whole to reverberate like a gong as starquakes jar the spinning sphere. Billions of years will pass, and little will change. Only a cooler, darker corpse of our sun will remain, decaying slowly into a *black dwarf* star, invisible in the night.

Larger stars live faster, and go to their deaths in a blaze of light and violence. In a star more than twice the mass of our sun, the fusion process does not end with the conversion of helium into carbon. Expanding over the course of their main-sequence life to ten times their original size, such stars climb rapidly into the red-giant branch. There they shift to helium carbon fusion, brighten for a time, and then settle still farther toward the red end of the spectrum. If a star is massive enough (about three times the sun's mass), it will pulsate during this period, as it passes through a region of the H-R diagram called the *Cepheid instability strip.* Expanding and contracting like a giant bellows, such a star can change its brightness by several magnitudes. Delta Cephei is the prototype of this class. In even more massive stars (around ten solar masses), the switch to helium fusion occurs before they even reach the red-giant stage. Such stars leap off the main sequence in less than 50 million years, traveling swiftly right across the top of the H-R diagram to become red supergiants. At their cores, temperatures rise beyond those required for helium fusion until carbon and helium fuse to form oxygen. Further contraction in even more massive stars pushes temperatures to 500 million degrees or higher, initiating reactions involving heavier and heavier elements, pushing temperatures higher still.

In the most massive of supergiant stars, such as eta Carinae, just before the star reaches the crisis point in its life, its internal structure can be something like this. Its outer layers have swollen to a diameter of a billion kilometers—more than three times the diameter of earth's orbit. This enormous envelope of hydrogen, enriched with molecules like titanium oxide, has grown so thin as to be practically a vacuum, so cool as to allow actual smoke and steam to form on its surface. Inside this dull red atmosphere, a thin layer, only a few thousand kilometers deep, still fuses hydrogen into helium. Compared to the deeper layers, the temperatures produced by this reaction seem cool. Inside the hydrogen-burning layer, a layer of helium fusion produces carbon. And inside that, a shell fuses carbon and helium into oxygen. A still-deeper shell can form, fusing carbon into light metals such as magnesium. At temperatures over a billion degrees, neon nuclei fuse into oxygen and magnesium, oxygen nuclei form silicon, sulfur, and phosphorus, silicon fuses to nickel, and finally, at the center, fusion reactions can develop that produce iron.

The structure of shells within shells totters on an insecure foundation. At each successive layer, the temperatures required to sustain fusion become higher, yet as more and more massive nuclei become involved, the energy return decreases. The fusion of nickel into iron is the last fusion reaction that can produce energy. To fuse iron into a heavier element requires more energy than the reaction releases. Once nickel fusion appears in the core, the end is only a stellar heartbeat away, for the pace of this process has been accelerating all the while. In a massive star with a main-sequence life of a few million years, the helium fusion stage can last as little as a few hundred thousand years, carbon-fusion only a few thousand. The accelerating cycle of core burnout, contraction, and the ignition of a new, hotter kind of fusion races toward a catastrophic end.

What may happen at the end is still not clear. The switch to a new kind of fusion may take place in a core that has collapsed into degenerate matter, incapable of radiating away the renewed outburst of energy. The attempt to fuse iron nuclei may be a powerful drain on the star's energy. The number of neutrinos generated by the core reactions may climb suddenly, drawing off large amounts of energy. In any case, the star has run out of fuel. The furious pace of the innermost layer of fusion has depleted the fuel supply while the energy required to support the structure inexorably increases. At the same time, it is radiating energy in enormous quantities. When the reaction stops, the headlong flight of energy from the core leaves the shells above unsupported; the whole structure totters, and, in a matter of a few hours, falls. Twenty-five times the mass of the sun attempts to occupy the center of the star. The rebound from that collision blows away most of the star's material at speeds up to 9 million miles per hour. The star has gone *supernova.*

At peak brightness, a supernova explosion shines with an absolute magnitude of −20; a supernova within 30 light years of earth would rival the brightness of the sun. For a few days it shines brighter than an entire galaxy of normal stars. In that unimaginable release of energy, light nuclei fuse into elements heavier than iron, a process that demands far more energy than it releases. Here, there is energy to spare. Fortunately, catastrophes of this magnitude are rare, occurring only in stars four or more times the mass of the sun. They tend to happen in a galaxy of our size several times each century (we can observe them in galaxies beyond our own, where they shine so brightly as to be unmistakable, though the distance between us is tens of millions of light years). We have seen no supernova in our galaxy for several centuries, probably because they have been occurring out of sight, on the far side of the galactic hub.

The next supernova to appear in our skies will be a major event for astronomers, who will have a chance to observe close at hand a process we have only seen at intergalactic distances, or in theoretical models. This will be especially important, because what happens to a star *after* a supernova explosion is one of the strangest events predicted in modern physical theory. Shattered, incapable of producing energy, the core remnant collapses further, just as a white dwarf forms from the wreckage of a sun-sized star. But if the supernova remnant retains more than 1.4 times a solar mass, its collapse goes further than that of a white dwarf. The gravitational forces generated as the star collapses are more than the subatomic particles can bear. Protons and electrons are crushed together, forming a mass of neutrons condensed into a substance so dense that a quart of it would weigh a trillion tons.

If the remnant is less than three times the mass of the sun, the collapse ends there, with a body about 10 miles across, rotating several times a second, radiating the heat of its gravitational contraction. It has become a *neutron star.* As it rotates, its intense magnetic field emits beams of radiation from each magnetic pole. The magnetic pole of a neutron star, like earth's, is often not aligned with its rotational poles. The star's rotation can cause this beam to sweep like a searchlight through circles of space, crossing the celestial sphere several times each second. If the beam brushes past the earth, radio astronomers receive it as a rhythmic hissing, from which comes the neutron stars' other name, *pulsars.* The pulses of radio energy are so rapid and so precisely timed that, when they were first discovered, in 1967, astronomers jokingly dubbed them "LGMs", for "little green men." We now know that these regular signals are not the beacons of a star-faring civilization, but the death rattle of a star.

Like the white dwarf, a pulsar has a bleak future: only a slow deceleration in its rotational velocity, its radio beacon dimming as it slows. But what becomes of supernovae whose cores retain more than three solar masses? Such a mass, collapsing in the wake of a supernova's burst, develops gravitational forces so great that not even neutrons can withstand them. The fundamental particles of matter crush together, the entire star collapsing into a sphere of infinitesimal radius. The core develops gravitational forces so high that the escape velocity within a certain distance from the center (called the *Schwarzschild radius*) exceeds the speed of light. A light beam trying to escape from inside the Schwarzschild radius would rise up, reach a maximum altitude, and go into orbit around the core. It cannot escape. The star has become a *black hole.*

The collapse continues, but to what destination we cannot say. The classical, Newtonian physics we are accustomed to use in describing the world around us no longer applies. It is possible to describe some of the characteristics of a black hole in terms of those physics, but such a description can only be a simplification.

At a certain distance from the center of the hole there appears a boundary zone called the *event horizon.* This is the point at which the escape velocity equals the speed of light. If light, the most powerful means of communication we know, is trapped inside this zone, then all events transpiring within it can never become known on the outside. Meanwhile, inside the event horizon, the core remnant collapses to a point, with no height, no width, no depth, only mass. Such an impossible object is called a *singularity*—what you can know about its existence seems to be limited to the barest facts about the object that formed it: its mass, rotation, and electromagnetic charge.

The presence of a singularity, concentrating so much mass into a small region of space, sets up a gravitational field so intense it punches a hole in space. A spaceship, or anything else falling into a black hole, would be stretched out by tidal stress into a wire of infinite length and infinitesimal thickness, falling forever down a bottomless well. At the boundaries of the hole, the laws of physics also stretch to their breaking points. Waves of light appear to bend, trapped in a tight orbit around the event horizon. Time and distance distort, so that a falling object seems to fall forever without progressing. The vicinity of the hole is full of the images of objects that passed into it long ago, the last light waves emitted from them hovering like ghosts above their grave.

And the stellar core itself? In a very real sense, it has ceased to exist in this universe. Oddly, this invisible object is not impossible to find. So powerful are the gravitational forces around a black hole that any matter falling onto the event horizon gives off enormous quantities of energy as it is ripped to shreds. The region of a black hole shines brightly in the most energetic range of the electromagnetic spectrum, especially if the hole was once a star orbiting closely around another. Matter from the surviving star is ripped from its atmosphere, and churned greedily down a maelstrom of glowing gases until it disappears, emitting powerful X rays as its temperature rises to millions of degrees. Several powerful sources of X rays in our galaxy appear to be the sites of such activity. One of them, *Cygnus X-1,* lies 10,000 light years away in the constellation Cygnus.

These *stellar endstates*—white dwarf, neutron star, black hole—are simply the most spectacular of the many ways a star can die. By far the most common fate is the red dwarfs', so quiet it is barely discernible from their life. With one-half or less of the sun's mass, red dwarf stars burn their hydrogen fuel more slowly, and live through the few stages of their lives unspectacularly. They also live an enormously long time; their lives may be so long that no red dwarf has died of old age since the universe was formed. A star of half a solar mass is still starting its life when giant stars formed at the same time have long since vanished. Red dwarfs typically take 100 million years to contract to the main sequence. Once there, such a star burns steadily at perhaps a few percent of the solar luminosity, but it burns exceedingly long. Two hundred billion years may not suffice for a red dwarf to leave the main sequence. When one does, it will drift quietly down, not exploding, not even shedding its outer layers, to become a black dwarf, a burnt-out cinder of a star. There it joins the larger, cold remains of what were once white dwarfs that long since preceded it to the stellar graveyard.

Black dwarfs vanish forever, but a strange resurrection awaits the stars that die most violently. Many supernovae occur in gaseous, dusty regions of space, because these are the environments most likely to produce the most massive stars. When a star explodes into such a cloud, the shock wave can set the cloud's molecules into motion. One hydrogen atom nudges up against another, and their tiny gravitational attraction keeps them together. Their combined pull snares another atom nudged close by, and then another, until the cloud is thick with small lumps of matter that combine, and under the combined pull of every atom in the cloud on every other atom these lumps coalesce, and merge finally into a solar nebula, contracting slowly to form another star rising like a phoenix from the ashes of the first (Fig. 26.4).

Evidence gathered in our solar system suggests that this is how our sun was born. The stars visible in our galaxy can be divided into two distinct populations. *Population I* stars are young, frequently hot stars, containing large quantities of metals such as iron and nickel. The sun is such a star. *Population II* stars (IInd, because they were discovered later) are far older, as much as 10 billion years old, and poor in metals. The difference in their chemical composition tells much about their origins. The stars of population II formed from virgin clouds of gas when the galaxy was young, and no

26.4 Three Herbig-Haro objects. Young stars, still shedding the dusty clouds in which they formed. (Lick Observatory photograph)

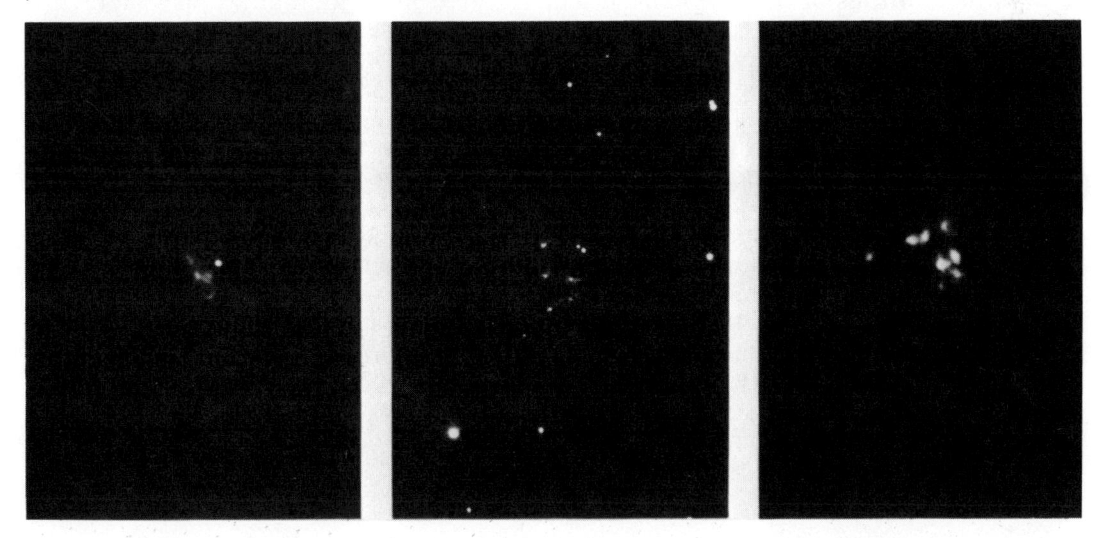

stars had grown so old as to transmute their hydrogen into metal. But as Population II stars aged, they flared briefly as supernovae, seeding the remaining clouds of gas with new stars; population I was born.

The older generation bequeathed to the galaxy not merely a new generation, but, as Carl Sagan has pointed out, a richer one. They gave us the carbon in our DNA, the calcium in our bones, the iron that tints the sands of Mars and gives our blood its bright red color. They also made those heavier elements whose synthesis requires the enormous amounts of energy available only in such moments of titanic waste: gold, silver, uranium. From the flesh, blood, and bone of a newborn infant, to the gold of ancient Egypt, the marble monuments of Rome—the warheads of our weapons—we are the heirs of ancient stars.

OBSERVATION

The observation of stellar evolution, the types and populations of the stars, involves searching out and identifying the individual types, rather than tracking the progress of a single star. You have already learned to recognize the visual sign of the stellar types, their colors. A later chapter goes into more detail on the periods—early and late—when a star's light is variable. There are two more aspects of a star's career that you can observe firsthand, however. One requires some special equipment; the other involves nothing more than persistence.

You can make a more detailed study of stellar spectra by purchasing a simple spectroscope (Edmund Scientific offers one). For visual use at the eyepiece of a telescope, these devices use either a glass prism or a *diffraction grating*. The latter is a sheet of optically planed glass on which are engraved extremely fine, parallel lines, to produce spectra by diffraction rather than refraction. In either case, you can expect to spend in excess of $100 for a suitable instrument (building a spectroscope is also well within the abilities of most advanced amateurs). Unless you have a large-aperture telescope mounted for astrophotography, you can't expect to do detailed spectroscopy of dim objects. But a simple hand-held spectroscope *will* show differences between the spectra of the brighter stars, and that alone is sufficient for some amateurs.

A more accessible mode of observing stellar behavior is hunting for novae and supernovae. Once considered two classes of the same object, it is now believed that a nova, unlike a supernova, is a temporary outburst of a white dwarf, caused by the rapid collapse of material onto the star's outer layers. In some cases, when the dwarf star is a member of a close double star system, material from the remaining star is drawn continuously into a shell suspended above the dwarf's surface. When this material accumulates into a sufficiently massive layer, it suddenly overcomes the forces holding it up, and crashes to the surface of the dwarf star. This impact releases enormous quantities of energy, and the star can brighten tens of thousands of times, reaching an absolute magnitude as high as −9—brighter than the most luminous supergiant stars. It reaches maximum brightness in less than a day, subsiding to its normal luminosity over a year later.

There are several classes of novae: *rapid novae, slow novae,* and *recurrent novae.* The last, considered as a kind of variable star, are discussed in chapter 27; there are many known examples, which you can watch for an outburst. In this class, matter accumulates continually and relatively rapidly, resulting in explosions several times a century. The mechanism behind the other two classes may be the same, but for some reason works more slowly, so centuries or millennia can pass between explosions. These events are important to astrophysicists trying to determine exactly how they work, so the discovery of a nova is an important event—one that amateurs can take part in.

Searching for novae (and supernovae: the techniques are the same) is similar to searching for comets, except that the object you are hunting for is a distinct point source of light rather than a diffuse cometary head. The brightness of novae at maximum also reduces the need for equipment. Binoculars are the most that you need, and worthwhile searches can be done with the naked eye. What's more important is a thorough knowledge of a limited area of the sky. Start by choosing a region, usually a single constellation (less if the constellation is large, more if it's small). With the aid of a good set of charts, learn the arrangement of its stars by heart. The Tirion atlas, with its coverage to eighth magnitude, is good for this. It represents perhaps the limit that the nova hunter should ask of him or herself. Start by memorizing only the naked-eye stars, and add a half-magnitude of coverage at a time until you reach eighth magnitude. You will find that the difficulty of the task grows exponentially as the stars dim below naked-eye visibility.

Once you know your territory, a nova hunt becomes simply a matter of comparing the sky on each clear night with the pattern of stars on the map (the one on paper, or the one in your memory). The more conscientiously you follow this practice, the better your chances of eventual success. There are on average 25 novae per year in a galaxy the size of the Milky Way; once every 2 weeks the opportunity is there. Of course, fewer than this number become visible from earth. They may be too distant to reach eighth magnitude even at maximum, or be hidden on the far side of the galaxy, or in regions obscured by the sun, or in the south circumpolar skies. You can increase your chances by observing on the plane of the Milky Way, where most novae occur. The constellation Aquila, for some reason, seems to be the most fertile region. The other constellations of the summer Milky Way are also productive.

If you sight a nova, you need merely watch it for a few minutes to be certain that it is not moving across the background stars in any way. Once you have a confirmed sighting, inform the Smithsonian Astrophysical Center (see chapter 23). An early notice can conceivably get major observatories onto your discovery before it reaches maximum light, yielding rare and valuable information about nova processes. In any case, you will want to keep some records of the nova's brightness. Chapter 28 gives more information on estimating the brightness of a star by comparing it to stars of known magnitude, and the techniques described there are also useful in such a case. The main difference is that you will not have a prepared chart of comparison stars; you will have to make your own. To make such a chart, simply start comparing the nova's brightness to the stars in its vicinity. Later, you can look up the magnitudes of the stars you used for your comparisons and arrive at a magnitude estimate for the nova. In addition to the time, each set of records should contain a note as to sky conditions (clear or hazy, moonlit or dark, and so forth) and an estimate of the field's altitude above the horizon. You should update these notes at regular intervals, as often as the nova's changing brightness requires.

You can also search for supernovae, using the same techniques of memorization. Because of the comparative rarity of supernovae, some amateurs have turned to patrolling other galaxies, using photographs taken at regular intervals in their search. This is an ongoing project for advanced amateurs, which has been promoted by the West Coast amateur, Ben Mayer, under the name *PROBLICOM.*

Multiple Stars

27

When you look at the Big Dipper on a clear night, you can see that Mizar (ζUMa) is clearly a double star. But when you look elsewhere, to the naked eye it seems that examples of such *binary star systems* are rare. Alpha Capricornii and a few others are all that appear. Through the telescope, however, we find that a startling new picture of the stellar population emerges (Fig. 27.1). Fully half of the sun's near neighbors in space, a total of two dozen stars, are *multiple star systems.* In such a system, two, three, and sometimes as many as five stars are gravitationally bound, orbiting around their common center of gravity. Thorough surveys of the heavens have found that multiple star systems are the rule, not the exception. Perhaps half the stars in our galaxy belong to such systems.

There are several classes of *binary star* ("binary," a term that properly applies only to double stars, is often used for all multiple systems). The different kinds are distinguished primarily by the methods used to detect them. In all classes but one, the physical phenomena involved are the same. The most interesting to the amateur is the *visual binary.* This is a system in which both stars are bright enough, and the *separation* between them is great enough, that both *components* of the system appear in a telescope. A binary too close together will look like a single star, because the angle separating them is less than the size of the Airy disc of each star. Their discs overlap, concealing the presence of two stars in a single (sometimes slightly elongated) disc.

Binaries too close for visual observation can sometimes be detected with a spectroscope. *Spectroscopic binaries* are revealed by the Doppler

27.1 Cygni, a wide visual binary. This photograph shows the typical appearance of an easy binary in an amateur telescope.

shifts in their spectral lines. The members of a multiple star system orbit around their common center of gravity, the *barycenter*. Most systems show multiple sets of spectral lines, with at least one set blue-shifting and red-shifting, as the stars' orbits take them toward and away from us. In some systems, however, the mass of the invisible companion is too small to cause the larger star to move appreciably, and its own spectral lines are too dim to appear in a spectrograph. Such systems can still be detected, however, if their proper motion is large enough.

Photographs of stars that have a large proper motion, taken at regular intervals, accumulated over decades, will show their motion across the field of background stars. Barnard's star, a dim red dwarf in Ophiuchus, has the highest proper motion of any known star (10.25 seconds per year). Long-term studies of its proper motion show that its path is not straight, but wavy. The wavy motion may be caused by the presence of two large planets, each about the size of Jupiter, pulling the star slightly off its straight course as they orbit. Binary stars detected in this way are called *astrometric binary* stars.

Finally, a class of otherwise undetectable binary can reveal itself when the orbits of the stars line up so that one star occasionally passes between earth and its companion, causing a stellar eclipse. Because such an eclipse appears as a dimming of the total light we see from the binary, this class of *eclipsing binaries* is generally considered as a variable star, and is discussed in more detail in the next chapter.

There is one more class of binary star that is not actually a binary at all. These are the *optical doubles,* in which two stars that are actually nowhere near each other in space appear side by side, simply because they lie along the same line of sight when seen from earth. Because of the proper motions of the stars (theirs, and the sun's), such optical doubles are always one-shot affairs. They draw near each other for a time, and then depart, the gap between them widening forever.

It is the close gravitational association of the stars in a multiple system that makes them a true stellar system. We recognize this association primarily through observing the orbital motion of the stars involved. Stars, like planets, follow Kepler's laws, and from applying those laws to the orbits of binaries astronomers are able to deduce the masses of the stars involved. Since mass is so important in stellar evolution, binaries have much to tell us about stars in general.

Like the planets, stars orbit around the common center of gravity of the two bodies. If one star is much larger than the other, the orbit will look as if the smaller star is actually revolving around the other. This is the case in the solar system, where the orbiting bodies are so much smaller than the sun that the system's barycenter lies near the center of the sun. If two stars are the same size, they will travel in large ellipses around a point midway between them. In some cases, these ellipses can be almost circular; usually, they are more eccentric. Because most visual binaries have separations of dozens of astronomical units, their orbits usually take scores of years. Some well-known binaries have orbits so long that they have not completed a single circuit since their discovery. But binaries found spectroscopically generally have small orbits, with shorter periods. In some cases, stars revolve around each other so close as to touch.

Special terms describe multiple star systems. They help to identify the stars involved in what is frequently a bewilderingly complex act. These terms also describe the stars' orbital behavior. The most massive star in the system is usually called the *primary.* It is commonly designated by the letter *A* added to the star's full name, such as "Alpha Centauri A". The smaller member of a pair is usually called the *secondary,* or *B,* as in "Alpha Centauri B." If other stars belong to the system, they are given letters *C, D, E,* and so on. Many older records do not distinguish between actual and optical companions; they use the term *comes* indiscriminately to refer to any dim star seen in the field of a brighter one.

The two key terms which describe multiple star systems are *position angle* and *separation.* Position angle measures the angle between any line in the telescope's field and the north point, moving through the east (90 degrees), to the south (180 degrees), west (270 degrees), and back to north (0 degrees) again. In the reversing field of a telescope, this progression will appear to be counterclockwise, with north at the bottom. The position angle of a double star is the direction in which the secondary lies, relative to the primary. For example, if the position angle (abbreviated PA) of a binary is listed as 45 degrees, the secondary member of the pair will appear northeast of the primary. The *separation* of a binary system is the angular distance, usually in seconds of arc, between the primary and the secondary, or other members, of a system (Fig. 27.2).

The full description of a double star in a catalog will usually consist of a string of numbers. First, the position, in right ascension and declination. Then the magnitudes of the components, usually separated by a comma. The separation will come next, and then the position angle: 1458 +22.1; 3.2, 8.5; 4.7″, 225°. These figures describe a binary star system at 14 hours, 58 minutes right ascension, north declination 22.1 degrees. Their magnitudes are 3.2 for the primary, 8.5 for the secondary. They appear 4.7 seconds of arc

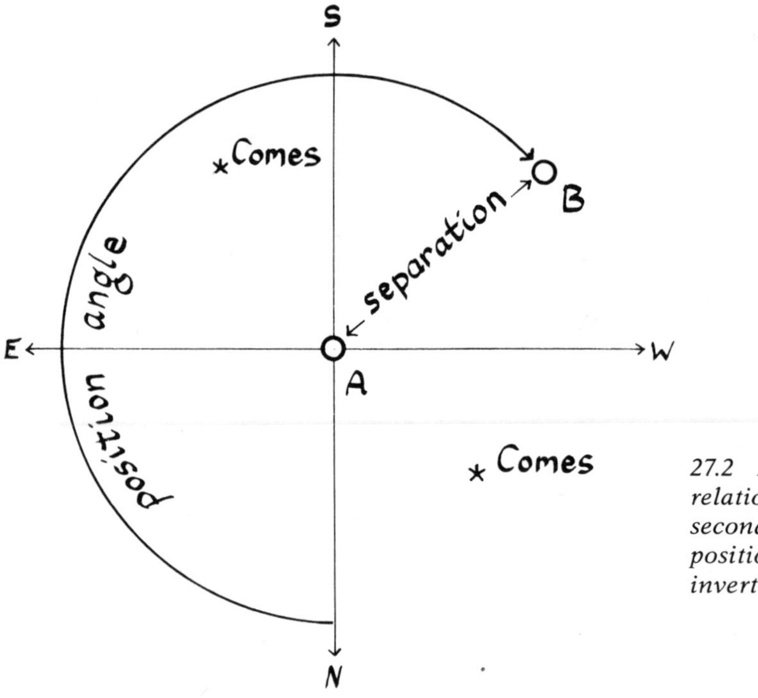

27.2 A binary star, showing the relationship of A, *primary, and* B, *secondary, components, separation, position angle, and comes. Note inverted, reversed field.*

apart, with the secondary southwest of the primary. In many listings, the date of these figures and the pair's orbital period are also included, so you can adjust for any changes that may have occurred, especially in position angle, since the binary was last measured.

Some systems require hundreds of years to complete one orbit, and have changed little since the heyday of astrometry in the last century. Others, with periods of a hundred years or less, change their position angle by several degrees each year. The elongated orbits of many binaries, combined with the skewed angles at which we view them, make the separations subject to change as well. Separation is usually greatest when the secondary is at the point in its orbit farthest from its primary; this point is called *apastrion.* Separation decreases after apastrion, reaching its minimum around the time of closest approach, or *periastrion.* Usually, unless the system's orbit is perpendicular or parallel to our line of sight, the times of greatest and least separation do not exactly coincide with apastrion and periastrion. The orbit we see in most cases is like the silhouette of the actual orbit; the actual orbit is called the *true relative orbit.* A perfectly circular orbit, for instance, seen from an angle of 30 degrees above its plane, would appear as an elongated ellipse. From measurements of the orbital velocities of the members of a binary, it is possible to determine the actual distances between them, and from there it is often possible to calculate the actual shape of the orbit.

One of the most important contributions of binary star research to astronomy is that it permits us to make an actual measurement of the masses of the stars involved, using Kepler's and Newton's laws. Once actual measurements of the binary's motions have given the true shape and size of its orbit, we can use a simple equation to find the total mass of the stars involved. Kepler's third law states that, for two bodies in orbit (such as the earth around the sun, or two stars orbiting the barycenter of a binary star system), the sum of their masses times the square of the orbital period is proportional to the cube of the semi-major axis of the orbit. Newton's discovery of the gravitational constant allowed him to turn that proportionality into an equation, one we can simplify by expressing mass in units of solar masses and orbital periods in units of sidereal years. In these terms, the total mass of the bodies in an orbital system (M) is equal to the cube of their average distance from the barycenter (d) divided by the square of the orbital period (P):

$$M = \frac{d^3}{p^2}$$

Once we know the total mass, determining how much of it belongs to each member is a simple matter of observing the relative distances of the two stars from the barycenter. The distance of each star from the system's barycenter is inversely proportional to its mass. This relatively straightforward pair of calculations has been enormously useful in confirming our theoretical knowledge of how stars shine.

In the case of two stars of unequal mass, for instance, comparing the stars' positions on the H-R diagram has helped to refine our knowledge of the relationship between mass and speed of evolution. In some complex cases, giant and supergiant stars have been observed to lose mass from their distended outer atmospheres as earlier-type companions actually siphon it off. Such a situation is the closest astronomers can come to running a controlled experiment on the relationship between mass and stellar aging. An extreme example of such a process has recently been proposed as an explanation for the planetary nebula stage of stellar evolution. The discovery that some planetary nebulae are actually very close double stars has led some astronomers to theorize that it is actually an extreme process of gravitational mass loss that throws off the glowing shell of the nebula. Close examination of planetary nebula cores may force a reevaluation of the model of planetary nebula formation.

OBSERVATION

Observing double stars can be one of the most demanding tasks you will ask of a telescope. To see a double well, you must clearly separate its components, showing dark space in between the Airy discs of the two stars. The close separations between many of the visual binaries require a telescope with the following qualities:

1. Resolution. The ability to separate a close pair, so that its components become visible as two distinct stars rather than an elongated Airy disc, is a direct test of your telescope's resolution. Naturally, the greater the aperture of your telescope, the more double stars you will be able to resolve. On most nights, however, instability in the atmosphere can impose a lower limit of 1 arc-second on your telescope's resolving power.

2. Definition. Poor definition, such as that produced by diffraction in the telescope's light path, or turbulent air, can blur the image so much that it never reaches the limiting resolution of its aperture. Refractors therefore perform better than reflectors on double stars, because of their unobstructed light paths.

3. Magnification. Assuming clear air and a low limiting resolution, splitting a close double requires the highest magnification your system can stand. Because low-contrast details, such as those on the surfaces of planets, are not involved, the magnifications used in double-star observation usually exceed the theoretical limiting magnification by a factor of two.

4. Mounting. Usually ignored in discussions of double star observing, the mounting may be the most crucial element of all. At those high magnifications, any instability in the mount will expand to a monstrous trembling projected on the sky. In addition, at magnifications of 500× or more, the earth's diurnal motion speeds up by the same amount. The sky seems to whip through the telescope's field of view at 125 degrees per minute, making smooth tracking essential and a clock drive highly desirable. If you don't have money for a professionally-machined, clock-driven mount, you may find that the Dobsonian mount is the best for this purpose. Its complete freedom from shake or backlash more than compensates for the need to move it on two axes.

The magnifications used on double stars can go as high as 100× per inches of aperture. Such magnifications are only rarely useful, however. More often the optimum magnification is about as follows:

Aperture Diameter (inches)	*Magnification*	*Aperture Diameter (inches)*	*Magnification*
2	200×	8	400×
2.4	220×	10	450×
3.1	250×	12.5	500×
4	280×	14	525×
6	350×	17	550×

As you can see, the amount of magnification per inch of aperture is not a constant figure. At 6 inches you can use nearly 60× per inch (the figure most commonly given for the upper limit of magnification). In 17-inch apertures, however, the limit is about 32× per inch. To achieve these magnifications, you need very short-focus eyepieces. For the 6-inch, *f*/5, a 2-millimeter eyepiece would be necessary—but even if you could find such a thing, you'd go blind trying to use it. A 6-millimeter ocular, with a 3× barlow, is a much better way of reaching this limit. Any time you try to stretch the capabilities of your system in this way, the quality of the eyepieces and barlow you use will make a vast difference. Quality Orthoscopics or Plössls will keep these extreme high powers from wrecking your telescope's performance. The barlow, especially, should be as good as you can afford.

Because instability in the atmosphere is most likely to limit the number of double stars you can separate, your first concern when hunting for binaries is to check the highly magnified image of a second- or third-magnitude star. If the Airy disc is small and distinct, with clear diffraction spikes or rings, the seeing will be good. If no diffraction rings appear, your chances of reaching the resolution limit for your aperture that night are poor. However, because turbulent air can calm down for brief periods, you still have a chance of moments of good seeing, even on turbulent nights. Waiting for those moments will require patience, but if your observing nights are hard to come by, you may find the waiting worthwhile. If nothing else, the time spent at the eyepiece allows your eyes to adapt to the dark and can give you practice at techniques such as averted vision.

There are several ways to increase the clarity of the image of a double star. The most important, perhaps, is to observe when the star is on the meridian, either at transit or upper culmination. A night of high winds will be hopeless, however, no matter how high the star rises. You can actually see atmospheric turbulence. With the telescope aimed at a bright star, remove the eyepiece from the drawtube, rack the tube all the way in, and set your eye at the focal plane. The patterns of light and shadow that flit across the objective are the shadows of waves of turbulent air high in the atmosphere. Try this on several nights, and you will soon be able to gauge the amount of turbulence from the appearance of these *shadow bands.*

You can also determine the relative height of the greatest turbulence by looking for shadow bands with the eyepiece racked out from the focused image of a bright star. Adjust the focus so that the bands appear sharpest. If the focal point is high above the focal point of the star (that is, if you have to move the drawtube far out in the focuser), the turbulence is close to the ground. Low-level turbulence is usually due to local cooling of the ground. It may clear up by midnight. High-level turbulence, indicated by shadow bands that focus close to the focal point of the star, is probably caused by a front moving into your region and will not improve much all night.

Take care that the immediate surroundings of your telescope are not contributing turbulence to the light path. Follow the instructions in chapter 12 on avoiding heat sources outside and inside the telescope. Your options for improving your observing conditions are not only a matter of avoiding trouble. You can easily build a simple accessory for your telescope that will help to minimize the effects of atmospheric turbulence. On some nights, you can almost eliminate it. In a reflector, this device can also reduce diffraction from obstructions in the tube.

The accessory is called an *aperture mask,* similar to the one sometimes used by solar observers. It is simply a sheet of cardboard (or plastic, or metal, or wood) placed over the aperture of the telescope. Light enters through a hole cut in the mask. This hole is smaller than the telescope's aperture. In a reflector, it is also off-center in the tube. Both the size and the position of the hole can improve the image. The smaller aperture is especially useful in telescopes larger than 5 inches, because on many nights atmospheric instability occurs in waves, and the wavelengths are longer than 5 inches. With an aperture larger than the waves, the image at any given moment will pass through two different waves, each refracting at a different angle. Result: blurring, as these two "lenses" superimpose their images. With a smaller aperture, the image passes through only one wave at a time. It will tend to swim, but will not break up and blur as much.

An off-center hole in an aperture mask can also improve the definition of any reflector by avoiding the diffraction caused by the secondary and spider. The size of the primary and secondary puts an upper limit on the size of this hole: the radius of the primary minus the radius of the secondary. In a 6-inch, with a 1.81-inch secondary, the upper size of the aperture mask that will avoid the secondary is about 2 inches. A 6-inch, *f*/5 masked in this way will behave exactly like a 2-inch, *f*/15 refractor—without the chromatic aberration. An aperture mask is more useful on larger-aperture reflectors, which are more vulnerable to bad seeing, and permit larger masks. A 6-inch telescope is not so sensitive to turbulence, but on nights when the atmosphere drops resolution below the limit of a 6-inch, a 2-inch mask may actually improve resolution.

In addition to turbulence, there is another limit to the minimum separation resolvable in your telescope. Dawes's limit (see chapter 6) applies to pairs of stars of equal magnitude, each brighter than magnitude 6. For pairs of dim stars, or stars of unequal brightness, the limiting separation is much greater. In practice, don't expect to reach the limit. Even if the atmosphere permits it, you need good optics and a few years of observing experience before you can routinely resolve stars to the theoretical limit. A more reasonable expectation, even with ideal conditions, is about twice that value. For stars with dim companions (within two to three magnitudes of the scope's limit), the separation increases by as much as tenfold, because the brightness of the primary can drown out a close, dim companion.

Whenever you're confronted with a binary like this, try installing an *occulting bar* in the eyepiece. This is simply a small strip of any flat, opaque material (thin black plastic works well) installed at the focal plane inside the eyepiece. This is easiest with the Kellner design, in which the focal plane is easily accessible through the back of the ocular. The bar should be about m316-inch across, and long enough to reach across the inner diameter of the eyepiece when it is in place. You can cut one to fit more easily if you first cut out a circle of the material with a diameter equal to the inner diameter of the ocular's barrel. Then cut the bar from that disc, choosing a slice that crosses the disc slightly to one side of center. Install the bar so that it cuts across the ocular's field and appears more or less in focus when you look through the eyepiece. It's helpful to hold the eyepiece above a brightly illuminated background as you install it; holding the bar with tweezers helps, too. One of the instant-setting glues will help to fix it quickly.

An alternative method of occulting a bright binary star is to set up your telescope so that the pair's primary appears hidden behind some nearby obstruction. The discoverer of Sirius B used the wall of a building for this. The success of this method depends on the position of the dim companion relative to the obstruction. For doubles with position angles greater than 180 degrees, the companion will appear west of the primary, preceding it across the sky. Such stars will appear in the field of a telescope just before the primary bursts into view. For position angles less than 180 degrees, the dim secondary will linger in the field for a moment after the primary fades behind the wall.

Another method used by some amateurs for ferreting out the close companions of bright stars takes advantage of the *diffraction spikes* produced by the spider of a Newtonian. You see these four bright points radiating from bright stars on most nights; they are caused by light diffracting around the spider's arms. Some of the light in the Airy rings of a bright star is drawn off by the spikes, diminishing the radius of the rings between the spikes. By rotating the telescope so that the companion lies between the diffraction spikes, you can sometimes pick up very close dim stars. A variation on this technique uses a special kind of aperture mask. This, too, is a cardboard or similar sheet placed over the aperture, but instead of a small, round opening, it has a full-sized opening with a polygonal circumference, usually in the shape of a hexagon. The sharp angles cause exaggerated diffraction spikes around a bright star. Rotating the mask causes the spikes to rotate, which is sometimes enough to reveal the close companion when it is between the spikes.

Such techniques, and the extremely high magnifications requiring expensive eyepieces, are necessary only when looking for "difficult" doubles —those of very narrow separation, or very unequal brightness. Luckily, there are a great many "easy" doubles, bright pairs widely separated, easy to find in the sky. The cataloging of double stars was one of the prime pursuits of nineteenth-century astronomy, a time when many professional instruments were not much larger than those available to amateurs today. There now exist several dozen such catalogs, listing thousands of doubles. You will find excerpts from them in many of the observing guides listed in the Bibliography. With the exception of stars carrying Bayer or Flamsteed designations, double stars are usually identified by their number in one of these catalogs, such as "a1532" for the 1,532nd star in the catalog compiled by R. G. Aitken. This is one of the most comprehensive catalogs, containing most of the stars accessible to an amateur. Other catalogs include those by F. Struve, Otto Struve, J. Herschel, S. W. Burnham, and J. Dunlop. The usual shorthand for references to these catalogs is the compiler's initial (usually a Greek letter) followed by the star's number in that catalog. A star in the Struve listing is denoted by a capital sigma (Σ) followed by a number; the sign for Otto Struve's catalog is OΣ; Herschel's is a lower-case h; S. W. Burnham's is a B; J. Dunlop's is a Δ. Many of the stars in these catalogs are extremely challenging, having been discovered by extraordinarily talented observers who pushed their equipment to its limits. You will find enough material in their lists to last a lifetime.

The following four doubles are good objects to start out with. They are all visible to the naked eye as single stars, prominently placed so as to make them easy to locate with a telescope. Each system has sufficient separation and components bright enough to make its identification easy to the first-time double-star observer. Use them as practice objects, to prepare for more challenging systems. Once you have observed them at powers adequate to show them to greatest advantage, experiment with *lower* powers. How little magnification do you need to detect a gap between them? Make your task even more difficult by defocusing the eyepiece slightly, or installing an aperture mask to cut your resolution. Using these familiar stars, and these techniques for increasing the difficulty of detecting their double nature, you will find it easier to detect the more difficult doubles that crowd the sky.

1. *Beta Cygni* (Albireo; mags. 3.1, 5.11; types K and B; sep. 34.3 arc-seconds; PA 55°) Albireo is actually a triple star system: the primary is a spectroscopic binary star.

2. *Gamma Andromedae* (Almach; 2.5, 5.0; types K and B; sep. 10.0 arc-seconds, PA 63°) The secondary is actually a triple star.
3. *Eta Cassiopeia* (3.5, 7.2; types G and M; sep. 12 arc-seconds, PA 308°) Orbit about 500 years long; separation slowly increasing.
4. *Gamma Virginis* (3.5, 3.5; types F and F; sep. 3.9 arc-seconds, PA 297°) Orbit 170 years; separation decreasing rapidly; position angle decreasing.

Once you have these, you may want to try the stars listed in the following table. They are listed in order of increasing difficulty, to allow you to hone your skills as you work down. This table was compiled by the late Dr. Joseph Ashbrook, longtime editor of *Sky & Telescope*. It appeared originally in the November 1980 issue of that magazine.

· SELECTED DOUBLE STARS ·

Object	*Separation*	*Magnitude*	*Right Ascension (1950)*	*Declination*	*Notes*
16, 17 Draconis	91″	5.2 5.6	16h 35.0m	+53° 01′	Opera-glass pair
ν^1, ν^2 Draconis	62	5.0 5.0	17 31.2	+55 13	Twin white stars
δ Cephei	41	4.0 7.5	22 27.3	+58 10	Primary variable
61 Cygni	29	5.5 6.4	21 04.7	+38 30	
α Canum Venaticorum	20	2.9 5.4	12 53.7	+38 35	*Cor Caroli*
ζ Ursae Majoris	14	2.4 4.0	13 21.9	+55 11	*Mizar*
κ Bootis	13	4.6 6.6	14 11.7	+52 01	
γ Arietis	8.2	4.8 4.8	1 50.8	+19 03	
ξ Cephei	8.0	4.6 6.6	22 02.3	+64 23	
ζ Coronae Borealis	6.3	5.1 6.0	15 37.5	+36 48	
γ Leonis	4.3	2.1 3.4	10 17.2	+20 06	
δ Serpentis	3.9	4.2 5.2	15 32.4	+10 42	
ϵ^1 Lyrae	2.7	5.4 6.5	18 42.7	+39 37	The "double double"
ϵ^2 Lyrae	2.3	5.1 5.3	18 42.7	+39 34	Test for 3-inch
ζ Aquarii	1.8	4.3 4.5	22 26.3	−0 17	
12 Lyncis	1.7	5.4 6.0	6 41.8	+59 30	
η Orionis	1.5	3.7 5.1	5 22.0	−2 26	
ε Equulei	1.1	5.9 6.2	20 56.6	+4 06	
4 Aquarii	1.0	6.4 7.2	20 48.8	−5 49	
32 Orionis	0.9	4.6 5.9	5 28.1	+5 55	
48 Cassiopeiae	0.7	4.8 6.5	1 57.8	+70 40	Slowly widening
14 Orionis	0.7	5.9 6.6	5 05.2	+8 26	
7 Tauri	0.6	6.6 6.7	3 31.5	+24 18	Hard in 8-inch
ω Leonis	0.5	6.0 6.7	9 25.8	+9 17	
η Ophiuchi	0.4	3.2 3.5	17 07.5	−15 40	12-inch test
14 Lyncis	0.4	5.9 7.1	6 48.7	+59 31	

Variable Stars

28

As they age, stars pass through a period of radical instability. This is the time when stars that have evolved off the main sequence into the red giant branch head across the upper margins of the Hertzsprung-Russell diagram toward the final ejection of their outer layers.

A star of large enough mass will pass through this phase several times in its career, as its source of energy shifts from helium to the fusion of heavier elements. Stars in this stage of their careers, called *Cepheid variables,* expand and contract by as much as ten percent of their entire diameters, and their brightness changes with their girths. There are about a thousand variables of this type known in our galaxy. They are typically such large, bright stars that we can also detect them in other galaxies, and can clock their pulsations over millions of light years. Because there is a strict relationship between the size and brightness of a Cepheid variable and the time it takes to pulsate, astronomers have used such observations to determine accurately the distances to many galaxies.

But the Cepheids are only one kind of variable star; almost two dozen types in all are recognized. Many of these are bright enough, and their variations large enough, to be detectable in amateur telescopes. Traditionally the area in which amateur astronomers have been able to contribute to scientific astronomy, the observation of variable stars is also one of the few realms of stellar astronomy in which we are able to *see* astrophysical processes in action over relatively short spans of time—hours to years. There are three major classes of variables: the *pulsating variable,* to which the Cepheids belong; the *eruptive variable,* of which the nova is member; and the *eclipsing binary variable,* of which the star Algol (βPersei) is an example. The mechanism causing the star's brightness to vary is different in each class. This difference is usually detectable, however, only by studying the *light curve* of each. The light curve is a graph showing the brightness of the star over a period of time. Figure 28.1 shows the light curve of a Cepheid variable.

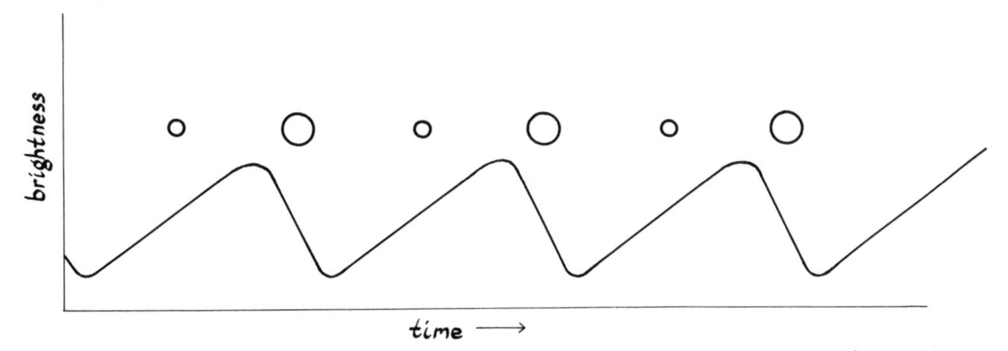

28.1 Light curve of a Cepheid variable, showing change in diameter (scale exaggerated) with brightness.

Cepheid variables, and their allied types, the *RR Lyrae* and *W Virginis* variables, are the most important members of the pulsating class. Other members of this class include the *long-period, semiregular,* and *irregular* variables. The pulsating action comes about by a combination of the processes involved in the growth of a star off the main sequence, as its outer layers respond to changes in the core. Like a young star settling down onto the main sequence, the outer layers of a variable try to maintain an equilibrium. The outward force of radiation and gas pressure from the core struggles to support the weight of the star's outer layers.

The efficiency with which the star's matter transmits energy is the crucial factor in achieving this balance. In a main-sequence star, much of the matter is ionized; ionized gases are relatively good insulators, and contain the star's heat. In a red giant star, the surface layers expand so much that the heat welling out from within is no longer sufficient to keep the gases ionized. The star starts to lose heat faster than it can generate it. Pressures in the interior drop, and the outer layers collapse, just as a hot-air balloon starts to collapse when its burners blow out. This collapse usually stops after the star has shrunk by about one-tenth. At that point, the gases have absorbed enough energy from the interior to heat up, ionize, and become efficient insulators again. With the energy-leakage plugged, the star starts to reinflate, returning to its old size. The star has completed one cycle of its variation.

The pulsation continues, however, because the star does not stop expanding until it has overshot its balance point. The momentum of the expansion swells the outer layers so far that they cool too much again, neutralize, and the whole process repeats. Oddly, the maximum brightness in a Cepheid's pulsation does not come when it is largest or smallest. At its largest, the surface grows so cool that it shines dimly. At its smallest, the internal layers are too opaque to radiate heat to the surface. It takes time for the increased radiation generated by the higher pressures to radiate through the contracted inner layers, so maximum brightness occurs while the star is expanding, before it reaches its greatest size.

For any individual Cepheid, the period from the star's *maximum* light to its *minimum* is remarkably regular. As a class, the periods of Cepheids range from 3 to 50 days. RR Lyrae variables have a much briefer period, usually less than a day. W Virginis stars fall in about the middle of that range, waxing and waning over periods of from 10 to 30 days' duration. For each variety, however, there is a strict *period-luminosity relationship*. The longer the period, the brighter the star's median absolute magnitude. (The median is the star's brightness at the middle of its range, between minimum and maximum.) Cepheids of 3-day periods, for example, have median absolute magnitudes of about −1.5. The 50-day Cepheids are truly ponderous: their absolute magnitudes of −5 puts them firmly in the supergiant luminosity class. Their brightness can vary by 0.1 magnitude to 2 magnitudes.

These figures hold for the "classical Cepheids," the first of which to be discovered was δ Cephei. These are all supergiant stars of spectral classes F to G. The W Virginis class is sometimes called the "type II Cepheid," because they are Cepheid variables belonging to the old, metal-poor stars of the galaxy's population II. Like the type I Cepheids, they are F or G supergiants, and vary by 0.1 to 2 magnitudes. They average about 1.5 magnitudes dimmer than the classical variety. The RR Lyrae stars are also population II Cepheids, sometimes known as "cluster variables" because they are common among the ancient stars of the globular clusters. Unlike the other types, they are white-to-yellow giant stars, spectral types A to F, and vary much more rapidly. Fainter by far than the classical Cepheids, they are still up to a hundred times brighter than the sun, shining with a median absolute magnitude of +1 to 0 magnitude. Their range of variation is from less than 1, to 2 magnitudes, over periods less than a day long. They are by far the most numerous of the Cepheids, being the second-most numerous of all the classes of variable star. More than 4,500 examples are known.

The period-luminosity relationship tells us a star's absolute magnitude simply by timing its pulsations. Being able to determine a star's absolute magnitude, of course, allows an astronomer to calculate its distance. The discovery of the period-luminosity relationship among Cepheid variables has been one of the milestones in our understanding of the true size of the universe. Early in this century, the detection of Cepheids in the Great Galaxy in Andromeda gave us the first clue to the real distances separating the Milky Way Galaxy from the other islands in our cosmic archipelago. Studies of RR Lyrae variables helped to determine the true size of our own galaxy, by allowing astronomers to determine the distance to the globular clusters that swarm around the galaxy's center.

It was in the course of these latter studies that astronomers realized the difference between the "classical" and type II Cepheids. No one had imagined that there might be a separate class of Cepheids, with absolute magnitudes too low for their periods. Thinking that all Cepheids shared the same period-luminosity relation, astronomers observing variables in distant clusters confused the type II Cepheids they found there with the brighter classical variety. The confusion caused them to underrate the absolute magnitudes of the latter. This in turn led them to drastically underestimate the distance of the classical Cepheids in the Andromeda galaxy. Working from that estimate (short by a factor of two), they could not understand why RR Lyrae variables failed to show up there as well. Over the distances thought to prevail, stars of RR Lyrae brightness should have been visible in the Mount Palomar telescope. The astronomers eventually realized their mistake, and the difference between the classical and type II Cepheids became plain. The distance modulus of the Cepheids observed in the Andromeda Galaxy was recomputed, giving a new figure (2.2 million light years) twice the distance of the previous estimate. At that range, RR Lyrae variables are too dim to detect.

The long-period and semiregular variables are M giants and supergiants, pulsating over much longer periods. Long-period fluctuations range from 80 to 600 days. The periods of semiregular variables are from 30 to 2,000 days long. There is a correspondingly large range of variation in brightness. The star Mira (*o* Ceti), has given its name to the entire class of long-period variables. Mira ranges from second to ninth magnitude, its size varying enormously as well. The mechanism responsible for such a monstrous change in a star is still poorly understood, but apparently involves large exchanges of energy, perhaps by shock waves traveling through the interior. One theory holds that these shocks are caused by the changeover in the star's core from hydrogen to helium fusion. As the more violent helium-burning process gets under way, shocks rack the star's interior, preventing the stability needed for the new form of fusion to settle down.

The long-period variables are the most numerous variable class, numbering well over 4,500 members. To give you some idea of the tremendous bulk involved in the pulsation of such a star, consider the case of Mira. Its diameter at maximum reaches as much as 435 million miles—five hundred times the size of the sun. Were Mira to occupy the center of the solar system, at maximum it would extend almost to the orbit of Jupiter. Yet its mass is no more than twice the sun's. Distended to such a horrible girth, this relatively modest star cools to temperatures so low that at minimum its spectral energy curve peaks in the infrared. Molecules of dust and carbon compounds and even water vapor breathe from its surface, enveloping it in haze. It is in stars like Mira that the many different organic molecules radio astronomers have detected among the stars may have their origins. And if such clouds were, as some theories hold, the source of the organic molecules that gave rise to life on earth, stars like Mira may be one of the ultimate sources of life.

In general, Mira-type variables do not show as much regularity in their light curves as the Cepheids. One maximum can be brighter or dimmer than the preceding by a magnitude or more, and the period can shorten or lengthen by weeks. There is some relationship between the length of the period and the range of brightness, but the strict period-luminosity relationship of the Cepheids does not hold. Most stars of this class have maximum absolute magnitudes of -1 to -3; semiregular and irregular variables show even greater range. One of the most familiar examples of the latter is Betelgeuse (α Orionis). It varies most noticeably over a period averaging 5.7 years, its brightness changing by almost a full magnitude. But there are shorter, smaller variations also present in this cycle, making its brightness at any given moment difficult to predict. Its average absolute magnitude is -5.5, brighter than the brightest Cepheids. At maximum, it reaches a radius possibly as large as 4.3 astronomical units. If it were to replace the sun, it would engulf all of the terrestrial planets. At minimum, its radius may shrink to less than 2 astronomical units. A much hotter star than Mira, Betelgeuse, like most other members of this class, is a true M-type supergiant. More than 2,000 are known.

The Cepheid, long-period and other classes mentioned above do not exhaust the number and variety of the pulsating variables, which include the δ Scuti and β Canis Majoris stars. Their variations are so small and rapid as to require electronic photometers to measure them. There are also the RV Tauri stars, pulsating red giants with periods as long as 5 months, and other subclasses and variations on the above themes. Because pulsating variables may represent important turning points in the evolutionary history of the stars, and many of the mechanisms driving them are still poorly understood, they will continue to be of interest to astronomers for some time, and more subclasses and variations will probably continue to be defined.

Eruptive variables are a class of unpredictable, violent stars, whose brightness changes radically over short periods of time. The nova is the most notable example of this poorly understood class. Novae are apparently capable of repeated outbursts, so long as there is a nearby source (as in a large, companion star) of fuel. These are called the *recurrent novae;* their outbursts typically occur decades apart. Another class, the *dwarf novae,* or *SS Cygni stars,* undergo much smaller eruptions, of magnitudes from 2 to 6; periods from 3 weeks to 2 years can elapse between these events. Apparently, some build-up process is at work here, as well, for the longer the wait between maxima (Fig. 28.2), the brighter the outburst. In a related class, the *Z Camelopardalis stars,* the same pattern occurs, except that the star can fail to produce a maximum, shining at constant light for two or three cycles before picking up again. In a weird reversal of the nova cycle, *R Coronae Borealis stars* blink out, *dimming* by as much as nine magnitudes, and taking a week to almost a year to regain their normal brightness.

28.2 Nova Cygni, 1975. Top: *at maximum brightness (magnitude 2).* Bottom: *at minimum (magnitude 15).*

Almost all of the eruptive variables are stars that have wandered off the main sequence. Unlike the pulsating variables, however (most of which lie in the upper right of the H-R diagram), most eruptive variables are hot dwarfs, the ragged hulks of stars that have finished their evolutionary track. Their outbursts represent the sudden collapse of material that has rained down onto the surface of a white dwarf from a nearby companion star. A few exceptions to this rule are also interesting in their own right. The R Coronae

Borealis stars, for instance, are F or K supergiants. Their sudden blackouts seem to stem from changes in their cores. There is another class, the *flare stars,* consisting mainly of red dwarfs—M-type stars on the main sequence. In these variables, increases of brightness greater than eight magnitudes but lasting only a few minutes occur at irregular intervals. These may be nothing more than outbursts similar to those observed on the sun, where we call them "solar flares" (perhaps a better term would be "stellar flares"). The extremely dim normal light of a red dwarf makes such a flare seem far brighter by comparison, causing the great increase in the star's light.

The final exception to the hot-subdwarf rule for eruptive variables is also the most numerous example of the class, the *T Tauri variables.* These are main sequence and subgiant stars, occurring in almost all spectral types and luminosities represented on the H-R diagram. They show irregular, fast variations of as much as three magnitudes and are often found in dense clouds of interstellar gas. It seems that this kind of eruptive variable is not an old star at all, but a new one. T Tauri stars are apparently just descending toward the main sequence for the first time. Their light variations may be the result of instability in their core reactions as the new star seeks for the first time the essential balance between radiation and gravitational pressures. The recent discovery of cool, solid material in orbit around T Tauri itself suggests that such stars may, indeed, be solar systems in the making.

The changing brightness of the third class of variable star has nothing to do with the star's internal processes. Eclipsing variables, as their name suggests, change their brightness when one member of a multiple star system passes between us and another member of the system. During such an eclipse, light from the hidden star cannot reach us, and the total light from the system drops. Variables such as these are important less for what they tell us about the internal changes of stars than for what we can learn about the orbits of close binary systems. Timing the eclipses, for instance, tells us the orbital period of the system, allowing us to calculate the orbits and masses of star systems too close together to be resolved by other methods. If such a system contains a star in its red giant or supergiant phase, subtle changes in its period can be indicators of mass-loss from the star's outer layers.

The exact shape of the light curve of an eclipsing variable can tell much about the size, kind, relative brightness, and condition of the stars involved. In Algol, the classic example of this type, a sharp, deep minimum occurs every 69 hours or so; the total light of the system drops from its normal magnitude 2.1 to magnitude 3.4. This minimum lasts a total of 10 hours. About 34.5 hours after that first minimum—halfway through the complete cycle—a second, briefer, and much shallower minimum occurs. Thirty-four more hours later, a second sharp minimum, identical to the first, starts a new cycle. What exactly is happening?

From studying this curve, we can deduce the following. There are two stars present; one of them is bright, and about 2.6 million miles across. The other is dimmer, but about 3 million miles across. Every 69 hours, the larger, dimmer star passes between us and the brighter companion, blocking its light; the system's total brightness drops sharply. Half a revolution later, the bright star passes between us and its dimmer companion, dimming the light only slightly, for a slightly shorter time. (There is also, apparently, a third star in the system, but its orbit is not in the same plane, so it does not take part in these eclipses.) A closer look at the shape of the curve reveals more. The total light of the system, for instance, increases just before the secondary, mid-cycle, minimum. This is apparently the result of the reflection of the brighter star's light off the surface of the dimmer companion, which shines at us just like the full moon right before it is eclipsed (Fig. 28.3).

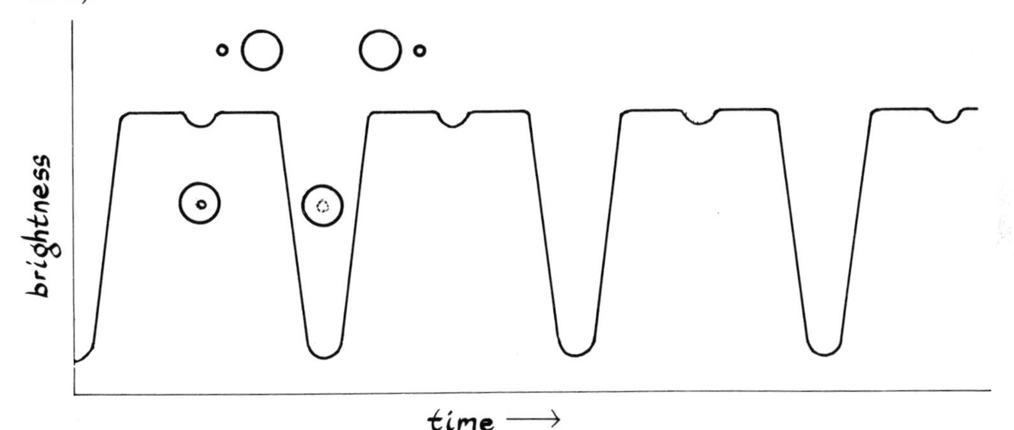

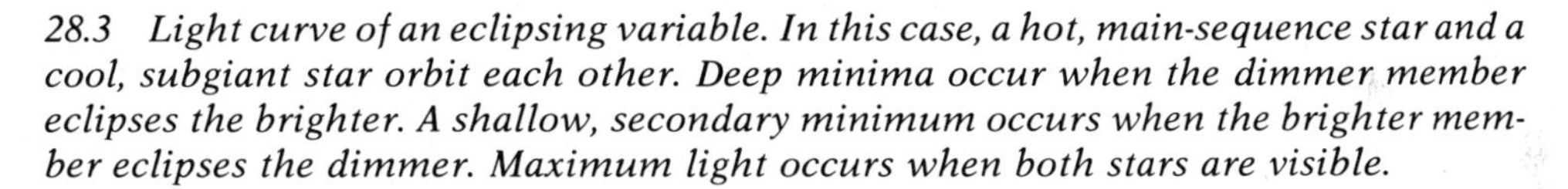

28.3 Light curve of an eclipsing variable. In this case, a hot, main-sequence star and a cool, subgiant star orbit each other. Deep minima occur when the dimmer member eclipses the brighter. A shallow, secondary minimum occurs when the brighter member eclipses the dimmer. Maximum light occurs when both stars are visible.

There are many eclipsing variables of the Algol type, showing a wide range of variations in the shape of the light curve. Generally, the longer the eclipse, the larger the eclipsing star. By the same token, the deeper the minimum, the brighter the star being eclipsed. For two stars of identical size and brightness, for instance, the primary and secondary minima will be of equal depth and duration. If the stars' orbits are not circular, but elongated, the secondary minimum will not take place in the middle of the cycle, but somewhat before or after, because the relative velocities of the stars slow down when they are near apastrion. The inclination of the orbits, relative to us, are also important in determining the precise shape of the curve.

In another type of eclipsing variable, the *β Lyrae* variables, the shape of the light curve reflects an even stranger situation. In such systems, the two stars are so close together that they almost touch. The enormous tidal stresses each exerts on the other change the stars from round to egg-shaped objects, rotating about their point of contact. In such a system, the light varies continuously, being greatest when the two stars lie side by side (from our point of view). At that time, the broadest face of their egg-shaped masses shines on us. But the maximum lasts only a moment, because the rotation of the system (which is fast, according to Kepler's laws, because the bodies are so close together) soon causes the stars not only to start eclipsing, but also to present us with their narrower silhouettes, showing less luminous surface in our direction. One subcategory of the β Lyrae class shows light variation from this reason alone: they are so close as to be egg-shaped, but do not actually eclipse. Both the Algol and β Lyrae variables usually involve stars much larger than the sun. One class similar to the β Lyrae type, but involving dwarf stars, is the W Ursae Majoris variable, or "dwarf eclipsing systems." They differ from the β Lyrae class only in the size of the stars involved, which are usually red dwarfs.

OBSERVATION

The key to observing variable stars—by which most people mean estimating their magnitudes over the course of one or many cycles—is finding *comparison stars.* A comparison star is a star of known magnitude close by the variable. By gauging the difference between the brightness of the variable and a number of comparison stars, you can arrive at a surprisingly accurate estimate of the variable's magnitude. The brightness and range of the variable determine the number of comparison stars needed, and the difficulty of making the comparison. Surprisingly, the brightest stars are the most difficult for an amateur to measure, because there are so few stars to compare them with. Such stars are so widely scattered that it is impossible to make comparisons with both stars in sight at the same time. For this reason, most measurements of bright stars are made with electronic instruments called *photometers,* which are extremely accurate versions of the light meter in a camera. With dimmer stars, it is easier to find a useful number of comparison stars, covering the magnitudes through which the variable ranges. With the dimmer stars, it is usually no problem to find such a selection within an ordinary telescopic field.

The method is simple in theory, and with a little practice becomes second nature. The essential observing aid is a *comparison chart,* several of which accompany this chapter. These are prepared by the American Association of Variable Star Observers (AAVSO), the central organization for collecting records of amateur observations of variables. Several examples of their charts appear at the end of this section, along with two lists of stars the Association recommends for observers with small telescopes. To learn more about the AAVSO program, and to obtain charts, you can write them at 187 Concord Avenue, Cambridge, Massachusetts 02138. Their catalog of available charts costs less than a dollar; individual charts even less.

Local organizations, such as astronomy clubs, may also have programs of variable-star observation under way. This is typically the first kind of research an amateur organization takes on. The reasons for the popularity of variable-star observation have to do with the large number of variables within the reach of amateur equipment, the simplicity of the technique, and the minimal requirements for equipment. For many amateurs, the chance to contribute useful data to scientific research is also irresistible.

You can practice variable-star observing with the naked eye. Binoculars will allow you to track some long-period variables through minimum, and a 6-inch, *f*/5 telescope is ideal for doing more ambitious work. The AAVSO recommends that beginners start with long-period and semiregular variables because of their large range of variation and slow, simple curves. Three to four observations per month are adequate for this kind of star. The most difficult part of observing such a star may be finding it in the first place. More on this follows, but with a long-period variable, which may be well below naked-eye visibility for much of its range, you may need some special information before you go out to look for it. The *AAVSO Bulletin*, published annually, gives the date of maximum and minimum for 700 long-period and semiregular variables. With this information in hand, you can start your observation of a long-period variable when it is at its brightest, making your first observing task much easier.

The two key factors in a telescope for variable-star observing are light-grasp and width of field. You need light-grasp to follow dim variables through their minima. A wide-angle, low-power ocular gives you a greater number of comparison stars in the same field as the variable. Any design of ocular, so long as it is achromatic and reasonably orthoscopic, will do. Kellners work well, though in focal lengths below 12 millimeters you may get better results with an Orthoscopic. The magnification you use will depend on the comparison chart for the star. The AAVSO charts come in four different sizes, with scales ranging from about 2 degrees to the inch to less than 5 minutes per inch. The *a* and *b* scales are large enough to help with finding the stars, while the smaller-scale, *c* and *d* charts should be used for making magnitude estimates. In the smaller-scale sizes, you will need moderate powers to distinguish the variable and its comparison stars. A wide-field design, such as an Erfle or Plössl, can be helpful here. The larger charts need less magnification. Generally, you want an eyepiece giving a field large enough to show all of the comparison stars on the chart.

A typical AAVSO chart will show the variable centered in the field, circled, and labeled with its name. Stars in the surrounding area are indicated, as on any star-chart, by small dots, the size of the dot corresponding to the magnitude. The limiting magnitude of the chart will depend on the range of the variable. A number of stars in the chart's field have been chosen, and their magnitudes carefully measured, to serve as comparison stars. Their magnitudes are usually written next to the star directly on the chart, given in tenths of a magnitude. So you won't mistake it for a star, the decimal point in the magnitude is omitted: magnitude 12.7, for instance, is given as simply "127". This figure is usually written immediately to the right of the star. In crowded fields where this is not possible, a short line connects the numbers to the star. Some stars of unusually wide range will have more than one chart, a large-scale and a small. The large-scale chart will show comparison stars for the brighter end of the variable's range, and the smaller chart will concentrate on the comparison stars needed to judge the minimum. For all telescopic variables, the chart shows the field inverted and reversed, as it appears in the eyepiece of a reflector.

In practice, observing a variable star is simply a matter of finding it in the telescope, identifying the comparison stars in the field, and estimating the variable's magnitude by comparison. Each of these simple steps, of course, can be baffling at first, so you should start out with one of the naked-eye stars listed below.

There are several cues to finding variables in the telescope. A long-period, semiregular, or irregular variable usually gives itself away by its deep red color. This tint usually deepens at minimum, diluting somewhat to orange near maximum. Such a star will usually be one of the reddest, if not the reddest, in the field. For stars not so marked, you must locate them by reference to the three brightest stars on the comparison chart. These may not necessarily be marked as comparison stars. Simply choose the three largest dots on the chart, and memorize their pattern and the location of the variable within this pattern.

Once you have found the approximate position of the field by star-hopping, use your finder or your lowest-power eyepiece to look for the pattern of the three bright stars. If the field is crowded, you may need to use as many as five stars forming a distinctive pattern. Don't try to use more than that—it's difficult for most people to memorize a pattern with six or more elements. One way to make this search easier is to cut down on the background clutter. Defocus the eyepiece slightly, so that the dimmer stars fade out of view. The brightest stars in the field will stand out clearly, making the pattern much easier to recognize. Once you have the pattern, center the field on the variable's location, refocus if necessary, and increase the magnification to the level needed to match the comparison chart. From there, you should be able to recognize the variable by its position among the handful of stars nearest to it on the comparison chart.

There are a number of ways to make that final recognition easier. Most important, the first time you try to find any variable, look for it when it is near maximum. In addition to the *AAVSO Bulletin, Sky & Telescope* lists in each issue several dozen variables that will reach maximum brightness in the next month or so, giving locations, magnitudes, and dates of maxima. As with all difficult objects, wait until it has risen well above the horizon haze (at least 30 degrees) before even trying to locate it. Until you are well practiced, try to choose stars with relatively uncluttered fields. This effectively rules out the Milky Way region, but there are so many variables that this shouldn't be much of a restriction. Finally, use your chart. Keep it with you, preferably on a clipboard, and study the pattern of bright stars as well as the immediate surroundings of the variable for a good, long time before you go to the eyepiece. Five minutes spent with the chart can save you a half hour of fruitless searching. You know you have it down when the pattern of stars appears before you when you close your eyes. You may find that doing a quick sketch of the bright star pattern before you look helps to fix it in your memory.

Once you have found the variable, you have to identify the comparison stars correctly. This is easier, of course, with the brighter comparison stars, but in the dimmer magnitudes the number of candidates multiplies, and the problem of identification increases. When in doubt, use the same techniques you used to locate the variable itself: pick out a distinctive pattern, and note the target's location in relation to it. It's almost always helpful to pick the pattern from stars slightly brighter, but not much, than the star you're looking for. It's also helpful if the pattern encloses the object; this keeps your gaze from wandering too far afield if you fail to identify the target on the first pass.

Finally, make your comparisons. There are several methods for doing this, but the easiest is the *interpolation method.* Take two comparison stars, one brighter, and one dimmer than the variable. It helps to imagine the two stars as occupying opposite ends of a linear scale of brightness. Your task is to place the variable in its correct location along that scale. If the variable seems to be closer in brightness to the dim star, it will be in the dimmer half of that scale. If it seems much closer, it will be very near the end. Naturally, the closer your two comparison stars are in brightness, the more accurate your estimate will be. As a rule of thumb, don't use comparison stars more than a magnitude's difference in brightness. Also, repeating the estimate with several pairs of comparison stars will help you to arrive at a more accurate figure. Finally, don't try to track variables with a range less than half a magnitude—you probably won't be able to achieve the accuracy necessary. With practice, however, you will find the interpolation method allows you to estimate the relative brightness of the three stars to a fifth of a magnitude; with practice, you can get it to a tenth (Fig. 28.4).

There are several possible sources of error in making your estimates. One of the most common is mistaking the comparison stars. I once spent a blissfully ignorant summer making painstaking measurements of what I thought was a long-period variable before I realized that the magnitude wasn't changing as much as it ought. I was carefully estimating the brightness of the wrong star entirely. I was also chagrined to find that, even though the star I had been tracking was not variable, I had been unconsciously fudging my results, underestimating its brightness because I was expecting it to dim. This is perhaps the most pernicious error in making measurements, similar to the Martian canal phenomenon. It takes a certain amount of discipline and disinterest to keep from seeing the results you hope to find. It helps me to remember that a star that *stops* varying (it's been known to happen, but very rarely) is an even more important discovery than one that continues to change according to expectations. It's also important to avoid looking at your previous records before making an observation. Since the AAVSO, for instance, doesn't require data more often than

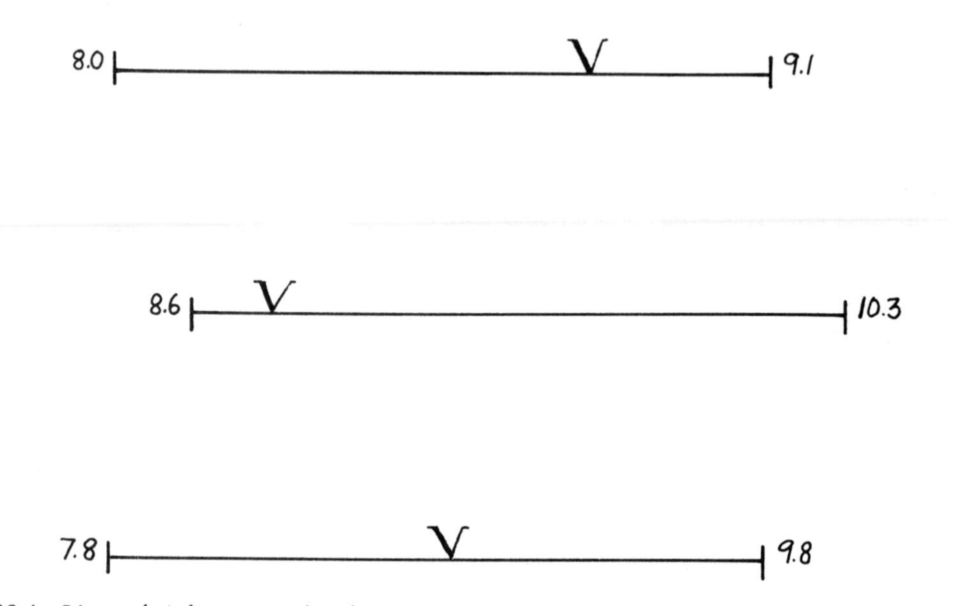

28.4 Linear brightness scales, for estimating the brightness of a variable relative to pairs of comparison stars.

monthly, there's no need to check your records more often than that. If you can keep yourself in the dark as to the star's behavior, you will be less likely to influence your observations.

In addition to psychological interference, there are also physiological conditions that affect your eye's sensitivity to light. Owing to the *Purkinje effect,* for instance, your retina becomes more sensitive to red light the longer you look at it. This causes some estimates of red stars to be brighter than they actually are. Make any estimates involving red stars (either the variable or comparison stars) as quickly as possible. You should also try not to compare stars of widely different color. The different sensitivities of different areas of your retina, responsible for the effectiveness of averted vision in detecting dim objects, also require that you try to observe each star in the same way. If you use averted vision to see any one of the three stars involved in the comparison process, use it on all of them. For some reason, stars tend to look brighter when they are in the lower region (that is, toward your chin) of any field. Try to keep stars near the center of the field, moving the telescope to do so if you must. The best way to compare two stars is to position them in a horizontal array across the field, equidistant from the center.

Naturally, if you go to the trouble of making brightness estimates, you will want to record them. I find the most convenient method is to draw up in advance a series of linear brightness scales, one scale for each pair of comparison stars. A series of parallel lines, with the magnitude of each star in large print at each end, is all you need. You may find it helpful to divide each scale into units of one-tenth magnitude. For comparison stars of magnitudes 8.9 and 9.6, for instance, you would divide the linear scale into seven equal parts. If you make up such scales, take care that the magnitude divisions do not match up from one scale to the next. The spot on the first scale for magnitude 9.2, for instance, should not be directly above the spot for magnitude 9.2 on the second scale. Scatter the scales right and left enough to keep from prejudicing your results. Under ordinary conditions, you can expect to find your estimates varying by several tenths of a magnitude, so be suspicious of yourself if you find all your estimates for the same star coming out the same.

There *is* such a thing as casual observation of variable stars, although it's a pretty closely held secret. There is no reason you should feel obligated to keep records, or even try for accurate estimates of the variables you observe. This is especially true if you are just starting out, when just identifying a variable is a challenge. There is a great deal of pleasure in simply watching a star brighten and dim in the field of a telescope, binoculars, or among the other bright stars in a constellation. The vivid red of a long-period variable flickering near minimum, the dramatic dimming of an eclipsing binary, or the violent outburst of an eruptive variable are rare opportunities to see with your own eyes changes taking place in the core of a distant star. The study of variable stars is important, and an enormously satisfying activity, but it's helpful at times to put the clipboard aside and simply stare, trying to imagine, without reference to any comparisons, what might be going on out there.

To make that imaginative leap, of course, you need to master some of the techniques of the variable star observer. To start off, here are a few stars you can practice on. The following section identifies a half dozen variables by (1) Bayer letter and constellation, (2) magnitude range, (3) length of period, in days, and (4) nearby comparison stars, also by Bayer designation, with magnitudes for each. There are AAVSO charts, supplied by that organization's director, Janet Mattei, for each. The first three stars are easily observable with the naked eye. The next three stars require some optical aid. As a general rule, you should not use a telescope on stars bright enough to see easily in binoculars, nor use binoculars for naked-eye stars.

A note on nomenclature. The first four of the following stars are atypical of the vast majority of variables in both their brightness and their names. The two differences are related. These stars are all bright enough during at least part of their cycles to have acquired Bayer letters. Most variables are not, and are usually designated by a special set of letters. The variables in any constellation are given upper-case roman letters, starting with *R,* as in "R Andromedae," and working through *Z,* as in "Z Camelopardalis." If a constellation has more than the nine variables, the lettering starts over with double initials. The last two stars that follow, R Leonis and SS Cygni, are labeled according to this system. Unlike the Bayer system, the order of the lettering does not indicate the relative brightness of the stars, but usually follows the order in which the variables were discovered (Fig. 28.5).

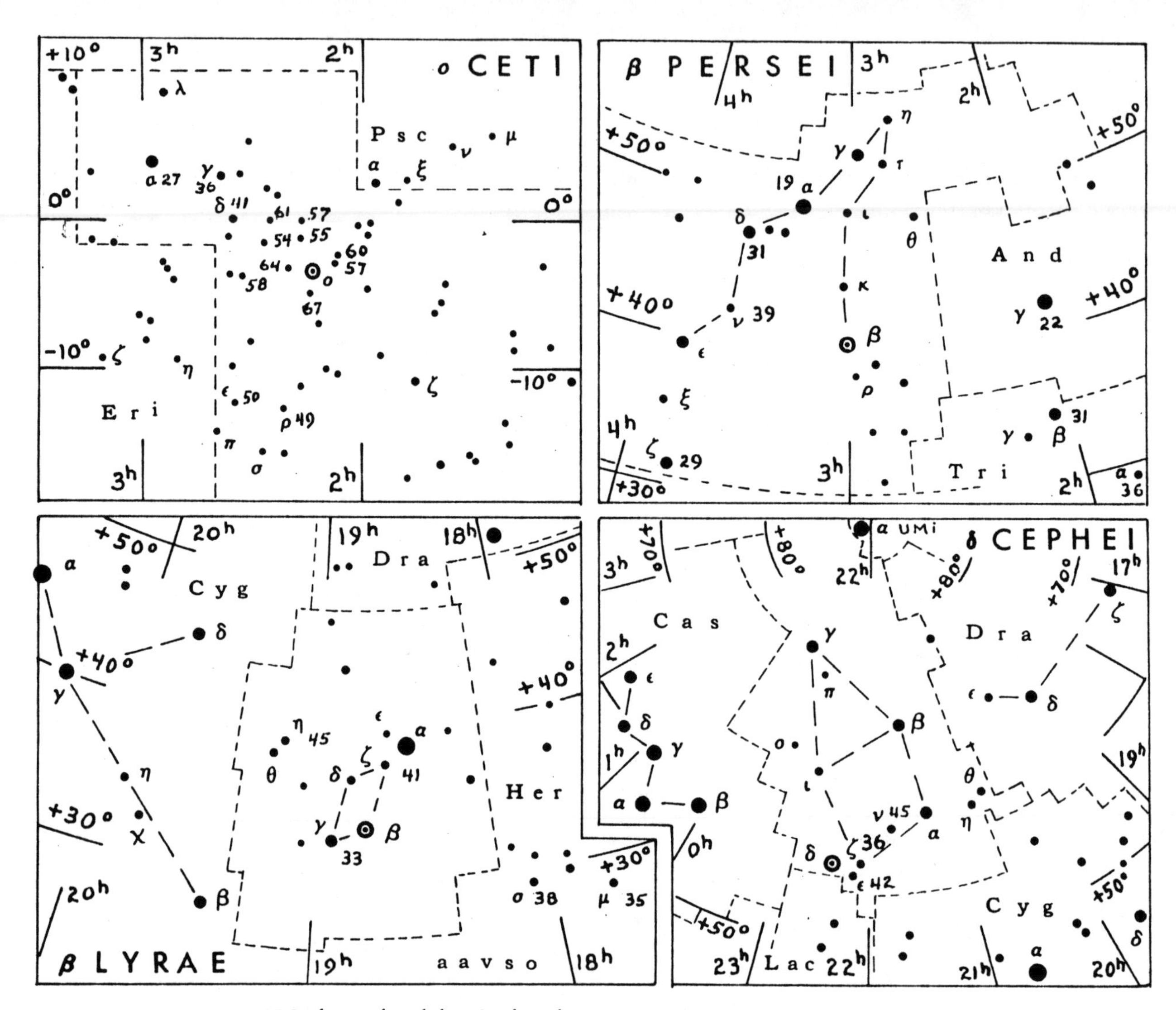

28.5 AAVSO charts for delta Cephei, beta Persei, beta Lyrae, and omicron Ceti.

1. *δ (delta) Cephei.* 3.6–4.3; 5.37 days; ϵ (epsilon) Cephei, 4.2, ζ (zeta) Cephei, 3.6, ν (nu) Cephei, 4.5. The prototype of the classical Cepheids, delta is 1,000 light years from earth.

2. *β(beta) Persei* (Algol). 2.2–3.5; 2.87 days; α (alpha) Per, 1.9, γ (gamma) And., 2.2, ζ (zeta) Per., 2.9, δ (delta) Per., 31., ν (nu) Per., 3.9. The prototype of the Algol type of eclipsing binary, Algol is 100 light years from earth. *Warning:* ρ (rho) Persei is also variable (range, 3.3–4.0; semiregular, period about 40 days).

3. *β* (beta) Lyrae. 3.4–4.3; 12.9 days; *γ* (gamma) Lyrae, 3.3, *ζ* (zeta) Lyrae, 4.1, *κ* (kappa) Lyrae, 4.3. One of the easiest variables to locate and follow, this is the prototype of the beta Lyrae class, about 1,000 light years distant.

4. *o (omicron) Ceti* (Mira). 3.4–9.2; 332 days; *α* (alpha) Ceti, 2.7, *γ* (gamma), 3.6, *δ* (delta), 4.1, *ρ* (rho), 4.9, *ε* (epsilon), 5.0. Maxima for the next several years:

1985: early April
1986: early March
1987: February, late December
1988: late November
1989: October
1990: September
1991: August
1992: July

5. *R Leonis.* (See Fig. 28.6 a and b for statistics.) A wide (10′) optical double with sixth-magnitude 18 Leonis, five degrees west of Regulus. Vivid red.

6. *SS Cygni.* (See Fig. 28.7 a and b for statistics.) Brightest example of the cataclysmic variables—the AAVSO would like to have your results. SS Cyg is a dwarf nova, displaying sudden outbursts in brightness at semiregular intervals. An extremely close binary star, in which the two components orbit within 200,000 kilometers of each other. A typical flareup of SS Cyg comes every 7 weeks, but intervals vary. Maxima last 10 days or less, after which the star rapidly drops back down to minimum.

For those interested in finding and following more variables, the AAVSO has prepared two lists of stars recommended for beginners. The first of these lists is of stars suitable for binocular- or spotting-scope observation. The second lists stars for small, mounted telescopes. They are all easy to find, because they are close to bright stars. The column of figures in each list is the variable's location, in hours and minutes of right ascension and degrees of declination; 072708, for instance, indicates a star at 7 hours, 27 minutes right ascension, 8 degrees north declination; italics indicate south declination.

Variables suitable for small, unmounted telescopes		***Variables suitable for small mounted telescopes***	
072708	S CMi	0214*03*	o Cet
094211	R Leo	023133	R Tri
103769	R UMa	043274	X Cam
141954	S Boo	074922	U Gem
154428a	R CrB	123160	T UMa
162119	U Her	123961	S UMa
1702*15*	R Oph	134440	R CVn
180531	T Her	142584	T Cam
1842*05*	R Sct	154615	R Ser
193449	R Cyg	163266	R Dra
230110	R Peg	194632	X Cyg
		205923	R Vul
		210868	T Cep
		213843	SS Cyg
		230759	V Cas

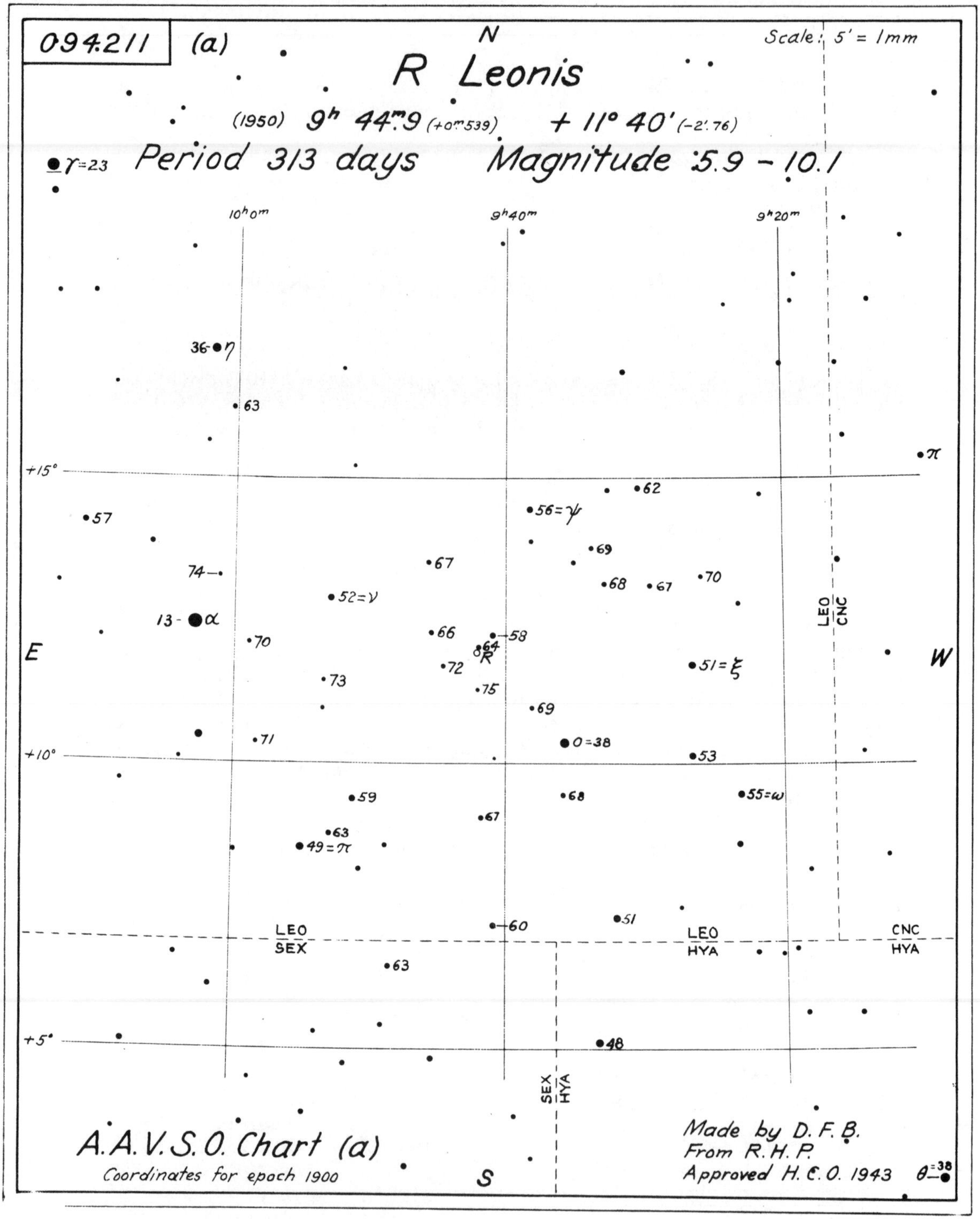

094211
(a)
N
Scale: 5' = 1mm
R Leonis
(1950) 9h 44m.9 (+0m539) + 11° 40' (-2'.76)
Period 313 days Magnitude 5.9 - 10.1
γ=23
10h0m
9h40m
9h20m
36-η
63
+15°
π
62
57
56=ψ
69
74-
67
68
67
70
52=ν
LEO CNC
13-α
66
58
70
E
64 R
W
72
51=ξ
73
75
69
71
O=38
+10°
53
59
68
55=ω
67
63
49=π
51
60
LEO SEX
LEO HYA
CNC HYA
63
+5°
48
SEX HYA
A.A.V.S.O. Chart (a)
Coordinates for epoch 1900
S
Made by D.F.B.
From R.H.P.
Approved H.C.O. 1943
θ=38

28.6a and b AAVSO chart for R Leonis.

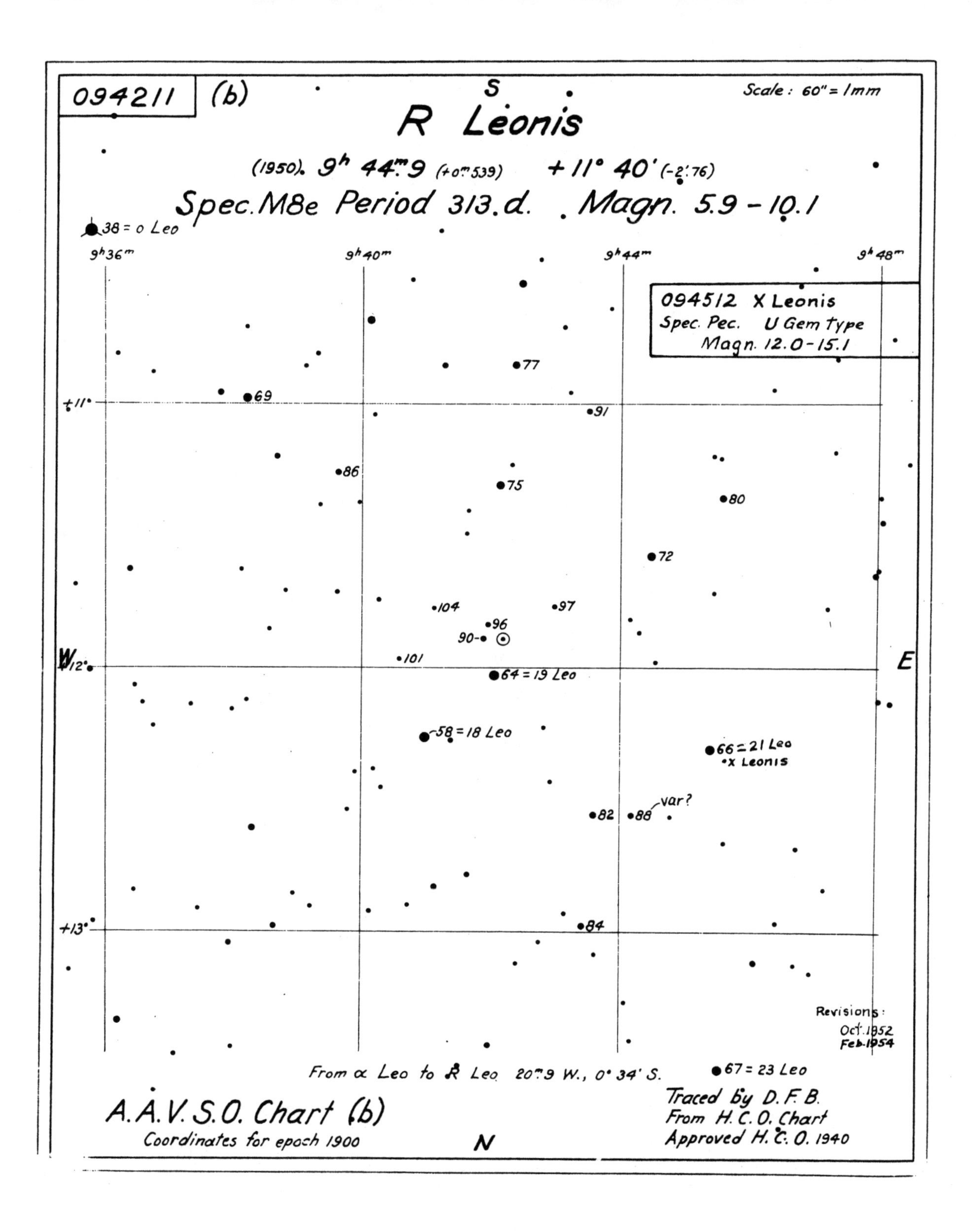

094211
(b)
S
Scale: 60" = 1mm
R Leonis
(1950). 9h 44m.9 (+0m.539) +11° 40' (-2'.76)
Spec. M8e Period 313.d. Magn. 5.9 - 10.1
38 = o Leo
9h36m
9h40m
9h44m
9h48m
094512 X Leonis
Spec. Pec. U Gem Type
Magn. 12.0 - 15.1
77
69
+11°
91
86
75
80
72
104
97
96
90-
101
W
+12°
E
64 = 19 Leo
58 = 18 Leo
66 = 21 Leo
X Leonis
var?
82
88
+13°
84
Revisions:
Oct. 1952
Feb. 1954
From α Leo to R Leo 20m.9 W., 0° 34' S.
67 = 23 Leo
A.A.V.S.O. Chart (b)
Coordinates for epoch 1900
N
Traced by D. F. B.
From H. C. O. Chart
Approved H. C. O. 1940

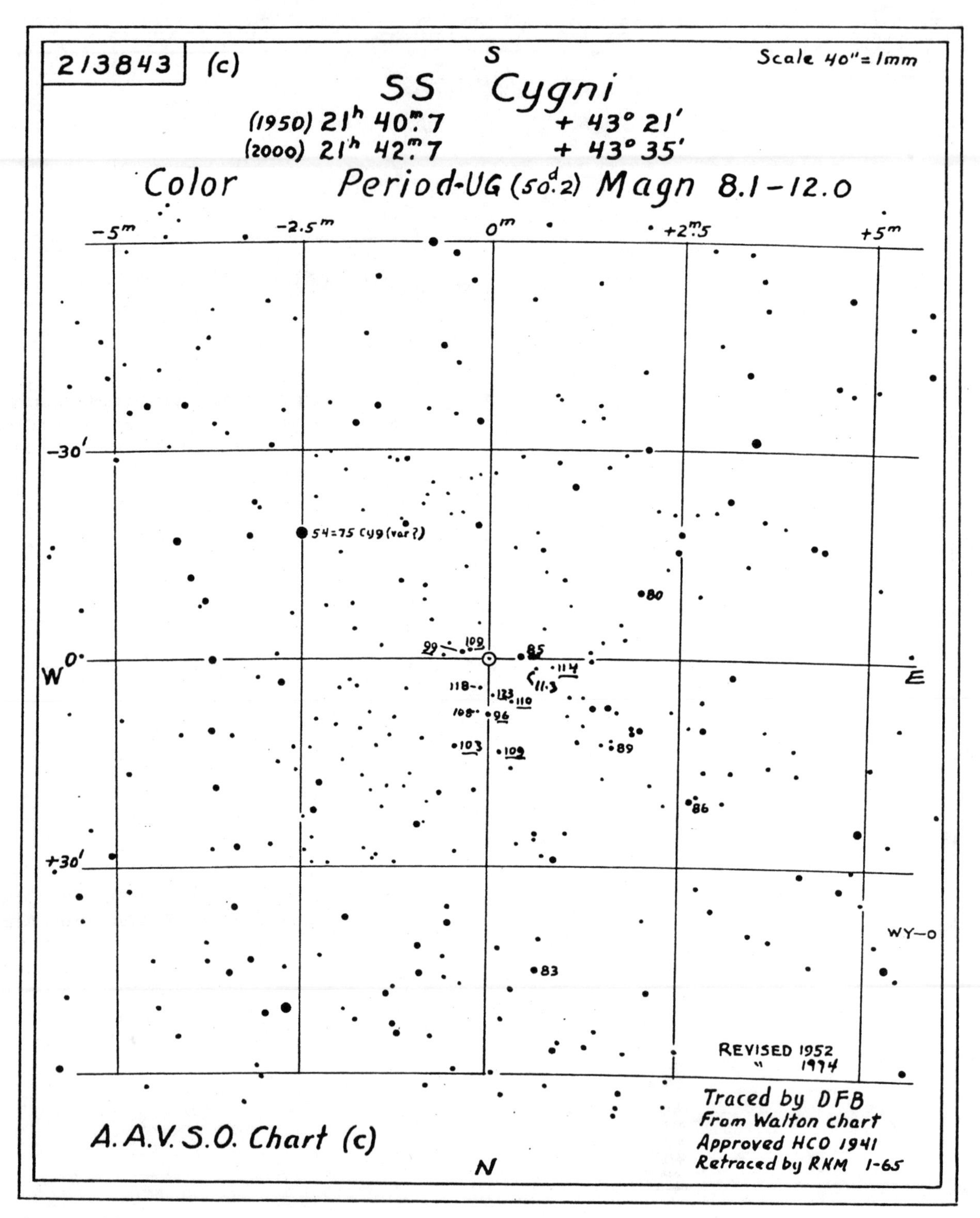

28.7a and b AAVSO chart for SS Cygni.

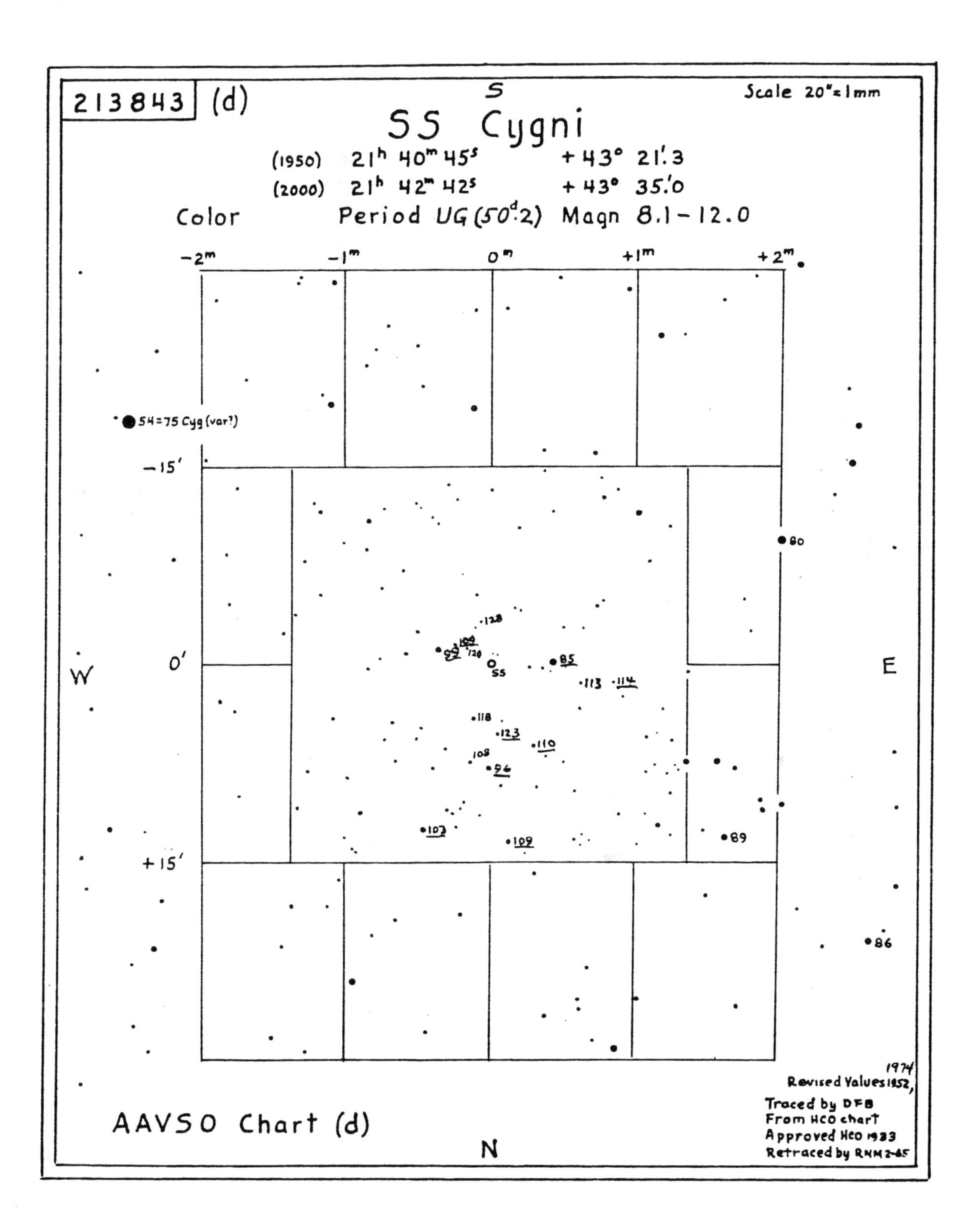
213843
(d)
S
Scale 20"=1mm
SS Cygni
(1950) 21h 40m 45s +43° 21'.3
(2000) 21h 42m 42s +43° 35'.0
Color
Period UG (50d.2)
Magn 8.1–12.0
−2m
−1m
0m
+1m
+2m
54=75 Cyg (var?)
−15'
0'
+15'
W
E
80
.128
120
SS
85
113
114
118
123
110
108
96
107
109
89
86
1974
Revised Values 1952,
Traced by DFB
From HCO chart
Approved HCO 1933
Retraced by RNM 2-85
AAVSO Chart (d)
N

29 Diffuse Nebulae

Many of the bright stars visible to the naked eye are hot, early supergiants, lying several thousand light years from the sun in the spiral arms of our galaxy. Because of their short life-spans, we know such stars must have formed recently. These arms are rich with clouds of gas and dust, from which stars form and into which dying stars pour out the elements they have formed during their lives. These clouds are called the *diffuse nebulae.*

You can see some diffuse nebulae with the naked eye. The Great Rift—the dark lane that divides the Milky Way from Cygnus to the southern horizon—is an enormous, dark cloud of gas and dust; it lies between us and the inner galaxy. The clouds of the Rift are *dark nebulae* (Fig. 29.1). There are two other classes of nebula, *emission nebulae* and *reflection nebulae.* One of the brightest of the emission nebulae is also visible to the naked eye, as the dim mist surrounding θ Orionis, the middle star in Orion's sword. Reflection nebulae are too dim to see with the naked eye, and even in amateur telescopes they are elusive. Dark nebulae appear only as silhouettes against the bright background stars. Emission nebulae, as their name implies, actually radiate light, by a mechanism similar to the one that lights a comet's tail, or an aurora. Reflection nebulae do not radiate, but passively mirror the light of bright nearby stars. In most cases, the composition of these clouds is the same.

Whether a nebula shines or not depends mainly on the stars around it. Dark nebulae are dark simply because no stars light them. They contain primarily hydrogen gas, dust, water, ammonia, methane, and many of the other compounds found in the atmospheres of the Jovian planets. The dust in

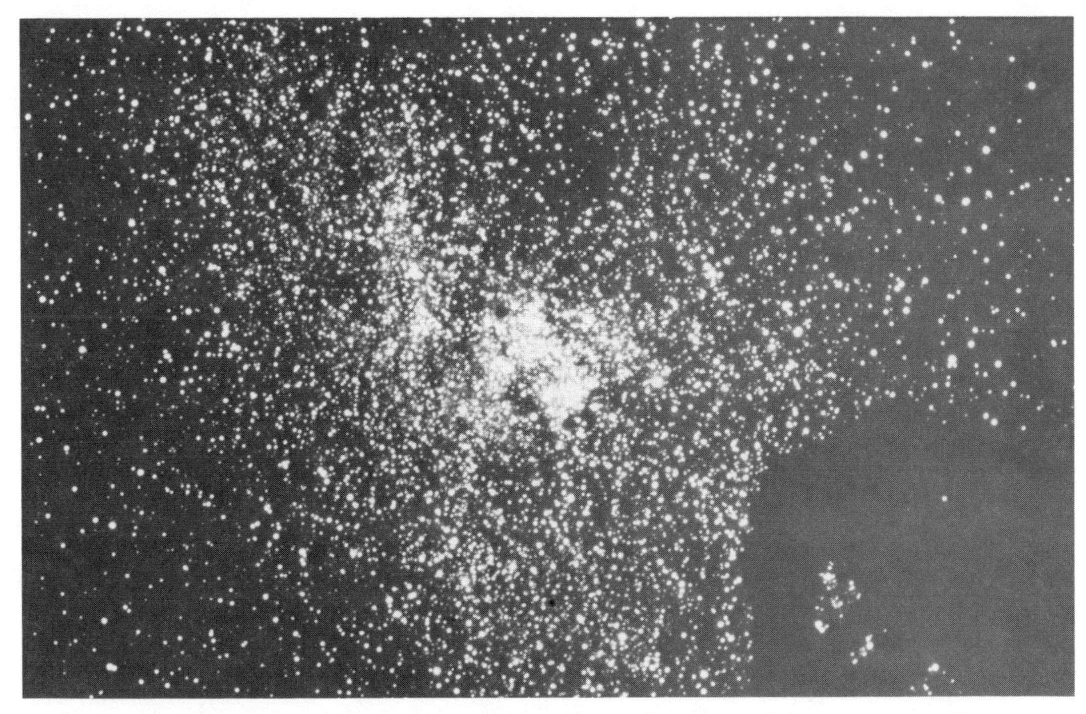

29.1 Star field with North American nebula. This rich region of the summer Milky Way contains many faint clouds of gas, ionized to shine by nearby hot stars. This field lies near Deneb, in Cygnus. (Official U.S. Navy photograph)

these nebulae is what blocks the light of the more distant stars. Dust blocks shorter wavelengths more effectively than longer ones, so these nebulae both dim and redden the light of any stars that do manage to shine through, just as the earth's atmosphere reddens sunlight at evening.

Emission nebulae shine because nearby stars are powerful enough to pump up the electrons in the cloud's atoms of gas to excited energy levels. When the electrons drop to less energetic states, they emit photons, and shine. The red light of hydrogen predominates, although the greens and yellow of oxygen and other gases also appear on long-exposure photographs; visual observation shows them usually as gray, occasionally a faint green. Reflection nebulae are also close to bright stars, but contain more dust—microscopic granules of water ice, as well as sand and graphite. These grains reflect starlight, especially the bluer wavelengths. The light from reflection nebulae is usually blue, although, again, the color is not apparent to the visual observer.

Even in the thickest clouds there may be only 10,000,000 atoms of gas per quart of space, and only a few hundred thousand dust grains in a cubic mile. This is about one quadrillionth (10 raised to the minus fifteenth—10^{-15}) the density of the air in this room. But in space, where the quarts and cubic miles pile up beyond counting, this surgically clean, rarefied gas is enough to dim starlight to a ten-thousandth of its normal intensity.

For astronomers trying to estimate distances by the distance modulus method, such dimming presents a problem. It distorts their results by changing the apparent magnitude of the star. Luckily, a spectrograph of a dimmed star reveals its true spectral class, showing the star's true color index. The difference between the true color index, and the unnaturally reddened appearance of the star, tells how much blue light has been absorbed. From there, the astronomer can calculate how much of the star's total light has been dimmed, and thus estimate what the star's apparent magnitude would be if there had been no dimming at all.

There are two kinds of dark nebulae. If the hydrogen gas in the cloud is cool enough, it is electrically neutral. Neutral hydrogen is called HI; clouds of HI, or *HI regions,* have densities of 10 to 100 particles per cubic centimeter. They are widespread throughout the galactic arms. Other dark nebulae, however, are both denser and dustier than the HI regions, and usually contain molecules such as ammonia, water, and formaldehyde—simple compounds formed from the light elements, such as hydrogen, oxygen, carbon, and nitrogen. These nebulae are often called *molecular clouds.*

A dark nebula absorbs light. This absorption warms it very slightly above the near-absolute-zero cold of space. Typical temperatures are around 100 degrees Kelvin (−279 degrees Fahrenheit). What happens when a large, O-B star condenses out of such a cloud, and floods it with light? First, it warms, reaching temperatures as high as 10,000 degrees Kelvin. Then two independent processes start in reaction. As the temperatures in the cloud rise, the gaseous atoms in the cloud become so energetic that their electrons fly completely out of their orbits. Like the gases in the sun, the atoms are ionized. Ionized hydrogen is also called HII, and therefore a heated nebula like this is often called an *HII region.* Free electrons in the cloud are continually recaptured and cascade to lower energy levels, emitting photons as they fall. Most electrons fall so as to emit the red H α line of the spectrum. The dark nebula has become an emission nebula.

In an emission nebula, other atoms besides hydrogen, such as oxygen, nitrogen, neon, and helium, also become ionized, and emit their unique wavelengths of light. Some of these emissions are extremely rare on earth and are called the "forbidden lines." In the extreme conditions of interstellar space, the forbidden lines are common. The forbidden line of oxygen, with a wavelength of around 5,000 Angstroms, is the one you are most likely to see. In large (14-inch or more) amateur telescopes, it appears as a ghostly green glow.

The second process started by the heating of a cool nebula is expansion. New stars form from the cloud. They heat the region around them, increasing the pressure within that region. This region of heated gas expands, pushing against the cooler, surrounding regions. If it expands into a region of cooler gas, a region of intense activity forms, like a cold front blowing across the plains. Winds blow through these clouds at speeds over 80,000

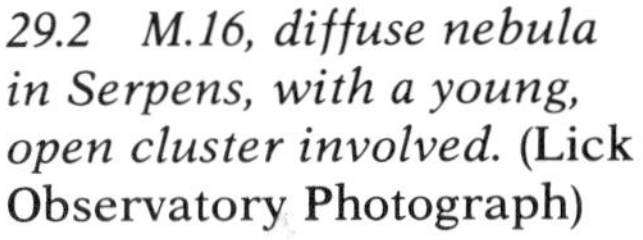

29.2 M.16, diffuse nebula in Serpens, with a young, open cluster involved. (Lick Observatory Photograph)

kilometers (50,000 miles) per hour. The increased activity of the expanding nebula may accelerate the formation of stars, especially if, as must happen, the blue supergiant stars exhaust their core supplies of fuel and explode in supernovae. The shock waves that expand through the nebula jar it further, and new stars settle out of the glowing cloud, like crystals precipitating from a supersaturated solution (Fig. 29.2).

OBSERVATION

"Nebula" was the term once applied to any telescopic object that was neither a star nor a planet. The word is Latin for "cloud," which was the appearance of almost any nonstellar object in the small-aperture telescopes of early astronomy. Today, the term properly refers only to those objects that actually are clouds—the diffuse nebulae of gas and dust.

Unlike any of the objects we have observed so far, all "nebulae," real or apparent, look *extended* and *diffuse.* These are observer's terms that become increasingly important with the objects described in this and the following chapters. An extended object is one that has an appreciable angular diameter; planets are extended, stars are not. A diffuse object is one without a distinct form, with boundaries that fade indiscernibly off into space; a galaxy is diffuse, a planet is not. All extended, diffuse objects, whether true gaseous nebulae or vastly different objects like globular clusters or galaxies, present a similar challenge to the amateur observer.

To understand this challenge, focus your telescope on any third-magnitude star, and slowly rack the eyepiece out of focus. As the area of the image increases, it becomes dimmer and dimmer. Eventually, it is indistinguishable from the background light of the sky. Even though the total brightness of a galaxy or gaseous nebula may be immense, that brightness is extended over a large area, making its *surface brightness* (luminosity per square arcsecond) very dim. The surface brightness of the extremely powerful Orion Nebula, for instance, is one nine-millionth of Jupiter's. In deep-sky observing, light-grasp is all.

Or almost all. Some nonstellar objects are less diffuse than others. Most planetary nebulae and galaxies are small enough to require medium-to-high magnifications. The stars in some clusters are so densely packed as to require moderate magnification to resolve them. But even with these objects, light grasp is essential. Most of them are quite dim; among deep-sky objects, eighth magnitude qualifies as bright, tenth magnitude as moderate, and twelfth magnitude as dim. Extremely bright objects like the Orion Nebula, or the Great Galaxy in Andromeda, visible to the naked eye under good conditions, are quite rare—only a dozen or so exist. Magnifying dim objects dilutes their light even further, as you spread it in the eyepiece to an even larger extent. In order to magnify dim objects, your telescope must offer adequate light-grasp.

It is in this realm that the large-aperture Dobsonian excels, but an ordinary 6-inch reflector is powerful enough to bring hundreds of every class of deep-sky object within your reach, even though you may have to exercise more skill to find and observe them. As with the previous observing sections in this book, the notes on observing that follow assume that you have access to a 6-inch scope. But they do not assume that you have developed the observing skills necessary to wring every last photon from it. The objects listed at the end of this chapter and those to come are for the most part among the brightest and most distinct of their class. Use them to learn the basic characteristics of each, and then go on to the more ambitious lists.

THE MESSIER AND NGC CATALOGS

This and the following chapters refer often to two catalogs of deep-sky objects visible in amateur telescopes. The *Messier Catalog* is named for Charles Messier, an eighteenth-century French astronomer. Messier was a passionate observer of comets, and compiled his list to keep track of the many nebulous objects that he sometimes mistook for comets. There are over a hundred items in the Messier list, including most of the bright diffuse objects in the sky. The list is a treasure trove for amateurs, including the Great Cluster in Hercules (M.13), the Great Galaxy in Andromeda (M.31), and the Orion nebula (M.42). At least one of every class of deep-sky object, excepting only the extreme oddities, like quasars and black holes, is represented on the list. Perhaps the best thing about the Messier catalog is that every object listed is visible to the skilled observer with a 6-inch telescope.

The *New General Catalog (NGC),* compiled around the turn of the century by the Danish astronomer John Dreyer, lists thousands of objects. The NGC and its supplemental *Index Catalogs (IC)* include almost every object within the reach of even the largest Dobsonian, and more. The NGC also contains many objects bright enough to rival the showpieces of the *Messier Catalog.* Hundreds of them are visible in a 6-inch telescope. Far more comprehensive than the Messier listing, the NGC contains numerous examples of every class of nebulous object. There are also other, specialized catalogs for particular kinds of deep-sky phenomena, such as the *Barnard Catalog* of dark nebulae; most of these, however, are limited to objects beyond the reach of amateur equipment.

There are several generally useful observing rules that apply to all diffuse, extended objects. Keep in mind the importance of surface brightness. Most lists of deep-sky objects give the integrated magnitude and angular size of an object; the ratio of magnitude over size tells you the actual apparent brightness. The exact ratio between size and brightness necessary to make an object "bright . . . easy . . . well-seen in a six-inch" (as the handbooks often say), or "what a waste of time!" (as the frustrated beginning observer says almost as often), depends on too many individual factors for any simple formula to predict. The best way to learn what's easy and what demands every bit of optical guile at your command is to start with the very brightest, large objects, and then work your way down through bright, small objects, then medium-bright, medium-sized objects, leaving the large, dim objects for last. This stepping-stone approach will keep you from arriving at a false, low opinion of your own capacities, or of your telescope's.

Any diffuse object, even a bright one, will be extraordinarily difficult to see against a sky lit up by the moon or by light pollution. Any haze in the air, natural or man-made, increases the problem. There are several solutions. First, get out of town: extended objects contrast better with the deep, dark skies. The moon needn't be new, but it must be below the horizon. Between new and full, you need to wait until moonset to observe; after full moon, you must get your observing in before moonrise.

If you can't get out of town, there are still a few tricks left in the tackle box. The LPR filters are the best thing that's happened to deep-sky observing since the great East Coast Blackout. They perform best on objects that radiate in the particular wavelengths they're designed to pass—HII regions in particular benefit from this kind of filtering, with planetary nebulae a close second. But even star clusters and galaxies, whose light spreads over the entire spectrum, benefit from the darker background these filters produce.

Take advantage of a fundamental characteristic of magnification: the sky background is always darker at higher magnifications, giving better contrast between sky and nebula. Finally, for the very largest nebulae, no matter what their surface brightness, you need to go after them with every weapon in your arsenal: RFT's functioning at optimal low magnification, with LPR filtering, dark, moonless skies, and averted vision. Some amateurs have found the 11 × 80 binoculars recently introduced by several manufacturers, equipped with LPR filters on each eyepiece, to be the most effective instrument for observing some of the more elusive, extended nebulae.

To locate the objects listed below, and those in the following chapters, see the star-charts in Appendices 5 and 6. Statistics on their size and brightness, as well as other information, is available in the listings of the Messier and selected NGC objects in the aforementioned Appendices.

1. *M.42 (NGC 1976): the Orion nebula* (Fig. 29.3). A faint gray haze enveloping θ^1 Orionis, the middle star of Orion's sword, this HII region is the brightest part of an extensive cloud of gas and dust extending throughout the constellation of Orion. It probably marks the leading edge of an enormous, gas-choked region about 2,000 light years away, where very massive O and B stars are currently forming. The diameter of the Orion nebula is dozens of light years. Fourth-magnitude. θ^1 Orionis is the star that causes the nebula to shine. It is a quadruple system, called the "Trapezium." Forming a lopsided rectangle about 20 by 10 seconds across, the primary of this complex multiple star system (Fig. 29.4) is the southernmost star of the group, an O star of fifth magnitude. Two of the other stars, the northern and western of the set, are eclipsing binaries.

2. *IC 349: The Merope Reflection nebula.* Lying within the boundaries of the Pleiades (M.45), the large, bright star cluster 15 degrees northwest of Aldebaran in Taurus, the position of this extremely elusive reflection nebula (Fig. 29.5) is so definitely marked that you can be sure of having it in your field whether you can see it or not. Try for it on very dark, clear nights. The southernmost of the half-dozen or so bright stars in the Pleiades is Merope, or 23 Tauri. This fourth-magnitude, B-type star marks the thickest part of the reflection nebula that extends throughout the Pleiades cluster. With averted vision, look to the south of the star, where the brightest portions of the nebulosity extend almost one-third of a degree. For the best view, hide Merope and the rest of the cluster off the edges of the field. On extremely good nights, you can see streaks of nebulosity across the entire cluster.

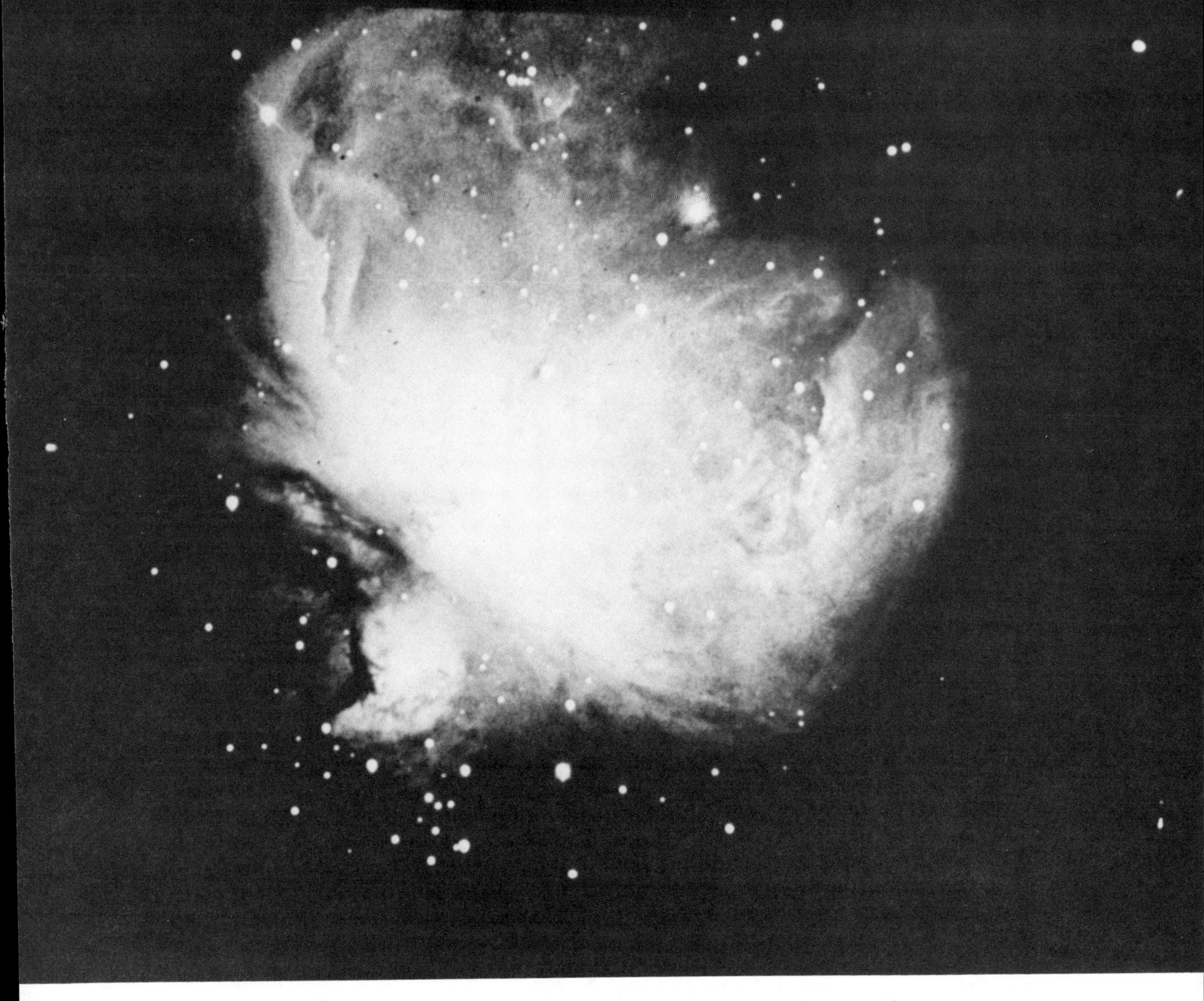

29.3 *M.42. The Great nebula in Orion.* (Official U.S. Navy photograph)

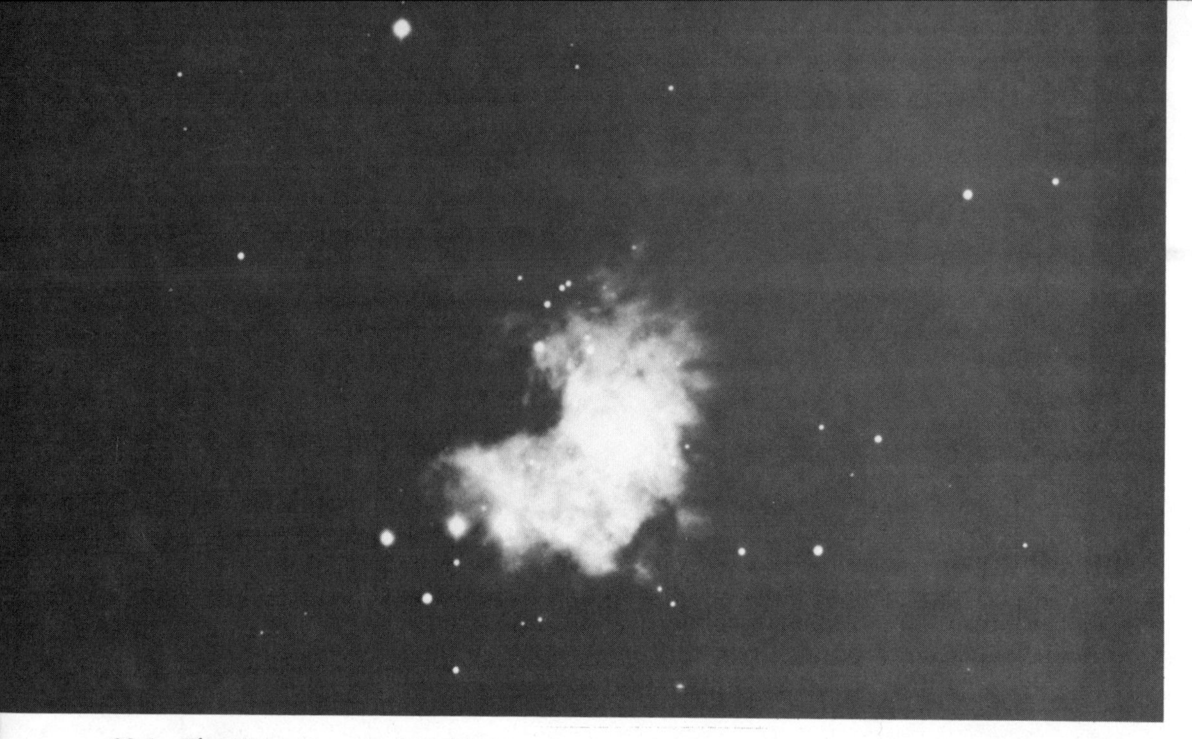

29.4 The Trapezium and Fish's Mouth, central regions of M.42. (Official U.S. Navy photograph)

29.5 M.45. The Pleiades, showing the Merope Reflection nebula. (Lick Observatory photograph)

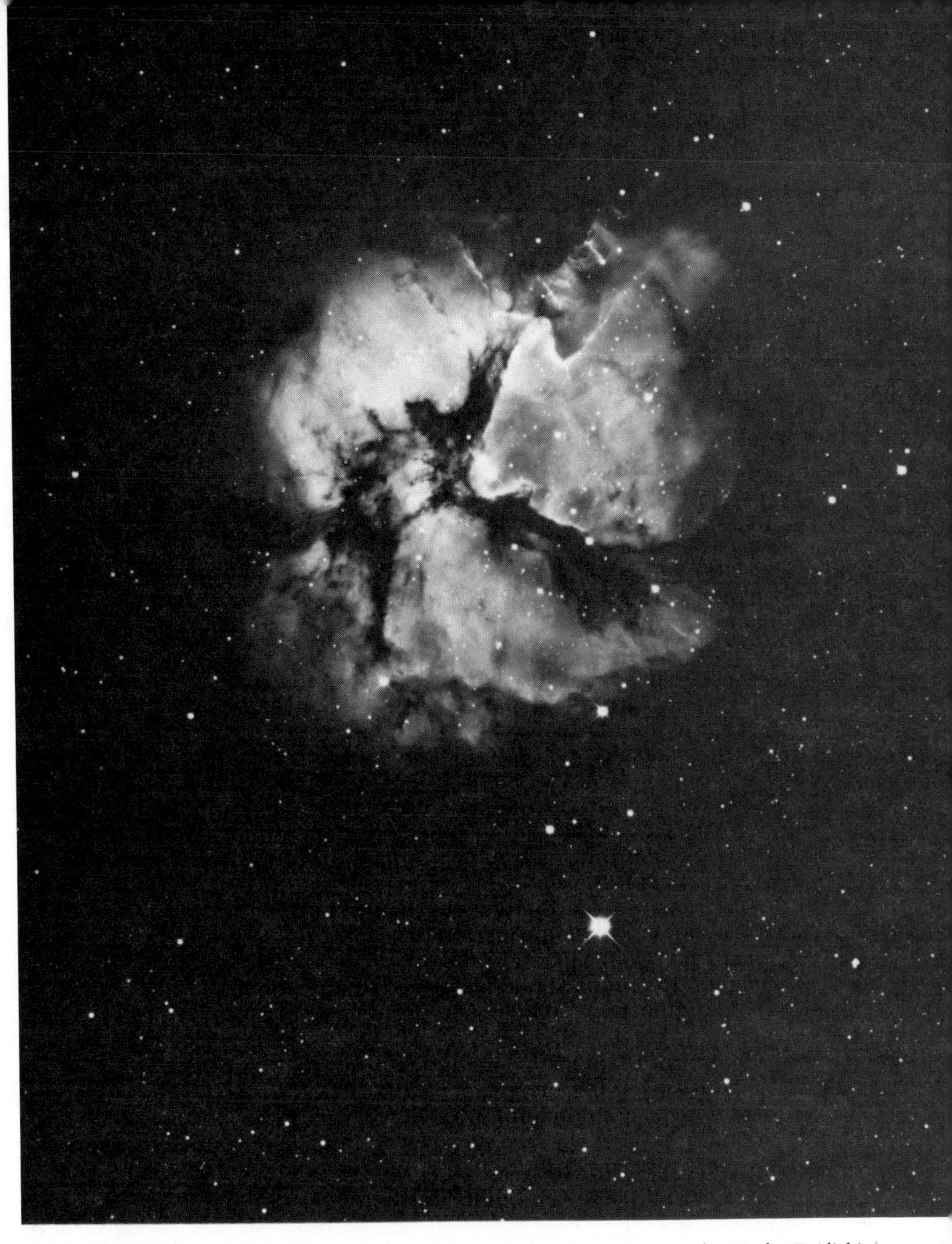

29.6 The Trifid nebula in Sagittarius. Only 1.5 degrees northwest of M.8, the Trifid is a dimmer object, although apparently lying at the same distance as the Lagoon; you can see it in a 6-inch telescope if your skies are dark and clear; 8 to 10 inches of aperture are necessary to see the dark lanes of obscuring matter. (Lick Observatory photograph)

29.7 M.17. The Omega nebula in Sagittarius, photographed in red light. (Palomar Observatory photograph)

3. *M.8: The Lagoon nebula.* A moon-sized condensation of the Milky Way (Fig. 13.5 on p. 115), it appears about 6 degrees north of γ Sagittarii, the star at the tip of the Teapot's spout. An HII region, with a small cluster of stars (NGC 6530) embedded in its eastern region, it is probably more than one hundred light years across. Its most striking feature is the dark lane crossing it from the northeast to the southwest—a dark nebula. The northeastern end of the dark lane contains two distinct dark spots, each about a minute or two across. These may be proto-stellar nebulae. The Trifid nebula (M.20, Fig. 29.6), appearing in the same low-power field just over a degree north of the Lagoon, is another ionized section of the same cloud. It also contains a very faint reflection nebula on its northern borders.

4. *M.17: the Omega nebula.* Star-hop north from the Lagoon by moving four low-power (2-degree) fields north, and then two fields east (Fig. 29.7). It lies about 2 degrees southwest of fifth-magnitude γ Scuti. Another HII region like the Lagoon, appearing about two-thirds the size of the full moon. An HI cloud fills the interior of the horseshoe shape, obscuring the background stars there.

Planetary Nebulae

30

If the diffuse nebulae represent areas where stars are born, then the *planetary nebulae* are at the opposite extreme of stellar evolution. These pale spheres of glowing gas are called "planetary" for their chance resemblance in small telescopes to a small, distant planet. Planetary nebulae form around most stars of solar mass or greater (up to four solar masses, and occasionally more). They form at the end of the red giant stage of their existence, after they have evaporated much of their original bulk. The formation of the planetary marks the end of fusion at the star's core, and the beginnings of its subsidence into the final, white-dwarf stellar endstate (Fig. 30.1). Far from being the swaddling clothes of a star, the glowing gas of a planetary is its winding sheet.

Yet in many respects, a diffuse nebula and a planetary are similar objects. Both are clouds of diffuse gas, mainly hydrogen, but with admixtures of heavier elements, and frequently large quantities of dust. Both shine by the same process, in which ultraviolet rays from a nearby star cause the gases in the nebula to fluoresce; in a planetary nebula, these rays emanate from the extremely hot surface of the white dwarf star at its center. There, however, the similarities end.

Planetaries form when the vastly distended outer atmospheres of red giant stars break away, and are then swept outward and energized by a shock wave released from the dying stellar core. This is an important process in stellar evolution, for it explains the relative rarity of supernova explosions. Only stars larger than several solar masses can go supernova. Such stars are not common in the galaxy, but they are far more common

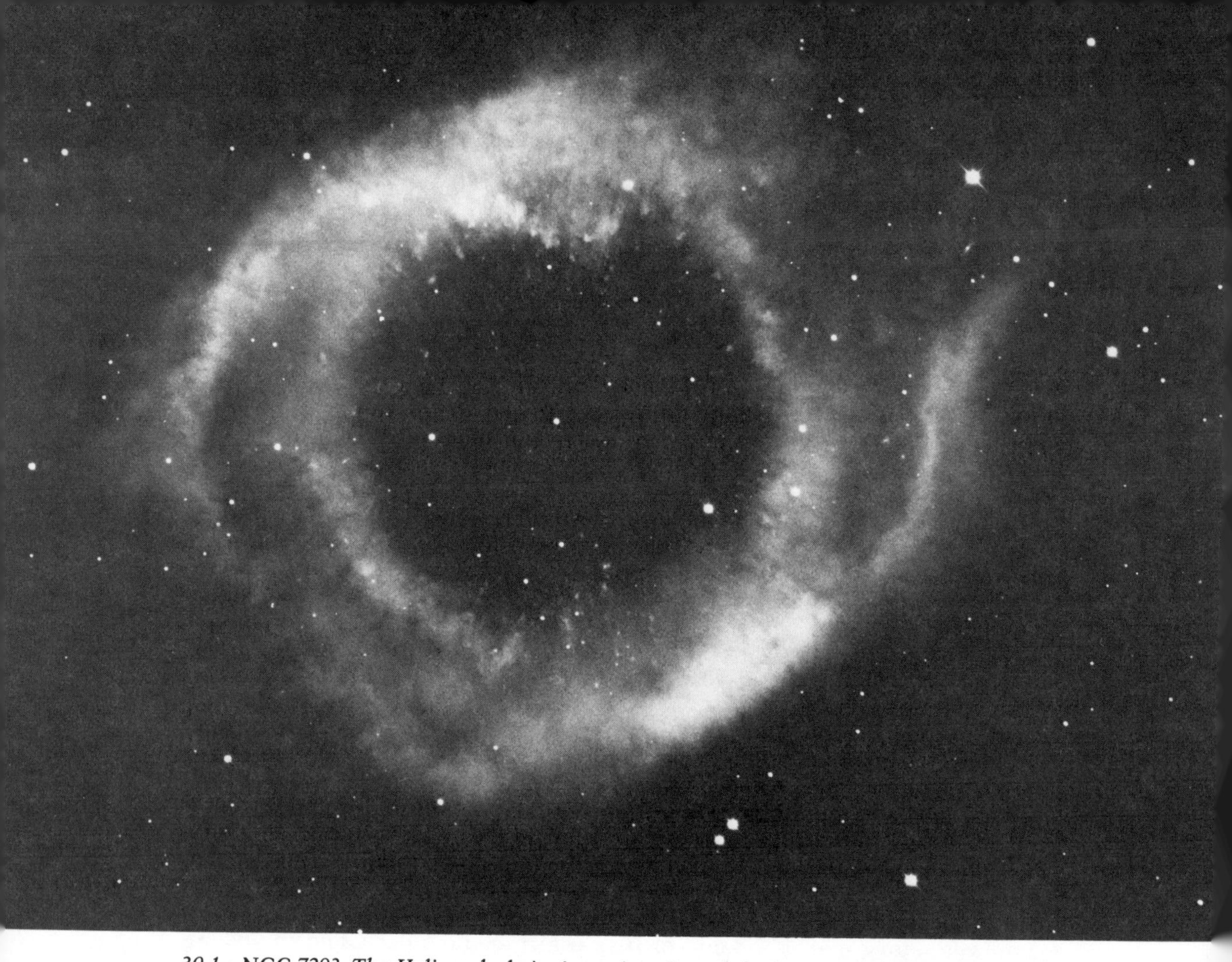

30.1 NGC 7293. *The Helix nebula in Aquarius. One of the largest of the planetary nebulae, the Helix is also one of the closest. Its large angular diameter (approximately 15 arc-minutes) and low surface brightness, combined with its location in the dim southern skies of fall, make this a difficult object to locate. It is visible as an extremely faint, gray ring, on dark, clear nights at low powers in a 6-inch telescope.* (Lick Observatory photograph).

than the observed frequency of supernova explosions would indicate. Something must be happening to prevent many massive stars from undergoing that final, cataclysmic flash. That something, apparently, is the process of *mass-loss* by which planetary nebulae are formed.

Astronomers have observed mass-loss going on around a variety of stars. Many red supergiants, such as Betelgeuse, have been observed spectroscopically to be surrounded by relatively cool clouds of gas, water vapor, and dust. A class of B stars, called "Be" (for "B-emission") stars, has also been observed to eject material from their outer layers. So do the larger long-period variables. When the outer layers of a star grow so far from the star's center of mass that its gravitational pull on them is too weak to hold them down, gases start to evaporate from the surface. In many cases, this process is enough to decrease the mass of the star below the danger point at which supernovae, neutron stars, and other disasters are likely to happen.

The exact process is still not entirely understood. In the current theory, a star that has been boiling off its mass for some time will be surrounded by a roughly spherical cloud of stellar "steam." The cloud contains primarily hydrogen, with significant quantities of helium and the other compounds that form in the core layers of a large star. Rates of mass-loss in very large stars can be enormous. Betelgeuse, for instance, is shedding its skin at about four solar masses every million years. A star losing mass at such a rate cannot remain a red giant very long. Eventually, a disturbance at the core sends a shock wave rippling outward, pushing the stellar atmosphere before it.

This wave travels much more quickly than the expanding cloud surrounding the star. As the leading edge of the wave overtakes the cloud, a number of things start to happen. First, the relatively cool, expanding gases in the cloud start to pile up at the leading edge, like snow in front of a snowplow. The increased density at the shock front, combined with the much higher energy levels of the gases in the wave, increase the pressure and temperature of the front. At the same time, the core of the star is exposed. Glowing with a temperature of up to half a million degrees Kelvin, it starts to flood the cloud of gas with ultraviolet light. As if at the flicking of a switch, the shell of gas plowed up by the shock wave ionizes and lights up. A planetary nebula has formed.

Expanding at speeds over 80,000 kilometers (50,000 miles) per hour, the glowing shell of a planetary will eventually reach a diameter of around 1 light year. At that point, the nebula is from 50,000 to 100,000 years old, and has reached the end of its life. Either the shell of gas dissipates, or the size of the shell is too great for the star at the center to irradiate. In either case, the planetary fades from view, the central star fading into an ordinary white dwarf as well.

There are over 1,000 planetaries known in our galaxy, from which astronomers calculate a total galactic population of about 10,000. To populate the galaxy so thickly, new planetary nebulae must form at the rate of five to ten each year to replace those that are constantly fading from view. The stars that form planetaries are remarkable objects, emitting very high-energy ultraviolet light from surfaces far hotter than those of ordinary stars. The coolest are at least 20,000 degrees Kelvin, with some shining ten times as hot. On closer examination, astronomers find that these stars, with their extremely high densities (typically about a solar mass, condensed into a much smaller volume), look remarkably like white dwarfs—very young white dwarfs, that have just started their long, slow slide into oblivion. The formation of a planetary nebula, then, also marks the end of the star's active life, and its transition into one of the stellar endstates. Over the estimated age of our galaxy, the rate of production estimated for planetaries indicates that as many as one billion white dwarfs must hide among the stars.

OBSERVATION

To the observer, planetaries differ from diffuse nebulae in one important way: their size. Few planetaries grow larger than 1 light year in diameter, and those that reach that size are often too dim to be visible by any means but long-exposure photography. Yet many planetaries are also extremely distant; the nearest members of the class are probably more than 1,000 light years away. They are often so small as to require significant magnification just to distinguish them from stars. Fortunately, there is an opposite effect working to the observer's advantage; the smaller planetaries frequently have much higher surface brightnesses than the diffuse nebulae. These two qualities can make planetaries hard to locate, but easier to see.

The problem of finding planetaries is primarily the need for high magnification: the field of view in a high-power eyepiece is often so small that you have to sweep around the sky for some time before you stumble across your target. The most reliable way to keep from wandering too far afield in such sweeping (and to slow yourself down in the process, lest you sweep right by your target too fast to see it) is to hop by field diameters, moving the telescope so that a single bright reference star that was on one edge of the field appears on the opposite edge. Stop, inspect the field, choose a reference star, and move on. Knowing the width of field in your eyepiece, you can keep track of how far you've swept in any one direction, and know when to turn back.

There are several tricks you can use to pull a small planetary out of a crowded field. With an LPR filter, you need only flick the filter back and forth between the eyepiece and your eye. This will make the planetary seem to blink on and off against the background stars, catching your eye as it does. Using this technique, you can use your wide-angle eyepiece, cutting down on the amount of sweeping necessary to find the object. A sheet of replica diffraction grating (available inexpensively from Edmund Scientific) held in the light path will spread the images of stars out into continuous spectra. A planetary, however, shines only at particular wavelengths, and so will break up into separate, distinct spots. The difference is not as eye-catching as the blinking effect of the LPR filter, but it will still allow you to use a low-power eyepiece.

If you have neither kind of filter, you will have to fall back on charts. The techniques described for finding variable stars can help here. And again, starting with the larger and brighter examples to familiarize yourself with the appearance of this class of object will be a big help. The following are four of the easier-to-find planetaries in the sky.

1. *M.57: The Ring nebula in Lyra.* At about 1 minute angular diameter, the Ring (Fig. 30.2) appears about twice the size of Jupiter, but with a much lower surface brightness. In a mid-sized amateur telescope, it will look like a small smoke-ring. The ring appearance comes about as we look at a bubble of glowing gas, which appears dimmest when we look straight through the thin shell, and brightest through the greater thickness of gas at its circumference. The central star, although exceptionally hot, is only magnitude 15. Between 1,000 and 2,000 light years from earth, the Ring's diameter is nearly 1,000 times that of the solar system. The apparent size of this system is less than photographs (or visual memory) make it; remember that it is less than one one-hundred-twentieth the size of a low-power field.

2. *M.27: The Dumbbell nebula in Vulpecula.* Brighter and much larger than the Ring, but far from any convenient guide stars, the Dumbbell is visible in a wide-field eyepiece (Fig. 30.3). Sweep 3 degrees north from γ Sagittae, the fourth-magnitude star at the head of the Arrow. The Dumbbell appears as a gray puff filling about one twenty-fifth of a 2-degree field. Its

30.2 M.57. The Ring nebula in Lyra. (Official U.S. Navy photograph)

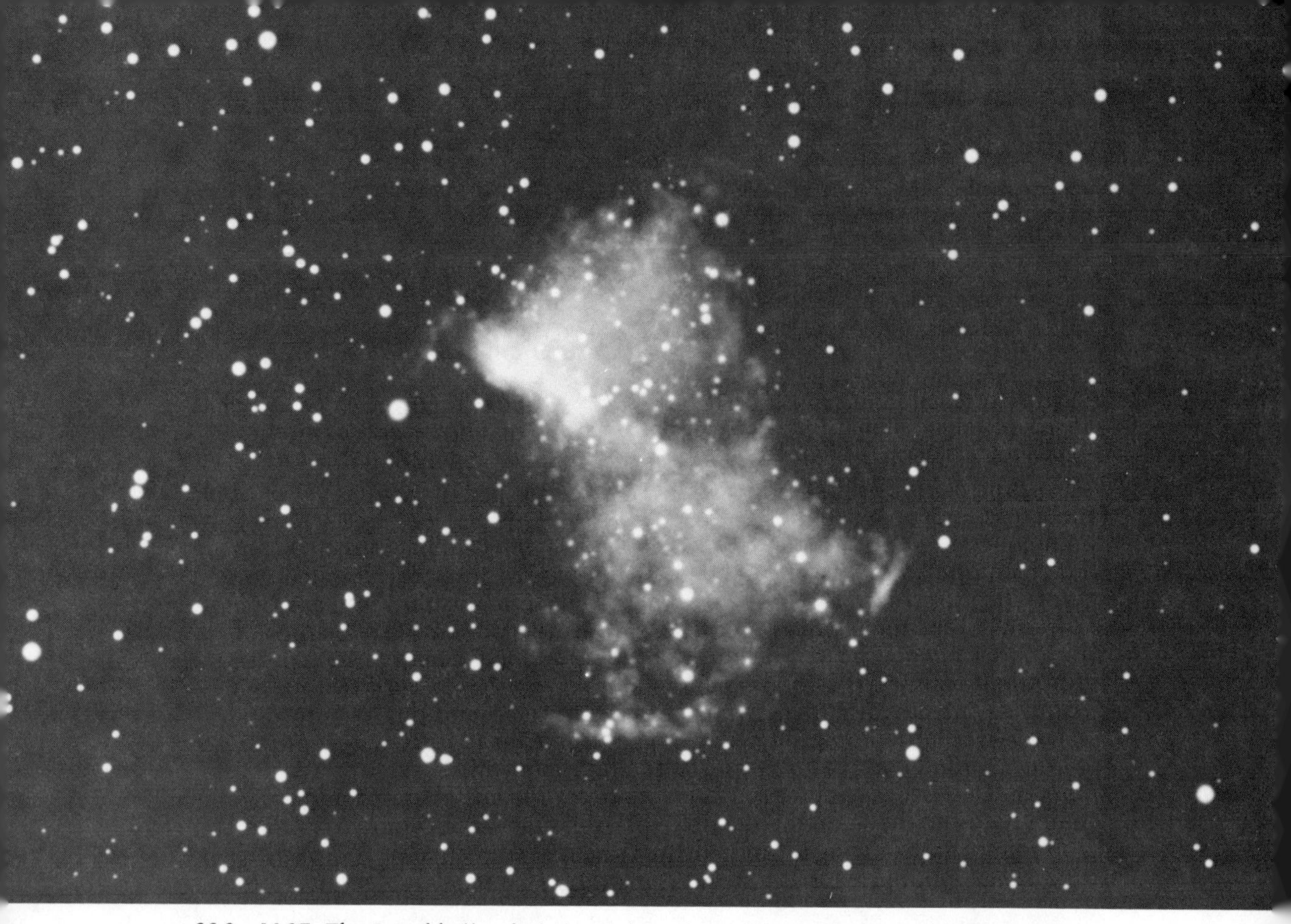

30.3 M.27. The Dumbbell nebula in Vulpecula. (Official U.S. Navy photograph)

integrated magnitude of 8 gives it a higher surface brightness than the Ring despite its vastly greater area. The dumbbell shape appears at 60×. Look closely on good nights for the thin wisps of nebulosity that extend from the broadened ends, and the mottling of light and dark sometimes visible across the broad disc. Probably one of the closest planetaries—certainly within 1,000 light years—the Dumbbell has a diameter of at least 2 light years, making it one of the largest of its class. The central star is several magnitudes brighter than that of the Ring, but still extremely difficult to detect.

3. *NGC 2392: The Eskimo nebula in Gemini.* A tiny, gray disc, about as big as Jupiter (Fig. 30.4). Starting at δ (delta) Gemini, jump 2 degrees east to fifth-magnitude 63 Gemini; the Eskimo forms a loose grouping with two sixth-magnitude stars on either side, less than a degree southeast of 63 Gemini. Surface brightness is relatively high, but so small as to require much magnification. The peculiar detail on the disc gives it in large telescopes an

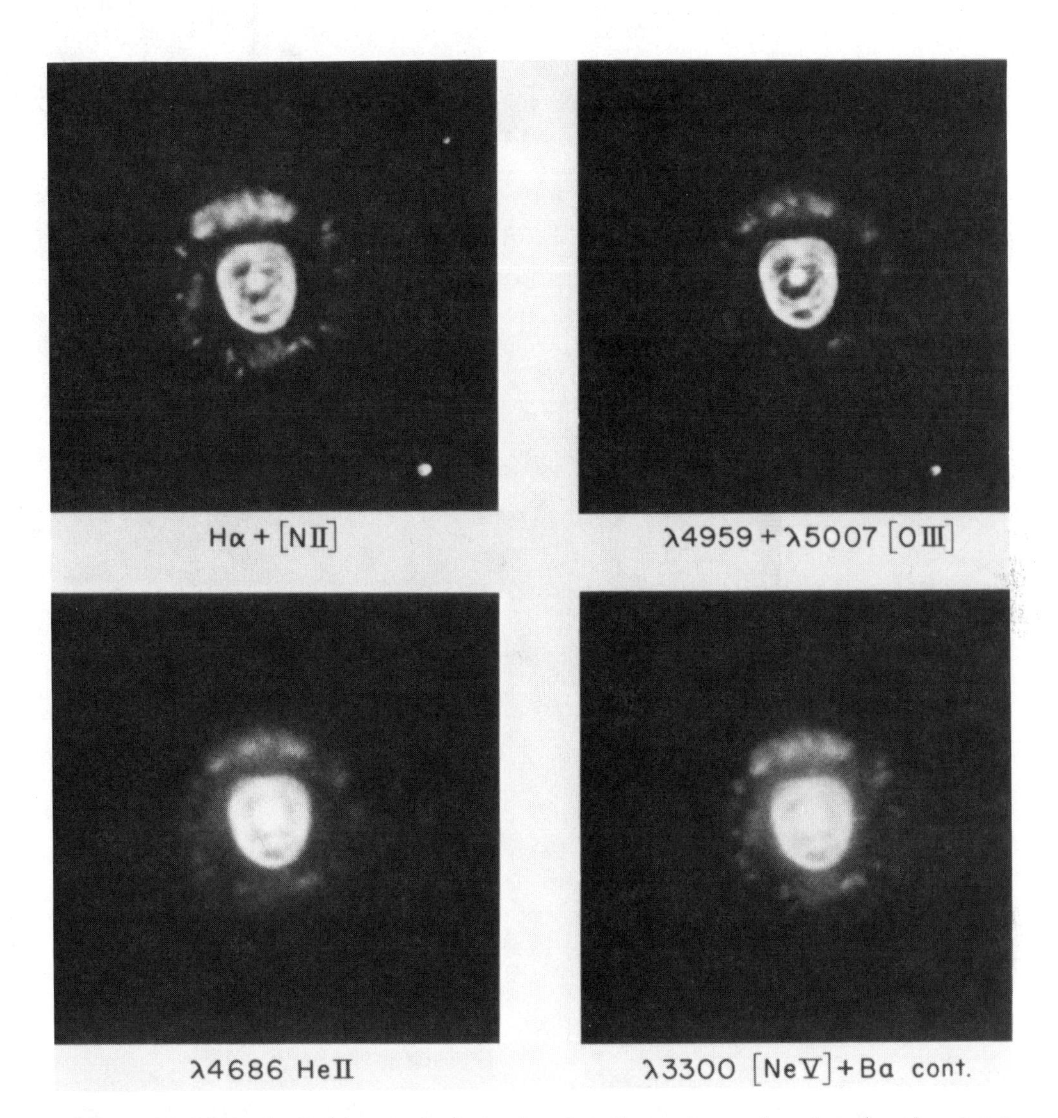

30.4 NGC *2392, the Eskimo nebula in Gemini. Four views, showing the planetary's appearance in red, yellow, violet, and ultraviolet light.* (Palomar Observatory photograph)

appearance that has been described as the face of an Eskimo peering out of a fur-rimmed hood. The "hood" is a semidetached, outer ring of ionized gas, perhaps the result of an early shock wave released by the central star before its final mass-loss. Distance estimates range from 1,300 to over 3,000 light years.

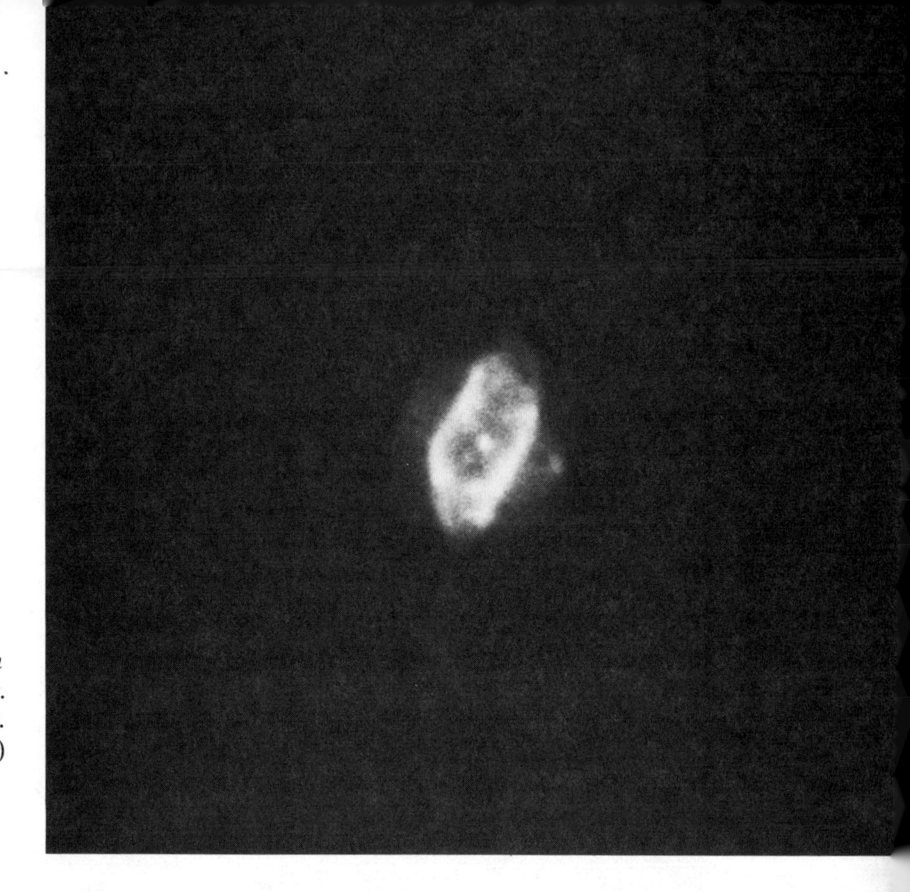

30.5 NGC *7009, the Saturn nebula in Aquarius. Photographed in red light.* (Lick Observatory photograph)

4. *NGC 7009: The Saturn nebula in Aquarius.* Takes its name from an odd pair of extensions (visible in a 12-inch) from the circular disc (Fig. 30.5). It is relatively bright, but small and hard to pick out. Center a low-power field on ν (nu) Aquarii. Move about 1.5 degrees west; the planetary should be centered in your field. The amount of detail visible in a mid-sized amateur instrument is limited, but you should be able to make out its elongated form.

Supernova Remnants

There are two more objects that don't properly belong to the class of planetary nebulae. Both are supernova remnants, the exploded shells of stars that failed to lose enough mass and suffered the catastrophic consequences. In a small telescope, however, they look remarkably similar to planetaries, and you may want to hunt for them using the same techniques you've practiced on planetaries. The first of these is also the most famous.

1. *M.1: The Crab nebula in Taurus.* Almost as large and bright as the Dumbbell, it is slightly more than a degree northwest of ζ Tauri. The Crab (Fig. 30.6) is all that remains of a supernova explosion that occurred about 7,000 years ago, 6,000 light years from us. The light from that blast reached us on A.D. 4 July, 1054. It was visible in daylight for over 3 weeks, and took

more than a year to fade from the night sky. In the millennium that has elapsed since then, the wreckage has expanded outward at 180,000 miles per hour, to form the present nebula, 6 light years across.

Radio astronomers have found the Crab to be one of the strongest sources of radio emission in the sky. X-ray observations have found it to emit powerfully in that region of the spectrum as well. Generating its magnetic fields, high-speed electrons, radio, and X-ray emissions, is the Crab pulsar, a rotating neutron star. Spinning at the fantastically high velocities of these superdense objects, it sweeps a powerful beam of radiation through the Crab each time it rotates: 30 times per second.

In an amateur's telescope, little sign of the fabulous astrophysics inside the Crab is visible, but its violent origins may be suggested by the twisted shape, the faint, "curdled" appearance of its surface, and (in large amateur instruments) the faint extensions visible around its periphery. The light from the Crab is a peculiar mix—the forbidden lines of oxygen, as well as the familiar lines of ionized hydrogen—and an undifferentiated blur of white light ("continuum emission"). An LPR filter will selectively brighten the emission areas, bringing out some detail in the Crab, but will not cause as dramatic an improvement of contrast as seen with planetaries. Likewise, the diffraction grating technique will not work so well, since the ionized re-

30.6 M.1, the Crab nebula in Taurus. (Official U.S. Navy photograph)

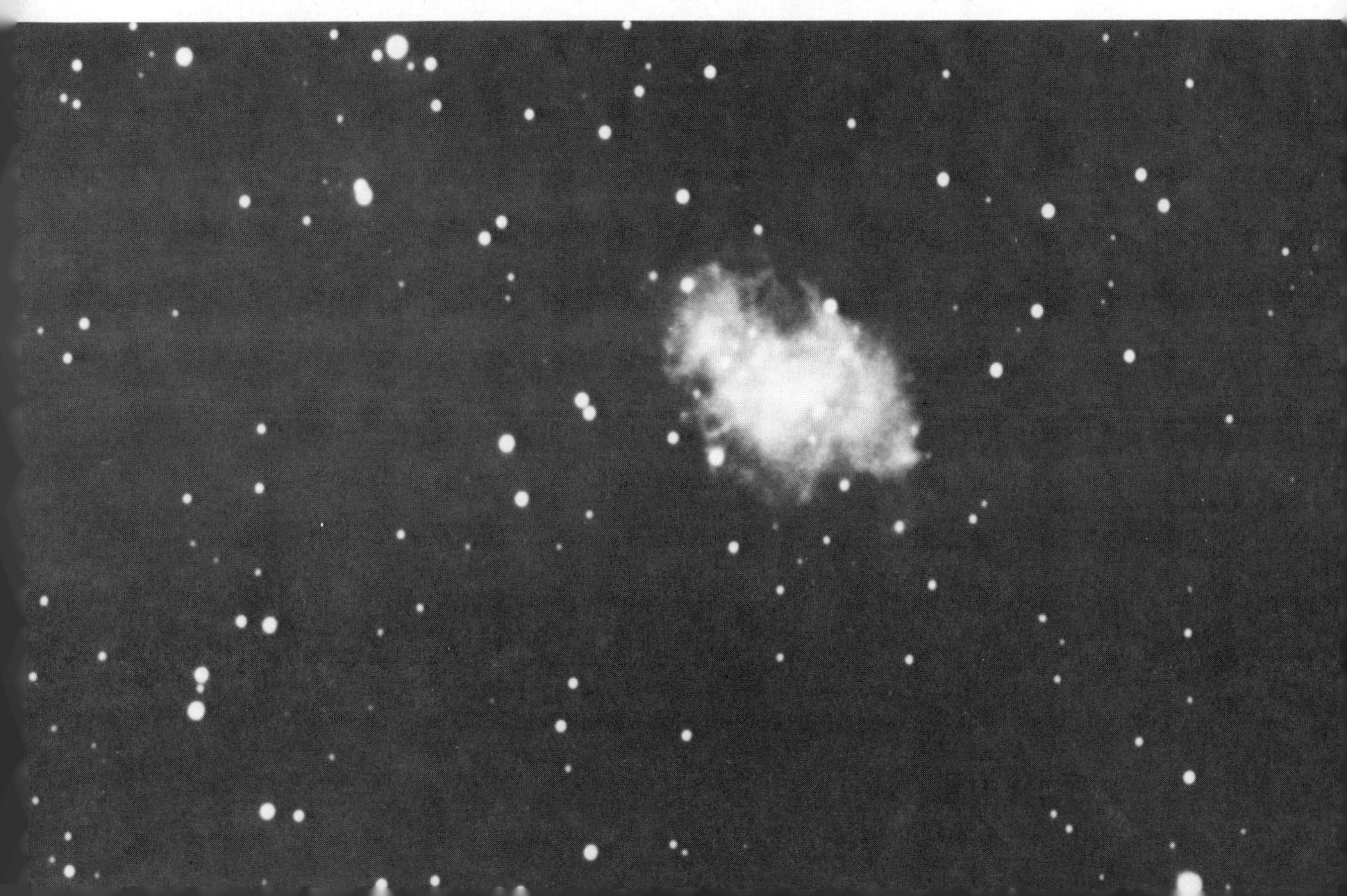

gions emit more dimly than the rest. Once you have found the Crab, use the highest magnifications it will stand (a good-quality barlow with a 9- to 12-millimeter eyepiece gives good results in the 6-inch, *f*/5).

2. *NGC 6960-6992: The Veil nebula in Cygnus.* An ancient remnant on the point of disappearing, it is one of the most elusive objects in the sky (Fig. 30.7). Use 52 Cygnus, about 3 degrees south of ϵ (epsilon) Cygnus, as a guide. Center 52 Cygnus, then inch the telescope south until it is just out of view. The brightest portion of the Veil's faint curve should be running north-south across the field, appearing as an extremely faint, narrow wisp of gray light. Don't move the telescope around to seek it out; 52 Cygnus lies in the middle of the Veil's 1.5-degree arc. Finding the nebula now is a matter of using averted vision. The other bright section of the nebula lies about 2.5 degrees east northeast of 52 Cygnus. The western section, *NGC* 6960, and the eastern (6992) are the brightest portions of a broken circle of nebulosity nearly 3 degrees across. Some 50,000 years ago, the center of the Cygnus Loop was the scene of a supernova explosion. This faint nebulosity is all that remains.

30.7 NGC *6992-5, the Veil nebula in Cygnus.* (Official U.S. Navy photograph)

31

Star Clusters

In the constellation Taurus, on the clear nights of winter you can easily see the asterism called the Pleiades. The Pleiades is one of the nearest and brightest examples of an *open cluster* of stars. Also called *galactic clusters,* these close associations of stars lie in the disc of the Milky Way, where they have formed relatively recently in astronomical terms. Open clusters contain anywhere from a few dozen to several thousand stars, usually of population I. The other class of star cluster is the *globular cluster*—each an immense gathering of up to half a million stars. Globulars orbit the center of the galaxy in a vast spherical shell more than 100,000 light years across. Metal-poor, extremely ancient, the population II stars in globulars are among the oldest in the galaxy.

Open and globular clusters are the two major kinds of close stellar groupings in our galaxy, but there are other ways in which stars congregate. *Moving groups* and *stellar associations* are looser aggregations of stars. The stars of the Big Dipper are an example of the former; most of its members show similar proper motions (shared, oddly, by Sirius). This indicates a common origin and a loose gravitational association. The bright O and B stars of the Orion association are an example of a *stellar association.* Unlike clusters, stellar associations are not gravitationally stable and disperse relatively quickly—in 1 to 2 billion years—after their members form.

There are more than 1,000 open clusters known in the Milky Way, of which hundreds can be seen in amateur telescopes. Usually about 30 light years across, a typical open cluster contains several hundred stars, most of which are held within the cluster by the gravitational attraction of the whole. The members of a cluster form together, probably in a large diffuse nebula similar to M.42. In some cases, such as the cluster embedded in the Lagoon nebula, you can see a cluster in the act of emerging from its original cloud. We know that the gravitational bond within an open cluster can be remarkably stable. The ages of some have been measured at 10 billion years—as old as the galaxy itself. Most open clusters, however, are among the younger stellar groupings in the galaxy.

The more massive a star is, the sooner it leaves the main sequence and becomes a red giant or supergiant. In a very young cluster, all of the stars lie on the main sequence; its brightest stars are hot O and B types. A few million years after the cluster forms, however, the most luminous stars have drifted across the top of the H-R diagram, into the red supergiant stage. A diagram of its stellar population will curve off the main sequence at its upper end, straying into the red supergiant branch. The point at which the cluster's population breaks off the main sequence gives us an accurate indication of the cluster's age. A cluster in which only the hottest, fastest-living stars have reached that point must be relatively young. A cluster in which the slow-burning stars like our sun have started to bloat into red giants must be 10 billion years old.

The typical open cluster is young. The Double cluster in Perseus, for instance, at only 2 million years, is younger than the human race. The 20-million-year-old Pleiades first shed their light on earth when our earliest hominid ancestors branched off from the rest of the primate family. The H-R diagram on the top in Figure 31.1 is typical of the open clusters. Studies of the globular clusters, however, tell a far different story. The stellar population of a cluster such as M.13 in Hercules (Fig. 31.1, *bottom*) leaves the main sequence in the F range. Moreover, some of the red giants have evolved into the region of Cepheid variability, indicating that they have been off the main sequence for a long time. Such a well-evolved population indicates an age for M.13 of about 10 billion years, typical of the globulars.

While open clusters stay within the Milky Way, globular clusters pass through the galactic disc and swing tens of thousands of light years out into the empty regions above or below the galactic plane. There they linger, in obedience to Kepler's laws, on orbits hundreds of millions of years long. They tend to concentrate near the galactic hub. Almost a third of the more than 100 known globulars occur in Sagittarius, where the center of the galaxy lies.

The density of stars in a globular cluster is hundreds of times that in the regions of the sun. Density at the center of a globular may rise extremely high. Recently, some astronomers have suggested that very massive black holes may lie at the centers of globulars. High-energy radiation has been observed emanating from the centers of some clusters, adding tantalizing support to the theory.

Measurements of the distances to globulars also show them to be much farther from us than the open clusters. The nearest open cluster, the Hyades

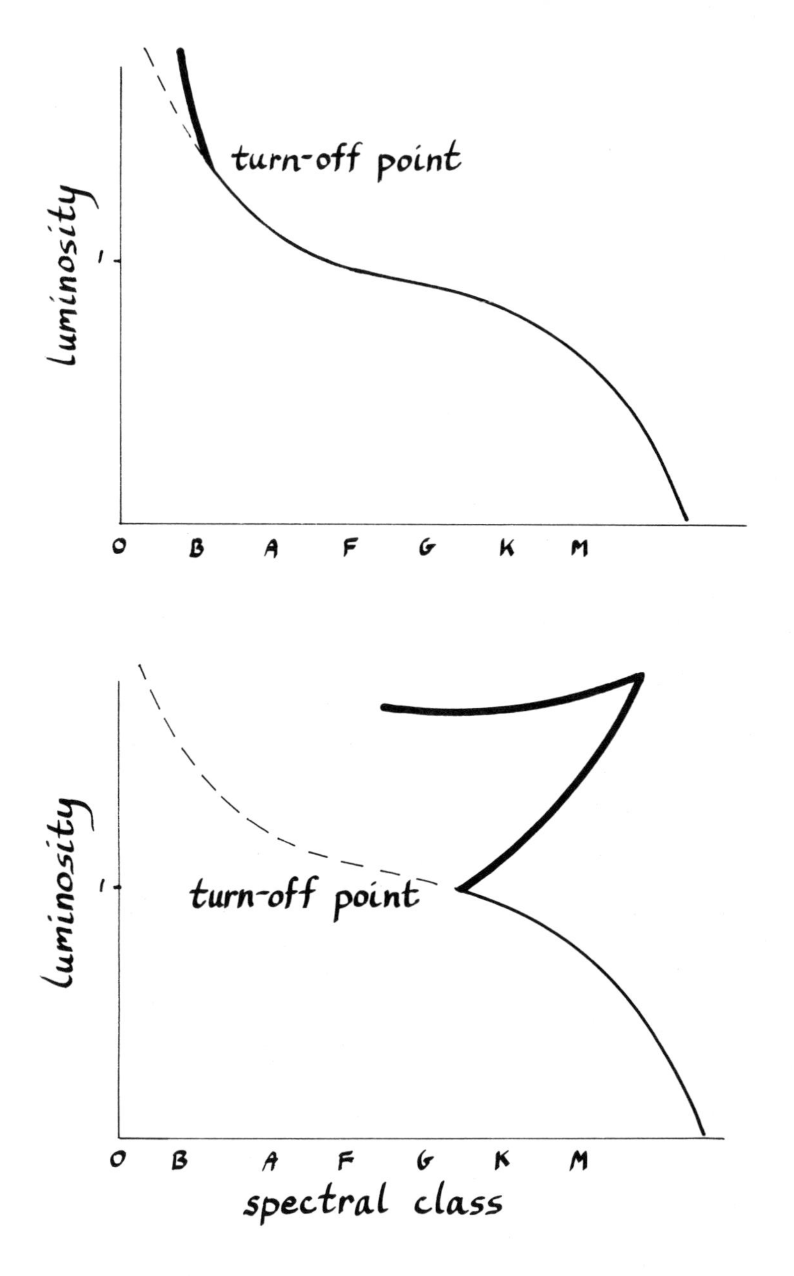

31.1 H-R diagrams of old and young clusters. Notice that the older cluster has no main sequence stars among the early spectral classes. The location of the turn-off point, where the population branches off from the main sequence, indicates the age of the cluster.

in Taurus, is about 130 light years away. The nearest globular, M.4 in Scorpius, is 8,000 light years away. The typical distance is in the tens of thousands of light years. The farthest known globular (in our galaxy) is NGC 2419 in Lynx. At 180,000 light years, it rides far out in the void of intergalactic space.

It was from the distribution of globulars in a spherical halo around the galaxy that astronomers learned the correct location of the sun within the

Milky Way. Before the twentieth century, they had assumed (half out of our geocentric instincts, and half from early observations of the density of stars in space around us) that the sun lay at the center of the Milky Way. Not until 1917, when Harlow Shapley surveyed the distribution of globular clusters across the celestial sphere, did the real picture emerge. Shapley found that the center of the globulars' distribution lay in Sagittarius, and deduced that the galaxy's center of gravity must lie there as well. His studies of population II Cepheids in the globulars told us the average distance to them, and thus the distance to the galactic center. The earth lies 30,000 light years, or about 60% of the galaxy's radius, from the hub.

OBSERVATION

Because of their great range in size and distance, clusters of stars can be everything from bright, large objects for which binoculars are the most suitable instrument (the Hyades, Pleiades, and Praesepe in Cancer), to dim, shrunken blurs, difficult to visualize in a 6-inch. Open clusters, appearing in the crowded star fields of the Milky Way, tend to blend into the background, especially when they are sparse or very far away. Globular clusters, far more compact and often isolated in thinly populated regions of the sky, can be easier to pick out. As with most deep-sky objects, you can find many of them listed in the Messier Catalog and the NGC. Other observers' handbooks, such as Burnham's and the Webb Society's, give descriptions of hundreds of the brighter clusters, detailing their visual appearance and physical charac-teristics. These descriptions can be an enormous help in observing.

When you use such handbooks, the descriptive terms can be bewildering at first. The important visual characteristics of a cluster are its brightness and size, as well as the density of its stellar population. The ratio between size and brightness is important for the same reasons as with the nebulae, but this can be complicated, especially for open clusters, by the population density. A large cluster with few stars is difficult to tell from the background—unless the members of the cluster are all conspicuously bright. The appearance of a cluster, then, depends on a combination of all three factors. The terms "sparse" and "compact" used above are typical expressions for clusters with thinly scattered or closely packed members, respectively. Synonyms for "sparse" include "poor," "little condensed," "low density," and so on. Compact clusters are also called "rich," "condensed," "populous," and so on.

There are also more formal systems, classified from poorest to richest clusters by letters or numbers—A to G for poorest to richest, I to XII for richest to poorest, and so on. In any case, the visibility of the cluster depends on its contrast with its background; the more bright stars that are packed into a smaller area, the more distinct the cluster will be. The interest in observing a cluster, however, often involves more subtle effects: the stars in a cluster can be of different colors and brightnesses, gathered densely or loosely strewn across the sky.

Most globular clusters are both very dense and distant. The apparent separation of the stars in a globular may be below the resolving power of your telescope, or the individual stars will be beyond the limiting magnitude. In that case, the cluster is often called "difficult," or "not resolved." It will still be visible, but as a pale blur of light like a diffuse nebula, without any of the pinpoint gleams of starlight that mark a resolved cluster. Also "difficult," of course, is any globular with an integrated magnitude near the limiting magnitude of your telescope. In either case, greater aperture is required to resolve or detect them. Globulars may be resolvable around their edges, where the stars thin out into space, but unresolvable at the center. Frequently, a globular will show another combination: well resolved at the edges, a nebulous blur at the center, but with bright points of light overlying the central blur. These are the resolved images of widely scattered, bright stars lying on the edge of the cluster that is closest to earth. To resolve a cluster "to the center," or "from edge to edge," then, can be two different things, depending on the density of the cluster at its true center. In the first case, there will be no blur of unresolved stars anywhere in the cluster. In the other, pinpoint images may appear across the cluster, but the center itself can still be unresolved.

The kind of instrument and magnification you bring to bear on a cluster should depend not only on its size and brightness but also on its density. To appreciate an open cluster's appearance, you need to see some of the sky around it, so that the ordinary stellar population provides a contrast. Generally, the field of view should be at least twice the diameter of the cluster. There are a very few clusters too big to see well in the field of an RFT. The Hyades, or a moving group like that surrounding **α** (alpha) Persei, will lap over the edges of any field smaller than that of binoculars or a finder. Apertures much larger than 6 inches will increase the resolution of some of the more distant clusters of either kind, and make some of the dimmer members of open clusters more visible. A 6-inch still offers sufficient light-grasp to put a large number of both kinds of cluster in reach. Magnification is generally more useful on globular clusters, most of which require at least 60× to resolve. The more distant ones will resist resolution in even very large amateur instruments. Generally, filters are not much help, except to increase contrast with the sky in light-polluted areas.

Finding clusters can be difficult. Photographs will lead you to expect something far brighter and more distinct than the very delicate shimmer of light you are likely to see at first. A good way to get some sense for the appearance of the dimmer clusters in a telescope is to look with the naked eye for the Beehive cluster in Cancer (M.44). At first, you won't see a cluster so much as a faint disturbance of light; one observer I know calls the effect "visual fizz."

The first time you try to locate one of the brighter telescopic clusters, scan the region first with binoculars, and use averted vision on anything that looks at all different from the background. Make a mental note of the location of that spot relative to bright nearby stars, and get your low-power eyepiece on it. Globulars are generally much easier to *see,* appearing almost stellar in binoculars or the finder. The challenge is distinguishing them from the ordinary stars in the field. In a low-power eyepiece, you can usually identify them by a subtle fuzziness; this will be more apparent with averted vision. For the dimmest objects, the techniques of star hopping and sweeping discussed in earlier chapters will work best.

Start by finding and observing, with naked eye and binoculars, the brightest and largest open clusters, the Pleiades (M.45) and Beehive (M.44). They are listed in the Appendix. For telescopic objects, try some of the following.

1. *The Double cluster in Perseus (NGC 869 and 884, also called* h *and* χ *(chi) Persei).* This closely spaced pair of very rich, bright clusters, each about ½ degree across, contains stars as bright as sixth magnitude, and appears in binoculars as a thick patch in the Milky Way a little more than halfway from α(alpha) Persei to δ (delta) Cassiopeia (Fig. 31.2). More than a degree across, these clusters are best seen in a 6-inch rich-field telescope at powers from 20× to 30×. About 8,000 light years from earth, the age of this system is estimated at only 2 million years.

31.2 The Double cluster in Perseus. (Lick Observatory photograph)

2. *M.35 in Gemini.* This large, rich, and compact cluster contains almost 100 stars bright enough to see in a 6-inch. About 2 degrees northwest of η (eta) Geminorum, 2,500 light years away. Use averted vision to examine the southwest border of the cluster. About 0.5 degrees from the center of M.35 is the open cluster NGC 2158, 5 arc-minutes wide, with an integrated magnitude of 11. Difficult in an 8-inch, it lies 13,500 light years beyond M.35.

3. *M.6 and M.7 in Scorpius.* Good binocular- and RFT-pair of open clusters. Use low power, from 18× to 30×. M6 is considerably smaller and dimmer. Its brighter members are B and A type stars, grouped together in a region less than 10 light years across. M.7, much larger, lies about 3 degrees southeast.

4. *M.11: The Wild Duck cluster in Scutum.* Very condensed, very rich open cluster, this is a triangle of several hundred stars, with a reddish, ninth-magnitude star at the apex. Use moderate magnification (60× or more, depending on aperture). Larger apertures show over 400 dimmer members. Use λ (lambda) Aquilae as guide, hop to β (beta) Scuti: M.11 is 2 degrees southeast. At a distance in excess of 5,500 light years, most of the stars visible in a 6-inch must be giants or supergiants.

5. *M.13: the Great Globular cluster in Hercules.* The brightest globular in the northern hemisphere, this cluster is visible in binoculars as a slightly hazy star on western edge of the Keystone of Hercules, one-third of the way from η (eta) to ζ (zeta). A 6-inch working at powers around 100× should be able to resolve it from edge to edge. Lower powers, or poor skies, will make resolution intermittent; the cluster changes from moment to moment, looking nebulous one second, then clearing into an array of bright points across its face. Estimates of the total population of the cluster range from 500,000 to 1 million population II stars, approaching 10 billion years' age. Twenty thousand light years away, about 100 light years across.

6. *M.22 in Sagittarius.* Brighter than M.13, considerably larger, harder to pick out owing to low altitude, crowded background. Use the guide star λ (lambda) Sagittarii, 2 degrees southwest. Distance, 10,000 light years. This cluster is easier to resolve than Hercules, except at center.

7. *M.15 in Pegasus.* Well-placed high in fall skies, 34,000 light years from earth, this cluster lies well above the disc of the Milky Way (Fig. 31.3). Its guide star, ϵ (epsilon) Pegasi is 4 degrees southeast. Smaller than M.13, with a similar diameter, this cluster is much more condensed. Compare its starlike core with M.13; it's a candidate for a black hole.

8. *M.5 in Serpens.* Well-placed high in spring sky, this cluster lies just over the border from Virgo (Fig. 31.4). Hop 11 degrees east from fourth-magnitude τ (tau) Virginis to 104 Virginis. Continue on that line through 110 Virginis. M.5 lies 4 degrees east of 110. It is a bright, large cluster, similar to M.13. On a clear night, binoculars can distinguish M.5 from any of its neighboring stars; haze or bright skies will require a low-power inspection. The

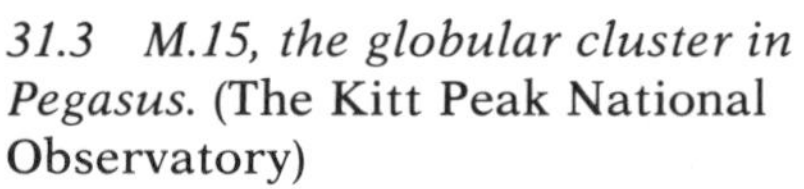

31.3 M.15, the globular cluster in Pegasus. (The Kitt Peak National Observatory)

31.4 M.5, the globular cluster in Serpens. (The Kitt Peak National Observatory)

third brightest globular (behind M.22 and M.13), M.5 is visible from mid-northern latitudes. It resolves at about the same magnifications needed for M.13. M.5 is slightly out-of-round, appearing definitely elliptical in large telescopes. With a distance of 25,000 light years, its age is estimated at 10 billion years, making it one of the most ancient clusters known.

The Realm of the Galaxies

32

Every object we have seen so far, from the moon to the farthest cluster, has been only a part of the Milky Way Galaxy. With a total population of 200 billion stars, the Milky Way is so huge and all-encompassing that until this century it was thought to be the entire universe. We now know that the universe is much larger, tens of billions of light years from edge (if there is one) to edge.* Billions of galaxies occupy the universe, separated by millions of light years. Observation of the galaxies takes the amateur observer through distances and spans of time that verge on the infinite.

If we could step outside our galaxy, the first feature we would notice would be the galactic disc, bright with stellar associations of blue supergiants and glowing HII regions. There would also be large dark regions of neutral hydrogen, and a rich scattering of main sequence population I stars. All of these gather most thickly in the *spiral arms.* The arms are the most characteristic feature of the Milky Way, forming a pinwheel pattern that spirals out from our galaxy's center. Each of the galaxy's several arms contains a high concentration of population I stars, gas, and dust. Between the arms, the galactic disc is sparsely populated.

At the center of the spiral, the *galactic bulge* is a brilliant mass of popu-

*When we deal with the universe on the scale of billions of light years, terrestrial geometry no longer applies. It's nonsense to speak of the universe having an edge, because every point in the universe is apparently its center. This concept may become clearer later in this chapter.

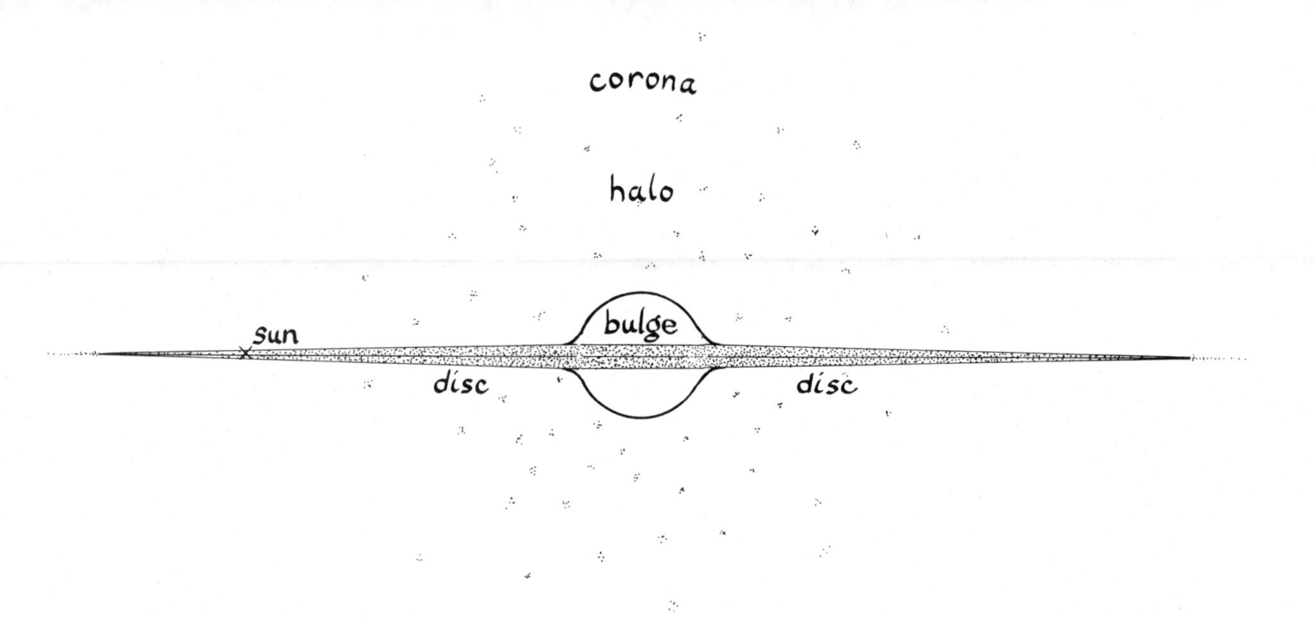

32.1 The Milky Way, showing its major dimensions.

lation II stars (Fig. 32.1). This is the galaxy's hub, a thickening in the galactic disc, distinct from the rest by the much redder light of its stars. And at the center of the hub, the *galactic nucleus* shines, looking from our imaginary vantage point above the disc like a supernally brilliant star. Here, in a region invisible from earth behind the thick clouds of the inner arms, something of awesome power beams out high-energy radiation, similar to the twin rays of a pulsar. Recent indications are that a black hole with the mass of a million suns may inhabit the center of our galaxy and others, consuming unimaginable quantities of matter and converting it into radiant energy.

Surrounding the galaxy is the *galactic halo* of globular clusters. These clusters are embedded in the *galactic corona,* an extremely thin cloud of gas interspersed with the corpses of ancient stars—white dwarfs, black dwarfs, and black holes. The gas in the corona circulates around the galactic center, fountained upward by supernova explosions.

If our imaginary vantage point were in the right place, we would see, standing off from the center of the Milky Way at distances up to 200,000 light years, almost a dozen *dwarf galaxies* and *irregular galaxies,* apparently satellites of the Milky Way. The two largest of these, the irregular galaxies known as the *Magellanic clouds,* are visible to observers in earth's southern hemisphere. To the naked eye, they look like two detached star clouds of the Milky Way.

This picture of our galaxy has been acquired only slowly by astronomers here on earth, because our present position blinds us to many of the more distant regions of our own galaxy. The great clouds of gas and dust that choke the spiral arms shroud the galaxy's hub and block our view of the Milky Way's far side. But not all wavelengths of light succumb to absorption in the diffuse nebulae. Like the red light that penetrates earth's atmosphere even at sunset, the longest wavelengths of the electromagnetic spectrum pass through the densest clouds, and are audible in radio telescopes on earth. Like a medical X ray, radio astronomy has given us a picture of the galaxy's internal structure that we could otherwise neither see nor photograph, and has allowed astronomers to assemble a reasonably complete picture of the Milky Way. The picture revealed, although still a hazy one, seems to show a system remarkably similar to many galaxies we can see and photograph, members of a class called the *spiral galaxies.*

A stack of three or four phonograph records, with a baseball at the middle, approximates the shape of a spiral galaxy. The Milky Way is one of the larger known spirals: 100,000 light years across, 1,000 thick, with a hub 13,000 light years across and 10,000 light years thick. The entire Galaxy rotates. Its spin takes the sun in the direction of the constellation Lyra, at a speed of about 800,000 kilometers per hour. Even at that enormous speed, our *cosmic year*—one revolution around the galactic center—is on the order of 200 million years.

Unlike the planets in the solar system, objects near the hub do not orbit at higher speeds. This suggests that a great deal of the Galaxy's total mass must orbit beyond the edge of the disc. This invisible mass, which must exist in the *galactic halo,* exerts its own gravitational force on the matter in the disc, radically changing the orbital behavior of the entire Galaxy. The halo, extending more than 300,000 light years from the center, may contain as much mass as the rest of the Galaxy—disc, hub, and corona—put together: the total mass of our Galaxy is currently estimated at one to two trillion times the mass of our sun.

With the aid of evidence gathered by radio and optical astronomy, maps of the Galaxy have been drawn showing much of the nearby structure. Such maps usually use the *galactic coordinate system.* In this system, the poles around which the Galaxy rotates are equivalent to the north and south celestial poles. These are the 90-degree points of *galactic latitude.* The north galactic pole lies in the constellation Coma Berenices, a few degrees west of β(beta). Its south pole is in Sculptor, about 10 degrees south of β Ceti. The *galactic equator* is the midline of the Milky Way, halfway between the poles. The equator is divided into 360 degrees of *galactic longitude.* The zero point of galactic longitude is the galactic center, in Sagittarius.

Maps of the Galaxy's local structure put the visual appearance of the night sky in a larger context. The sun and most of the stars we see with the naked eye lie in one spiral arm. It is called the *Orion-Cygnus* arm, and forms an arc with a radius of 30,000 light years around the galactic center. This arm, which is a few thousand light years thick, contains the bright stars of the Orion association, as well as the thick HI cloud of the Great Rift in the summer Milky Way. The Great Rift is a relatively nearby feature, but other such obscurations occur throughout the Milky Way, lying near the galactic equator. This band of dark nebulosity is typical of spiral galaxies. When we look out in the direction of the Summer Triangle, we are looking down the length of our spiral arm, in the direction where the Galaxy's rotation is carrying us. When we look out through the opposite side of the celestial sphere, we look back down the spiral arm, in the direction we have come.

About 5,000 light years outward from the sun, across a relatively empty region of space, lies the *Perseus arm.* This is the section of the Milky Way we see in the northern skies on fall evenings. It is apparently connected to the Orion arm. The spot where they join is about 15,000 light years from here, in the constellation Puppis, low in the winter sky. When we look at the southern half of the sky on summer evenings, we gaze inward, across another gap of 5,000 light years, toward the next inmost arm, the great *Sagittarius-Carina arm,* which winds between us and the bulge of the galactic center. The galactic bulge is visible as the Great Sagittarius star cloud.

The complicated branchings and rebranchings of the spiral arms, obscuring each other by their intertwinings as well as by the vast clouds of gas and dust inhabiting them, makes the full structure of the Milky Way difficult to piece together. We do know that there are only a few arms, tightly curled around the center so that one of the outer arms we see may be a continuation of one of the inner arms, after it has curled a full turn around the entire Milky Way. There may, in fact, be only two major spiral arms involved in the entire structure. The confused picture near the galactic hub, where the arms branch off, makes it difficult to trace them to their origins.

Equally obscure is the force responsible for the arms in the first place. It was early recognized that something was peculiar about the existence of a spiral pattern in a rotating galaxy. Unless the entire galactic disc rotates as a solid, the spiral pattern cannot be very long-lived. As long as different parts of the Galaxy rotate at different speeds, the survival of the spiral pattern is a mystery. We know from the large number of spiral galaxies observed elsewhere in the universe that the spiral pattern must be reasonably stable. Apparently, some complex mechanism is at work, continually imposing the spiral pattern on the differentially rotating disc.

Among the many processes proposed for this, one of the theories most favored at present holds that the spiral is a pattern formed by the combined elliptical orbits of all the stars in the Galaxy. In this model, these orbits precess, so that the major axis of one ellipse gets slightly out of line with the major axis of the ellipse next outward, and so on. Because the ellipse is longest at its major axis, this causes the stars at the ends of the major axes to jam up against the narrower part of the next ellipse outward. The result, in the regions where the ellipses touch, is a region of higher stellar density (Fig. 32.2).

The combination of many ellipses skewed out of line and touching in this fashion forms an extended region of increased stellar density, spiraling outward from the galactic center. The points where the ellipses touch are regions where the stars are bunched unnaturally close (and perhaps disturbed by this action into a burst of star formation, producing stellar associations of O-type supergiants): a typical spiral arm. The stars continue to move around the Galaxy on their elliptical orbits, but the spiral arms remain stationary; stars pass through the arms over the course of their orbits. This pat-

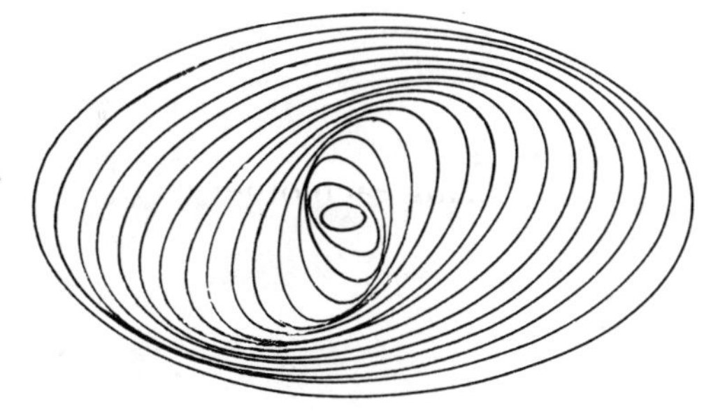

32.2 Formation of spiral structure by precessing elliptical orbits. Orbits of stars around the galactic center (ellipses), precess over time, forming regions of greater stellar density. These are the spiral arms.

tern is stable, and given the elliptical orbits of stars in galaxies, inevitable. Exactly what causes the ellipses to skew in the first place is still not clear.

The examples of the thousands of galaxies visible in space can also teach us much about our own. From surveys of these, astronomers early in the twentieth century were able to distinguish basic categories of galaxies. One of these is the class of spiral galaxies, to which the Milky Way belongs. Most of the brighter galaxies belong to this class, but this does not indicate their true percentage in the total population of galaxies in the universe. Spiral galaxies are the equivalent of the giant stars in the Milky Way. We see many more of them than we do red dwarfs, merely because the giants are bright enough to be visible over great distances.

The most common class of galaxy is the cosmic equivalent of the red dwarf, the *dwarf elliptical galaxy.* As its name suggests, this class of galaxy is a subcategory of the *elliptical galaxies,* which includes some of the brightest, most massive galaxies known (Fig. 32.3). Ellipticals vary from perfectly spherical to highly elongated egg shapes. They can range in size from the smallest dwarfs, which contain about a million solar masses, to the giant ellipticals, which can have as many as 10 trillion solar masses. The range of size in spirals is nowhere near so great. A small spiral still masses 1 billion stars; a large one, 400 billion. The volume of these galaxies is proportional to the masses: ellipticals range from 2,000 to 500,000 light years across, and perhaps larger; spirals, from 20,000 to 150,000.

A spiral galaxy has two kinds of stars, the hot, bright population I stars of the disc, and the older, slightly dimmer population II stars of the hub. The integrated spectral type of a spiral can range from A to K. An elliptical, how-

32.3 NGC *205, an elliptical galaxy.* (Official U.S. Navy photograph)

ever, consists almost entirely of population II stars. Its integrated spectral type ranges only from G to K. As in our own galaxy, population II stars appear with very little gas and dust—ellipticals seem to contain almost none of either. An elliptical galaxy, in fact, is very much like the core of our own galaxy, minus the disc. The predominance of population II stars in ellipticals suggests that, in these ancient galaxies, the processes of star formation have stopped.

There is also a third kind of galaxy, the *irregular galaxies.* They have the same dual stellar populations as spirals, but they lack any trace of regular structure, spiral or elliptical. They account for only about 3% of the galaxies visible in amateur telescopes. The Magellanic clouds belong to this class. There are also more extreme variant forms, the *peculiar galaxies* (Fig. 32.4), such as *ring galaxies,* which seem similar to spirals except that the spiral arms have merged into a single circular band. *Ring-tail galaxies* frequently occur in pairs, or as members of closely spaced clusters of galaxies. They, too, seem to be special cases of the spiral class, distinguished by one spiral arm that has been stretched out to several times the length of the entire galaxy. They may be survivors of collisions or near misses with other galaxies. Both of these classes are rare, and usually lie too far away to be visible in amateur telescopes.

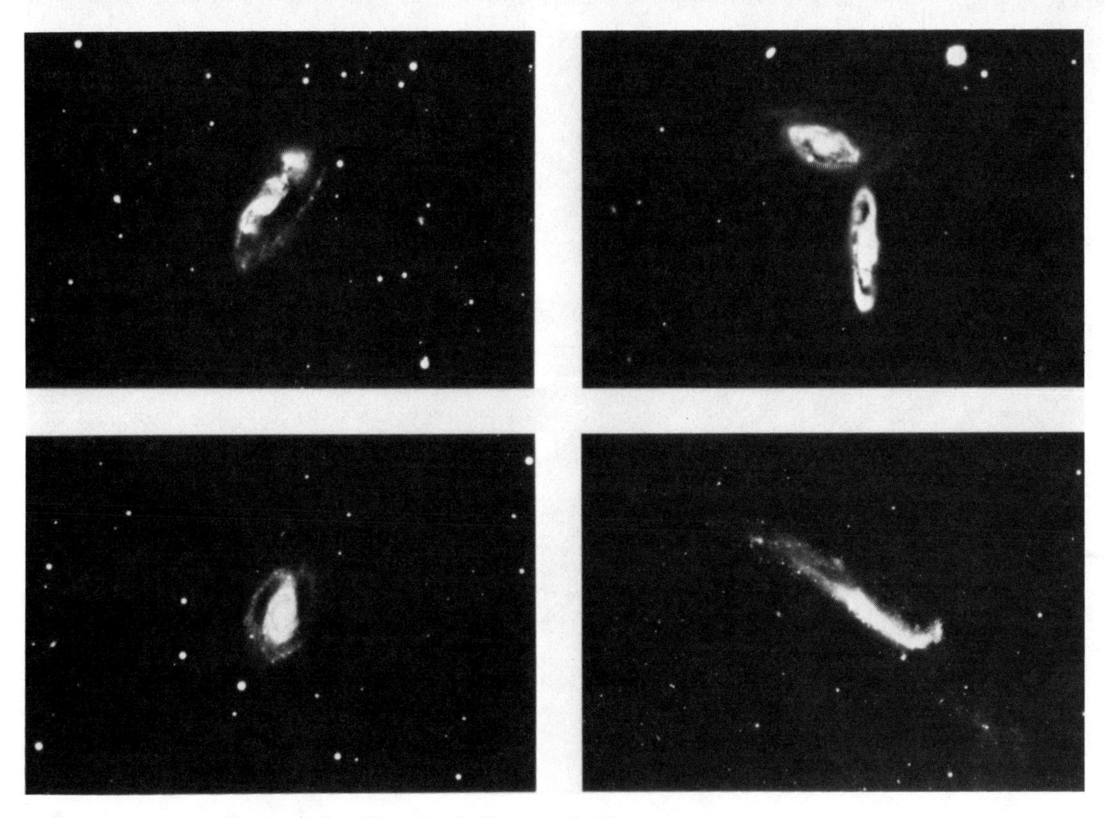

32.4 Peculiar galaxies. (The Kitt Peak National Observatory)

Another variant class, the *exploding galaxies* (Fig. 32.5), includes members bright enough to see with amateur equipment. A subgroup of this class includes objects such as the *Seyfert galaxies,* spiral galaxies in which something has happened at their cores to send extremely hot gases exploding outward at very high speeds. Other members of this subgroup include the rare *N-type galaxies* and *BL Lacertae objects.* Most of these poorly understood objects lie at vast distances from earth—tens and hundreds of millions of light years, where they appear as vague, small gleams on long-exposure plates taken by the great observatories. Lying at these distances and beyond, to the ends of the known universe, are the *quasars.* Appearing virtually starlike, radiating immense quantities of energy from regions only a few light years across, quasars may be the supremely brilliant nuclei of galaxies undergoing some kind of explosion.

Among the normal classes of galaxies, the spirals and ellipticals, there is yet a wide range of variation. Spiral galaxies (Fig. 32.6a–g) occur in two major subgroups, the ordinary spirals such as the Milky Way, and the *barred spiral galaxies.* In the latter, the two spiral arms branch out not from the nucleus, but from a straight bar with the nucleus at its center. The bar contains the same young stellar population and clouds of gas and dust as the arms of an ordinary spiral.

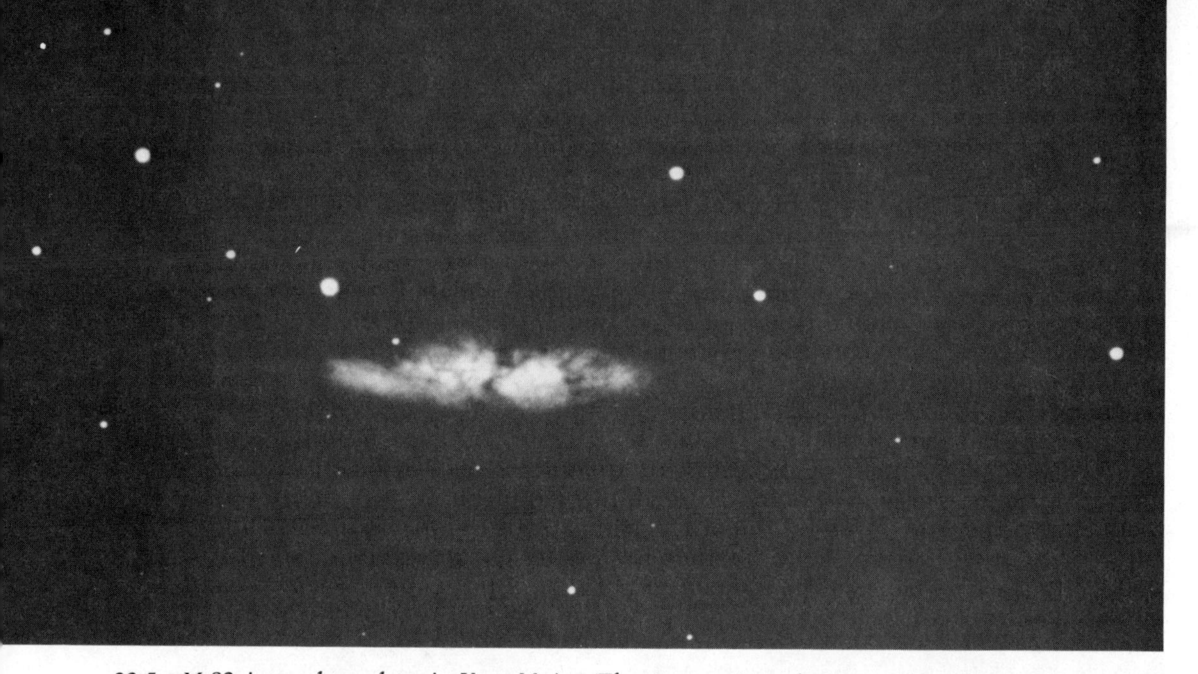

32.5 M.82, irregular galaxy in Ursa Major. The companion of M.81, and visible in a 6-inch telescope, this galaxy is apparently undergoing a violent explosion at its core. (Official U.S. Navy photograph)

►

32.6 Spiral galaxies. (a) *M.51, the Whirlpool galaxy in Canes Venatici; an Sc spiral. A 6-inch telescope will show its companion galaxy plainly, and on extremely clear nights, the spiral arm sweeping across its face.* (b) *M.33 in Triangulum, this Sc spiral is visible—with low powers, clear skies, and persistence—in a 6-inch telescope as an extremely pale gleam, about the size of the full moon.* (c) *M.66. This spiral galaxy, class Sb, is a member of the Virgo cluster. In a 6-inch telescope it appears as a faint, elongated gleam with a bright nucleus.* (d) *M.63, spiral galaxy in Canes Venatici. This compact Sb spiral appears about 5 degrees southwest of that galaxy, and appears in amateur telescopes as an oval gleam, 5 to 8 arc-minutes across, with a bright nucleus.* (e) NGC *4565. An edge on spiral galaxy in Coma Berenices, it appears in amateur telescopes as a thin sliver of light. This galaxy is a member of the Virgo cluster in an Sb system, similar to the Milky Way.* (f) NGC *7217, a spiral galaxy in Pegasus. This Sc spiral is visible in a 6-inch telescope as an eleventh magnitude nucleus, surrounded by an extremely faint, almost circular glow.* (g) NGC *5907, an edge on, spiral galaxy in Draco. This Sc spiral appears in a 6-inch scope as a featureless streak of light, at eleventh magnitude.* (a,c,d,e,f,g, Official U.S. Navy photograph; b, Lick Observatory photograph)

a

b

c

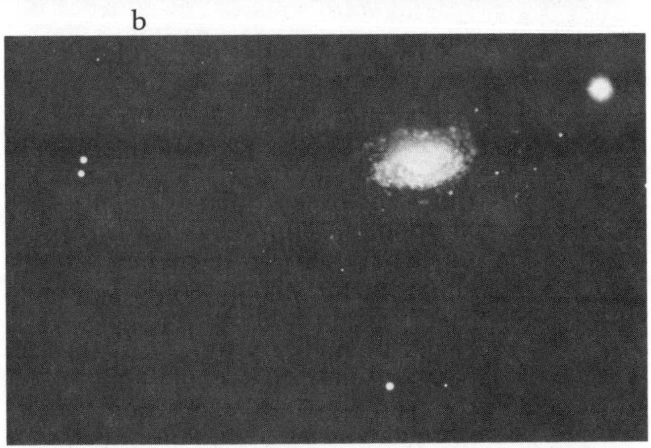

d

e

f

g

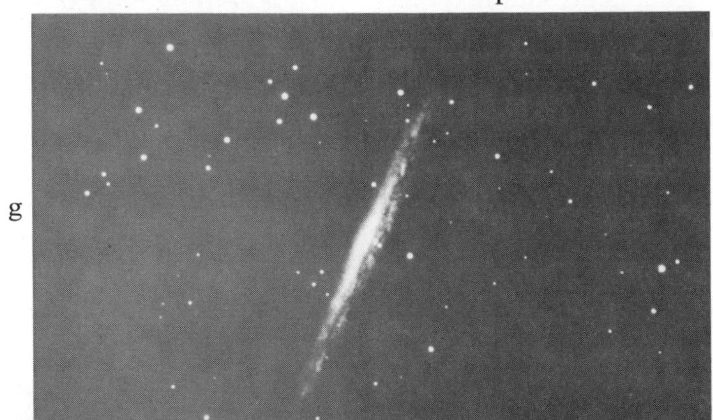

Astronomers have further classified the spirals by the tightness with which the arms coil around the hub, and the size of their nuclei. A spiral with arms tightly wound, so that it looks almost like a smooth disc, is an a-type spiral, usually written *Sa.* Sa galaxies tend to have large nuclei, and their spiral arms contain fewer HII regions and bright stellar associations. Spirals with moderately wrapped arms are called *Sb*; the Milky Way is an Sb galaxy. The most loosely wound of all, with very small nuclei, are *Sc.* The barred spirals show the same variety, and are classified similarly, with the addition of a B to their abbreviations to indicate the presence of a bar: *SBa, SBb,* and *SBc.* Ellipticals are classified similarly, except that numbers are used to indicate the degree of flattening of the ellipse. An E0 ellipse is almost spherical, and an E7 is the most elongated member. An intermediate form, the S0, is similar to the core of a spiral galaxy, without the disc; it is highly flattened, and lacks the band of obscuring dust clouds that marks the discs of true spirals.

This classification system, called the *Hubble sequence,* was invented by the astronomer who has contributed more to our knowledge of the galaxies than any other, Edwin Hubble. Hubble also conducted an important large-scale survey of the distribution of the galaxies in space, using a large telescope to photograph more than 1,000 selected areas of the sky. The survey uncovered a number of interesting facts about where and how galaxies appear in our skies. First, there is a large region of the celestial sphere, the *zone of avoidance,* where they do not appear at all. This is a long band running roughly north and south through the winter and summer skies, appearing thickest in the constellation Sagittarius. It is caused by our own galaxy, which blocks our view of other systems with its clouds of interstellar gas and dust.

When Hubble and later astronomers investigated the distribution of galaxies more closely, they found that, like stars, galaxies occur in clusters. These clusters can contain anywhere from a few dozen galaxies to thousands. The Milky Way belongs to a small cluster, called the *Local Group.* A cluster 65 million light years away in Virgo contains thousands. Clusters of galaxies clump together into *superclusters.* The Milky Way and the other galaxies in the Local Group, as well as the 2,500 galaxies of the Virgo cluster, all belong to the *local supercluster.*

Some 150 million light years across, the local supercluster contains about one quadrillion (10 to the 15th power—10^{15}) stars. The Virgo cluster (Fig. 32.7) apparently marks the center of the local supercluster. Other clusters have been detected, at distances of over 3 billion light years. Although none contains as many members as the Virgo cluster, some, such as one about 200 million light years away in Pegasus, are fabulously rich, with as many as 30 galaxies per million cubic light years—twenty times the density of the Local Group. Almost 3,000 clusters of galaxies have been found.

Within these clusters, many galaxies occur in double, triple, or even more complex combinations (Fig. 32.8). Members of close systems actually orbit around each other, like double stars. In some cases, the gravitational

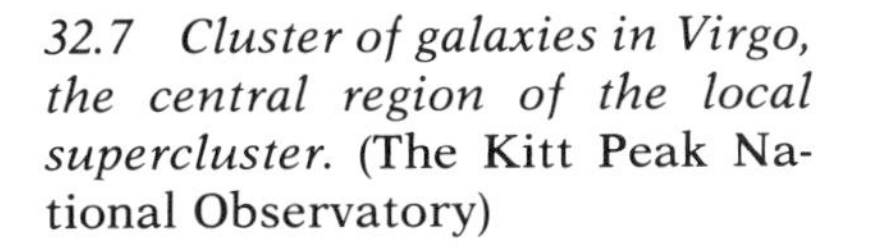

32.7 Cluster of galaxies in Virgo, the central region of the local supercluster. (The Kitt Peak National Observatory)

32.8 Cluster of galaxies in Coma Berenices. This cluster contains over 1,000 members, and has a high concentration of elliptical and So types. (The Kitt Peak National Observatory)

attraction within such a system can distort its members from their normal shape. The irregular shape of the Large Magellanic cloud may be owing to this cause.

This tendency of galaxies to occur very close together (Fig. 32.9) relative to their sizes also increases the likelihood of collisions among galaxies. Collisions of stars are rare, even in dense globular clusters, because the distances among them remain great, relative to their sizes. Compared to galaxies, stars are amazingly compact objects. Even in the heart of a globular cluster, the stellar population is still so dispersed that, if the stars were the size of human beings, they would be over 2,000 miles apart. On the same scale, the average distance between stars in the spiral arms of the Milky Way is more than 32,000 kilometers (20,000 miles). If galaxies were the size of human beings, however, the separation in close pairs drops to a few feet. At these close quarters, collisions become likely.

But if the Milky Way were colliding with another Sb spiral right now, no one but astronomers would notice. Thanks to the vast distances separating the majority of stars, in the course of an entire galactic collision, the odds are a trillion to one against a stellar collision. The effect on the entire galaxy, however, as the regular orbits of stars around the galactic center were all terribly disturbed, could be much greater. Computer simulations of such collisions suggest that the form assumed by colliding pairs of spiral galaxies might be something like a large, distorted egg—something like the giant elliptical galaxies common in large clusters of galaxies.

The other important feature of galaxies in clusters is that they all revolve around the cluster's center of gravity. Clusters in turn form gravitational bonds with other clusters, revolving around their common center. The Local Group, for instance, is apparently orbiting the center of the local supercluster, moving (relative to the supercluster's center) at about 900,000 miles per hour. At that speed, we will complete one orbit of the supercluster in about 300 billion years—20 times the life expectancy of our sun, or 15 times the present age of the universe.

For many years, astronomers have tried to deduce a scheme for the evolution of galaxies. With the large numbers of examples available, you would think that this task would be no more difficult than deducing the evolution of the stars. However, the scale of time and distance over which we must observe the galaxies is so great as to have defeated most of these efforts so far. Stars younger than 1 million years are all around us; the galaxies are apparently older than 10 billion years. Stars similar to our sun are available for study only 4 light years away; the nearest spiral galaxy is more than 2 million light years distant.

Operating under these handicaps, astronomers have been able to produce only the most general and speculative models for galactic evolution. Among the current candidates, one theory postulates that ellipticals may be the result of collisions between spirals. Other studies have found that ellipticals do not rotate around their geometric axes, as they must in order to be dynamically stable. What happens to ellipticals? Do they stabilize, or turn

32.9 Interacting spiral galaxies. (Lick Observatory photograph)

into some other form? The origins of galaxies are even more obscure, although recent studies of quasars suggests they may have a part to play. This issue involves one of the great scientific quests of all time, the search for the origins and ends of our universe.

This story has been told elsewhere in great detail, but a brief summary of the tale so far will help you to understand the final place of the Milky Way within the universe. It is a story that begins in the early years of this century, with the discovery of the true nature of the "spiral nebulae." Once the value of Cepheid variables as distance indicators became known, and the distances to the nearby galaxies were measured, our picture of the size and nature of the universe took a sudden leap toward infinity.

As the measurement of intergalactic distances reached outward, new tools were required. Cepheid variables are visible only over a limited distance. Beyond their range, other "standard candles" can give a rough idea of distance. A standard candle is any familiar class of object, such as supergiant O stars, for which one can determine a reliable average absolute magnitude. With absolute magnitudes brighter than the Cepheids, such stars are visible at distances where Cepheids fade below detection. The average brightness of the O supergiants visible in a distant galaxy, subtracted from the known average absolute magnitude of such objects, gives the distance modulus to the galaxy. The process has a larger margin of error than the Cepheid yardstick, but, on average, it is still reasonably accurate.

At very great distances, all of the stars in a galaxy merge into an unresolvable blur. Even globular clusters (average absolute magnitude −10), and large HII regions (−12) merge into the distant glow. What then?

By some clever statistical juggling, it is possible to gain usable estimates using entire galaxies as standard candles, but you can stretch such a technique only too far. Eventually, statistical uncertainties do you in. We need to go beyond the distance modulus, to some entirely new kind of yardstick.

Edwin Hubble, working in the 1920s, found that yardstick. Other astronomers had known for some time that many of the galaxies that were obviously far distant from us showed a peculiar feature in their spectra. Absorption lines in the combined spectra of all the stars in these galaxies were all shifted toward the red end of the spectrum. Their knowledge of the Doppler effect told astronomers that these galaxies were moving away from us. Hubble's work showed that there was a strong connection between the amount of the red shift observed in a galaxy's spectrum (that is, its speed away from us), and its distance (Fig. 32.10). Hubble's law states: The farther away a galaxy is, the faster it is moving away, or *receding.*

Further work has refined the exact relationship between distance and speed of recession. Much of this process has involved trying to find a precise value for the *Hubble recession constant:* the number you have to plug into the formula to translate red shift into distance. There is currently a heated debate over the exact value of this figure. One camp holds it to be around 50 kilometers (30 miles) per second per megaparsec, the other that it is about twice that figure. In either case, the significance of the figure is that, for every million parsecs you look out from earth, the speed at which the objects you find there are moving away from you increases by 50 to 100 kilometers (30 to 60 miles) per second. One million parsecs from here, the average galaxy is receding from us at a speed of 50 to 100 kilometers per second. A billion parsecs from here, the average speed at which objects are receding from us is 50,000 to 100,000 kilometers per second—almost half the speed of light. And so on.

This discovery had two implications. The immediately useful one, for astronomers trying to derive galactic distances, was that they had an elegant, simple yardstick. An ordinary spectrogram of a galaxy, measured for the Doppler shifting of its spectral lines, gives all the information necessary to tell its distance. Using this technique, objects have been measured at distances of 10 billion light years. Whether this represents the ends of the universe, however, only time—and larger telescopes—will tell.

The second implication of the cosmological red shift, however, makes the distances of things even more important. What does it mean that everything at great distances is moving away from us, no matter what direction we look? The entire universe is expanding. The direction of that observed expansion, however, does not mean that we happen to lie at the universe's center. To someone 10 billion light years away, *we* are the ones who seem to be rushing outward (which is why the center of the universe is at once everywhere and nowhere). The direction of the expansion depends on where you happen to be observing from. The most neutral way of stating the case is that the distance between everything is increasing.

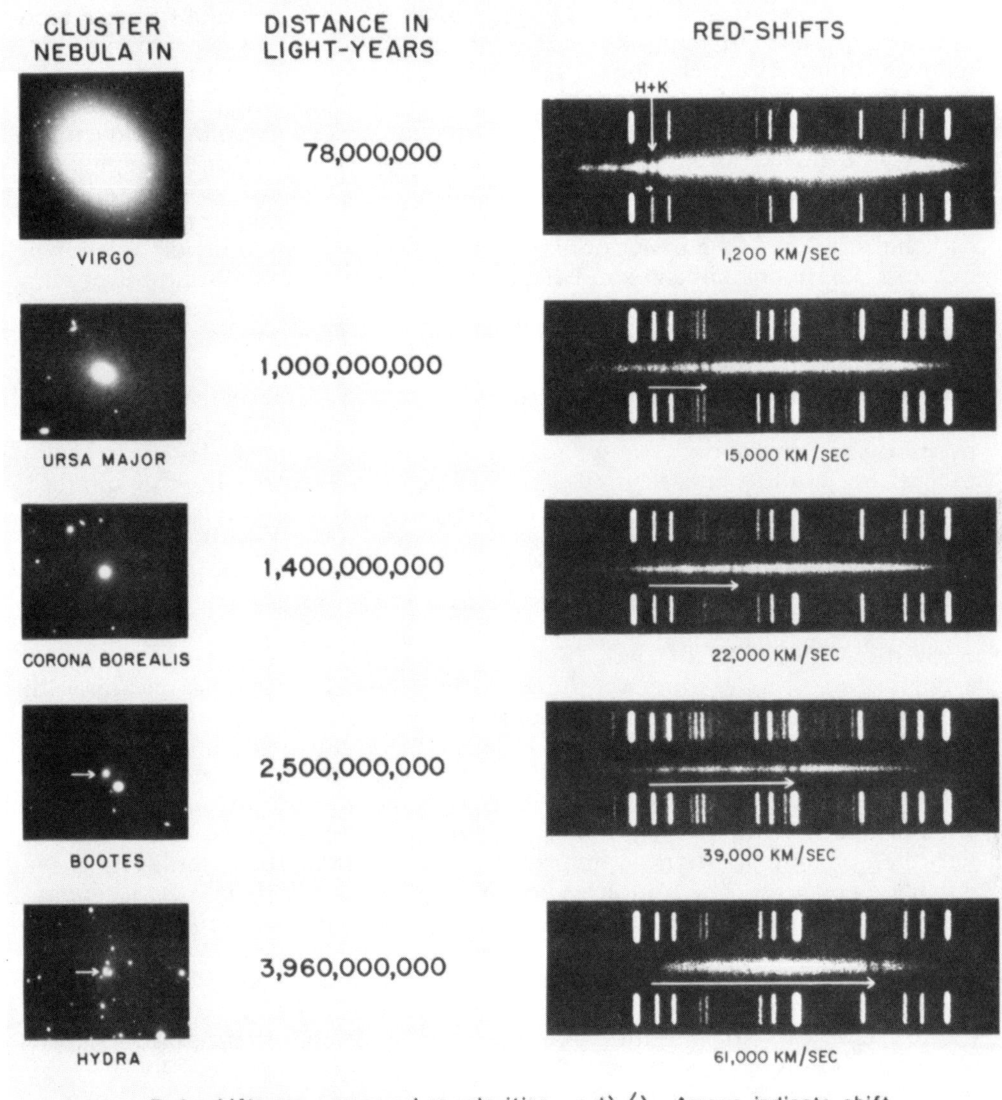

32.10 Red-shifted spectra of distant galaxies, showing relation between distance and shift of spectral lines. (Palomar Observatory photograph)

Once astronomers swallowed *this* notion, two enormous questions immediately arose: Where will it all end? Where did it all start? The second question is closer to an answer, simply because the fact of expansion points us in only one direction. If you follow the expansion process backward in time, it stands to reason that, long ago, everything was much closer together. Follow that reasoning as far as you can, and you find that, once upon a time, the entire universe all occupied a single point.

Something happened. Around 20 billion years ago, time began. The point that was to become the universe exploded, in the primeval fireball we call the *Big Bang.* The exact nature of this explosion may never be known, because it happened in a realm where little is certain, and conditions of temperature and density are so extreme as to beggar language. Recent theoretical breakthroughs, based on the principles of quantum theory, have suggested the following process.

Space, and the matter within it, looks "smooth" to us now: the forces governing the universe, such as gravity, follow systematic rules, and behave predictably. This is because we see things at a very coarse level of resolution. If we could observe matter and energy interacting on the smallest level—at resolutions of the *Planck scale,* 10 to the minus thirty-fifth of a meter (10^{-35} meter)—we would see that this smoothness is an illusion. Just as the cloudy appearance of the Milky Way masks the billions of its stars, the smoothness of the world masks an infinitesimal realm of utter chaos, where events happen randomly, in a state called *quantum weirdness.*

Before the universe began, this chaos was all there was. At some "time," a portion of this seething randomness happened to form a bubble, with a temperature something in excess of 10,000,000,000,000,000,000,000,000,000,000,000 (10^{34}) degrees Kelvin. Being that hot, naturally it expanded. After a few billion billionths of a second, it had grown to a size perhaps one trillionth the size of a proton. At that point, its temperature dropped low enough—10 to the twenty-eighth degrees Kelvin—for the fundamental particles of matter, such as quarks, to freeze out of the bubble. As these particles appeared, the bubble supercooled, and suddenly its rate of expansion increased enormously. For an extremely brief period—billionths of billionths of a second—it inflated.

That brief period of abnormally rapid growth is a recent refinement of the theory of the expanding universe. It describes events that occurred when the universe was still smaller than an atom, and when temperatures were still so high that none of the matter that may have been present in the original bubble could have survived: the universe was pure energy. The theory of the "inflationary universe" helps to explain certain odd phenomena, such as the extraordinarily even distribution of matter, the absence of antimatter, and the uncanny balance between the expanding force of the Big Bang, and the contracting force of the combined gravity of all the matter in the universe. All seem to be owing to the early instant of extremely rapid inflation, which was so fast that it smoothed out irregularities in matter distribution.

At the end of the period of inflation, the universe may have reached a diameter of a few centimeters; its temperature was still billions of billions of degrees. But it had cooled enough for particles of matter and antimatter to form—and instantly annihilate each other in yet another blaze of light. All that remained from that annihilation was a thin haze of matter—apparently because slightly more matter than antimatter was formed—and a fireball, spreading outward. That fireball, and the smoke of its burning, was the universe. It was now perhaps a trillionth of a second old.

The temperature of the expanding fireball dropped rapidly, cooling in a few minutes to a few billion degrees. Matter continued to condense out of energy: first protons and neutrons, then electrons, and finally neutrinos—all the stuff of which you and this book are made. After an hour or so, the temperature had dropped below a billion degrees, and protons and neutrons could combine, forming hydrogen, deuterium, helium.

To describe this event from a detached perspective, as if we were objective observers hovering somewhere outside the event, is, of course, nonsense. With Everywhere still enclosed in the point-universe, there would have been Nowhere for us to watch from, and of course Nothing for us to be. But if you had been hovering in Nothing, watching this process, you would have seen nothing. The outrushing fireball was too dense to emit radiation, and remained so for a million years, until its temperature dropped to a few thousands of degrees, and then it was transparent, and filled the universe with the light of its first day (but of course, it didn't *fill* the universe—it *was* the universe). In a billion years, this cloud of energy, light atoms, and neutrinos had cooled enough for galaxies to form. The expanding cloud cooled still further, until, today, its temperature is a couple of degrees above absolute zero. The echo of the Big Bang is still audible, a dim hissing heard at certain wavelengths in radio telescopes.

Where will the expansion cease? Will everything continue to recede from everything else forever? Or will there come an end, and will the universe gradually start to collapse into another primordial fireball, starting the whole process over again? The answer depends, oddly, on the mass of the universe. From the initial Bang, the universe attained a speed of expansion. If that speed is greater than the universe's own escape velocity (determined, like the escape velocity of a planet, by its mass), then the universe will not stop its expansion. Such a universe is said to be *open.* If the velocity of expansion is lower than the escape velocity, the universe will eventually reach the limit of its outward thrust, just as a ball thrown in the air comes to the top of its arc, slows, stops, and starts to fall. The crash of that long fall may be the Bang beginning another universe, as the fireball formed at the end of the contraction process leaps outward in another great expansion. Such a universe is said to be *closed,* and *pulsating.*

Astronomers are now attempting to learn the fate of the universe by measuring its average density. If that density turns out to be less than 5 × 10 to the minus thirtieth power (5×10^{-30}) grams per cubic centimeter, there is too little matter in the universe. The universe has achieved escape velocity. It will continue to expand forever, as the stars redden and die, until eternity will find it as a limitless empty haze, expanding infinitely into the night. This emptiness, as time goes on without end, will become even emptier, as the fundamental particles of matter age, and decay. As the years stretch on into infinity, nothing will remain but a faint warmth—the last vestiges of the Big Bang—and a few primitive atoms of positronium—positrons and electrons orbiting each other at distances of hundreds of astronomical units. These particles will spiral slowly toward each other until, touching, they will vanish in the last flash of light.

The visible universe contains not a tenth of the matter needed to stop the expansion. The search for "missing mass"—invisible matter such as may be present in the halos of galaxies—may turn up much more (the theory of the inflationary universe suggests it may be there), but the best estimates now are that the universe is open. One recent study indicates that the neutrino, now thought to be a massless subatomic particle, may actually have a tiny but measurable mass. The neutrino is so common that it can be said to fill the universe as air fills a room. If it does turn out to possess some mass after all, then that may be enough to fill the gap, giving the universe sufficient density to halt its expansion, and turn in upon itself, perhaps to recreate the universe in some far distant era.

When astronomers with the most powerful radio and optical telescopes look out at objects 10 billion light years away, they are seeing light that started on its journey 10 billion years ago. The most distant objects, then, give us a glimpse of the universe as it appeared in its youth, only a few billion years after the first galaxies started to condense out of the outrushing waves of the Big Bang. As the next generation of large telescopes sees first light, starting with the Hubble Space Telescope in the mid-1980s, our reach may go even farther, back into the earliest days of the galaxies. The observations of *quasars* made to date indicate that what we find there may be strange indeed (Fig. 32.11). Since their discovery in 1960, these *quasi-stellar objects,* or *QSO*s, have staggered most of our expectations for how matter ought to behave. Appearing as very dim, bluish stars (the brightest, 3C 273, appears about magnitude 12.8), their spectra show red shifts so high that they must be receding at speeds that are significant percentages of the speed of light. These are objects traveling from us at enormous speeds, at the edges of the universe.

To be visible at all at such a range, quasars must be extremely bright—the distance modulus tells us that their absolute magnitudes range up to −26. But quasars also vary in light. This is a simple trick for a star, but when something as powerful as a quasar does it, astrophysicists work overtime. An object's brightness cannot vary in a time less than light would take to cross it. Otherwise, light from its far side would still be reaching us at its previous level after light from its near side, at its new level, had arrived. The variation in light from such an object would be "smeared out," and very difficult to detect. An object that varies in a day, then, must be less than 1 light day across. The variation of any object therefore puts an absolute upper limit on its size.

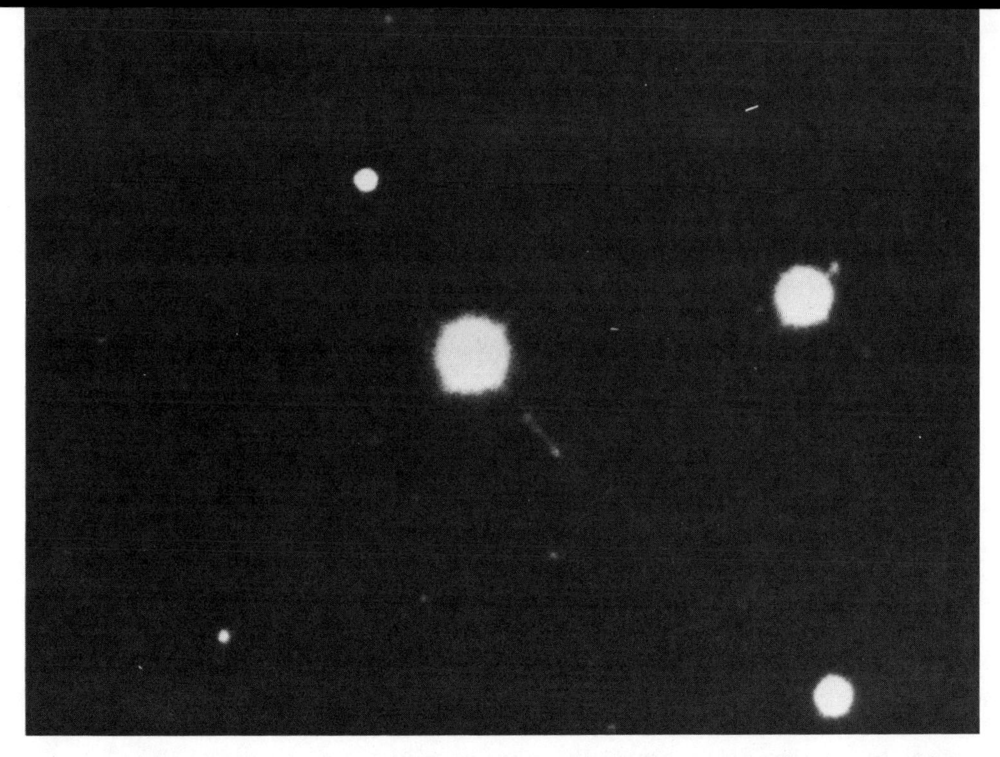

32.11 3C273, quasar in Virgo. The thin jet, extending toward the lower right, is 1.5 times as long as the Milky Way. The quasar itself emits energy from a region much smaller, hidden here behind the blaze of its Airy disc. (The Kitt Peak National Observatory)

Variations from quasars are distinct, and occur over a period of only a few weeks, indicating that they are no more than a few light weeks across. No one has yet come up with a satisfactory explanation of how an object the size of our solar system can produce more light than the largest elliptical galaxy. Recent evidence suggests that some of the nearer quasars may be active nuclei of galaxies that are so far away that the galaxy itself is invisible—only this supernally brilliant core shines clearly over the distance. It could be that a quasar is some unimaginably violent phase in the early evolution of a galaxy.

OBSERVATION

Galaxies offer an almost infinite range of challenges to the amateur, from the large, bright swath of our own Galaxy across the sky, to the dim glow of light rays millions of years old. It is even possible to glimpse one of the nearest quasars in a 6-inch telescope. Only the largest amateur telescopes show the finer structure of the galaxies, and it takes an instrument of observatory caliber to resolve their individual stars. However, it is entirely possible to see such details as the lanes of obscuring gas that mark the equators of spiral galaxies, close satellite galaxies similar to the Milky Way's Magellanic clouds, and the overall form of many of the nearby galaxies.

As with any nebulous object, the difficulty of detecting a galaxy depends on its integrated magnitude versus its angular diameter. Many galaxies appear in the telescope as vague and diffuse as a section of the Milky Way that has drifted from the main. Others, if they are intrinsically bright and compact, can be remarkably clear edged, showing distinct details of form. Any galaxy will be easier to find and see well through dark, clear, moonless skies. Light-pollution-rejection filters can be a great help, doubling or tripling the contrast between the faint white haze of the galaxy and the sky. Most of the galaxies described in the following section are bright enough to be visible on good nights from suburban locations, without special filtering.

Magnification of galaxies works very much like that on globular clusters, except that there is no chance of resolving a galaxy, so the need for high magnification is less. Use the lowest power possible to find galaxies, then shift upward until you reach a balance between the magnification necessary to bring out detail and the dimming of the magnified image. Some observers find that a long-focus eyepiece in a barlow lens gives better results than a short-focus eyepiece used alone: the larger exit pupil and longer eye relief make the image seem brighter.

Observing galaxies can be frustrating, because these fascinating objects appear at first so dim, vague, and nondescript—not at all like the pictures accompanying this chapter! Like most objects, they reveal their individual characteristics as your eye becomes more practiced. Use the following objects to get your bearings, and then go on to the more difficult ones listed in the Appendix. With experience, you will become sensitive to the subtle details that distinguish these distant island universes.

1. *The Milky Way.* Already covered in part in the last several chapters, this, the largest and brightest of the galaxies in our sky, is easy to overlook in our pursuit of more elusive game (Fig. 32.12). Take some time in the summer and winter seasons when it presents itself best, and scan the length of it for the details described in this and other chapters, trying to piece them into a larger picture. Instruments for such a project can be simple: under good, dark skies, the naked eye is a remarkably fast, wide-angle instrument; 7 × 50 binoculars are also good. If you have strong arms, or adequate support, 11 × 80 binoculars are wonderful, expecially if equipped with LPR filters.

2. *M.31: the Great galaxy in Andromeda.* The other massive galaxy of the Local Group, Andromeda (as most amateurs call it) is also a large, Sb spiral, about 2.2 million light years from earth, at the far end of the Local Group (Fig. 32.13). Find it by hopping from β (beta) to μ (mu) to ν (nu) Andromedae, where binoculars (and on good nights the naked eye) will show an elongated haze, brightening toward the center, about a degree west of ν (nu). Binoculars are the best instrument for tracing the full extent of the disc. On good nights, it can appear almost 4 degrees long, extended in position angle 320 degrees. With an integrated magnitude of 5, and an extension in a typical telescope of about 150 by 30 arc-minutes, this is the largest, brightest galaxy in the sky. From our point of view, we see it inclined 12 de-

32.12 The Sagittarius star cloud. Visible to the naked eye on clear nights, above the spout of the Teapot in Sagittarius, this haze of stars and dark nebulae marks the central bulge of the Milky Way, 30,000 light years away. (Official U.S. Navy photograph)

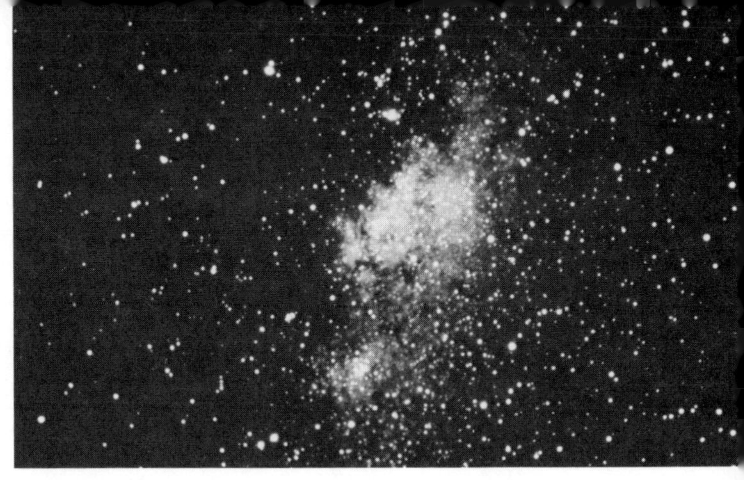

grees from edge on—similar to the rings of Saturn in a typical year; its northwest edge is closer to us. About 130,000 light years across, M.31 is slightly larger than the Milky Way, and contains perhaps 50% more stars.

At lowest power, in a 6-inch, the extremely faint outer edges of the spiral usually appear smooth and evenly lighted. On good nights this region looks mottled—larger (10-inch and up) apertures show this mottling to be the pattern of the spiral arms. The northwestern edge sometimes shows signs of

32.13 M.31, the Great galaxy in Andromeda. This near-twin of the Milky Way lies at a distance of 2.2 million light years. It is visible to the naked eye on clear, dark nights; amateur telescopes can reveal some of the details shown here, such as the band of dark nebulae around its equator. (Palomar Observatory photograph)

the cloud of obscuring dust that marks the galaxy's equator. Higher magnification brings out the very bright, small nucleus. About 30 minutes northwest of the nucleus is NGC 205, a satellite galaxy. It is an eleventh-magnitude E5 elliptical, appearing about 5 minutes across, and distinctly elongated. Much closer to the nucleus on its opposite side lies M.32, an E3 elliptical satellite galaxy. At ninth magnitude it is brighter and more distinct than NGC 205. Rounder and slightly smaller as well, it is a relatively easy target. NGC 205 is about twice as large as M.32, but both are dwarf systems, containing only a few billion stars each. These are by far the easiest dwarf elliptical galaxies to find and observe.

3. *M.81 and M.82 in Ursa Major.* This close pair of galaxies appear ½ degree apart, in the same field of a 6-inch telescope even at moderate powers (Fig. 32.14). M.81, the brighter of the two, is a relatively easy object, but much more challenging than the Andromeda galaxy, primarily because it is small, and not well marked by guide stars. Its surface brightness, however, makes it distinct against a dark background sky. Find this pair by first locating a small (1 degree) elongated triangle of fifth-magnitude stars, containing σ(sigma) and ρ(rho) UMa. This asterism marks the ears of the Bear, and lies about 8 degrees northeast of o (omicron) UMaj, the Bear's third-magnitude nose. M.81 lies 5 degrees east and 1 degree north of this triangle. In a low-power eyepiece, the separation between the pair of galaxies spans about a quarter of the field, and the galaxies themselves appear much smaller than Andromeda. M.81 lies about 7 million light years from earth, a good example of a tightly wound, Sa spiral. About 30,000 light years across, it contains almost as much mass as M.31.

Smaller than M.81, with a diameter of about 15,000 light years, M.82 is a peculiar galaxy. The thin, splinter shape is typical of a spiral galaxy seen edge on, but M.82's silhouette appears distorted. A 12-inch telescope shows that, unlike most edge on galaxies, with their typical dark lane of obscuring gas running lengthwise down their midline, dark lanes cut *across* M.82. Large telescopes have shown that M.82 has suffered, relatively recently, an enormous explosion. Clouds of gas with a total mass millions of times that of the sun are rushing outward from the center of this galaxy at speeds over 2 million miles per hour.

4. *M.104: the Sombrero galaxy in Virgo.* A distant member of the Virgo cluster, M.104 lies 37 million light years from earth, appearing almost edge on to us. In a 6-inch it looks like a distinct sliver of light, slightly bowed at the center, in a poorly marked region of the sky. Using low power, hop 10 degrees due west of Spica, and you should find the Sombrero in your field, looking about the same size and brightness as the Ring nebula.

Magnification will bring out the "sombrero" shape and the dark lane of dust that marks the rim of this tight spiral. An eighth-magnitude object, spanning over a distance of 40 million light years, this is one of the brightest and largest galaxies. As wide as M.31 and much more massive, the Sombrero probably contains six times as many stars as the Milky Way. The Sombre-

32.14 M.81 and M.82 in Ursa Major. (Palomar Observatory photograph)

32.15 Sombrero galaxy (Hale Observatories)

ro's red shift shows that it is receding from us at 2.5 million miles per hour.

5. *M.59 and M.60.* This pair of elliptical galaxies lies near the heart of the Virgo cluster, 4 degrees west and slightly north of ϵ (epsilon) Virginis. M.60 is by far the easier of the two to find; M.59 may not appear until after you have located the former, and then scan the region about ½ degree to its west with averted vision. One of the largest galaxies known, M.60 is a giant elliptical, elongated about halfway between the extreme E0 and E7 forms; its total mass is approximately five times that of the Milky Way. It is more distant than many of the members of the Virgo cluster, receding from us at almost 3 million miles per hour, at a distance of around 50 million light years. M.59, ½ degree west of M.60, appears as about half as bright as its companion, yet it is apparently much closer to earth, showing a recession velocity only one-fourth that of M.60. The two galaxies are apparently not a physical pair. M.59 is a much smaller system, too, yet it manages to pack more mass than the entire Milky Way into that smaller volume.

6. *3C273: Quasar in Virgo.* This is the nearest and brightest quasar known, and at magnitude 12.8 is within the reach of a 6-inch telescope on an extremely dark, clear night. Its position is marked on the Tirion star atlas at 3 degrees west and 4 degrees north of γ (gamma) Virginis. Use the AAVSO finder charts (Figs. 32.16 and 32.17) to pick out this extremely dim object from other stars in the field. Those interested in tracking its light variations will need at least an 8-inch, and preferably a larger, scope. I include it here not because it is an easy object, or for any intrinsic visual interest (although observers with large-aperture telescopes may want to track its variations), but because it is 3 billion light years away. Think of it.

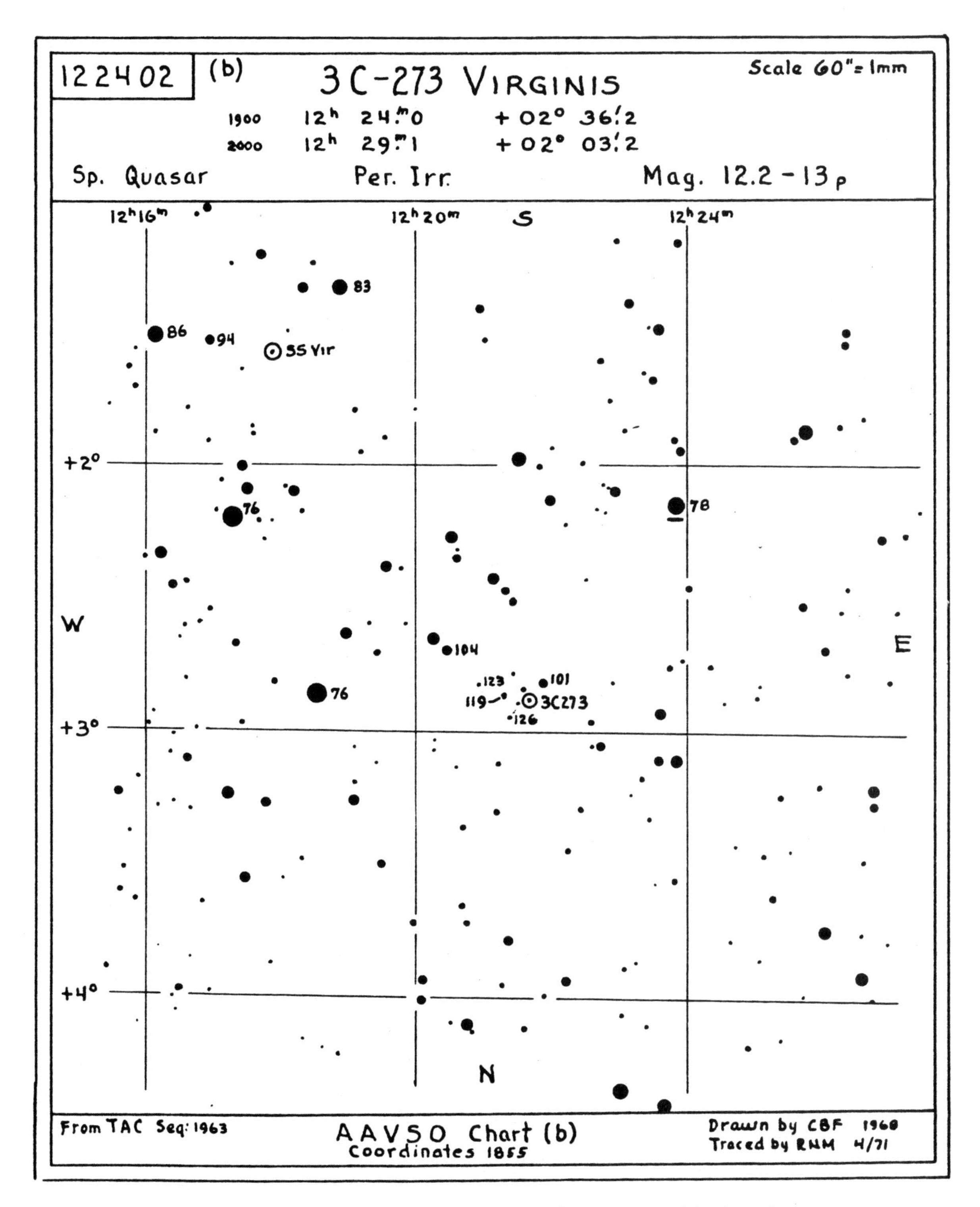

32.16 AAVSO finder chart for Quasar 3C 273 in Virgo. The region of this chart lies northeast of eta Virginis; the quasar itself is about 4 degrees northeast of that 4th magnitude star.

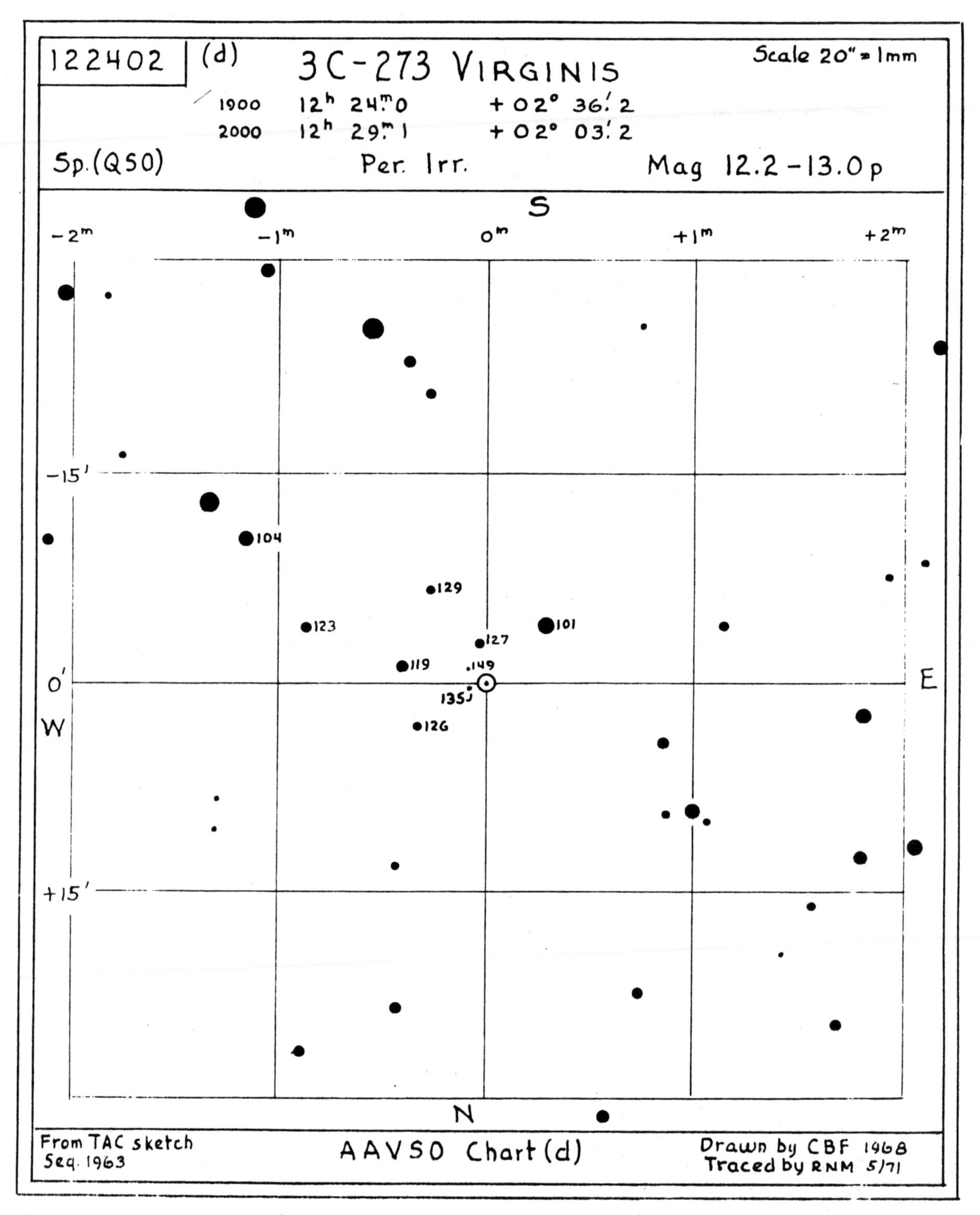

32.17 AAVSO comparison chart for Quasar 3C 273. Note the very fine scale: 1" = 8.5 arc-minutes. The diameter of the full moon on this scale is over three inches.

Building a Telescope

A 6″, *f*/5 Newtonian reflector is an easy afternoon project (although assembling the materials by mail order may take 6 to 8 weeks). The plans accompanying this appendix assume that you have:

- a 6″, *f*/5 mirror
- a 1.83″ major axis secondary
- a spider
- a diagonal holder
- a tube
- a primary mirror cell (to fit the tube)
- an eyepiece mount

A finder scope is optional; you may choose to substitute a simple peep sight, made by gluing golf tees to the outside of the tube.

PRELIMINARY DESIGN

Before purchasing the mechanical parts for your telescope, make a careful plan of its optical layout. This will help you to determine such factors as

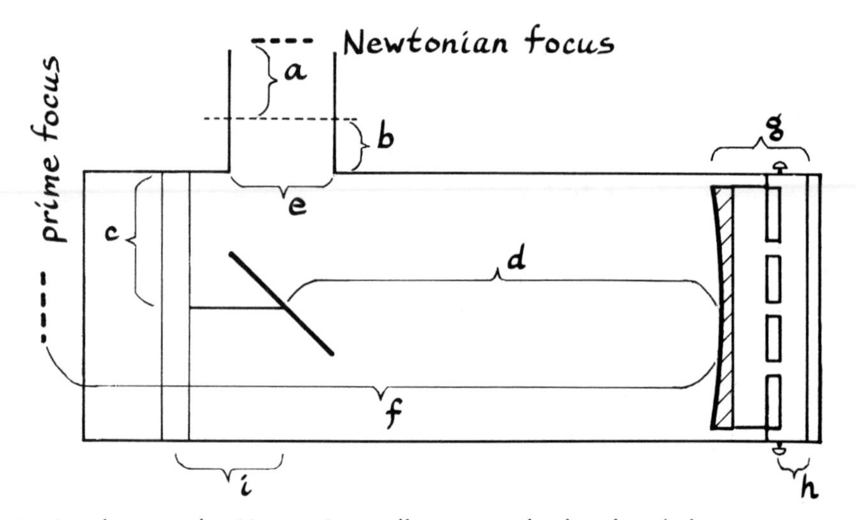

A1 Design layout of a Newtonian reflector: a, *the height of the Newtonian focus above the minimum height for the drawtube;* b, *the minimum height of the drawtube;* c, *the inner radius of the tube;* d, *the separation of the secondary and primary mirrors;* e, *the width of the drawtube opening;* f, *is the focal length of the primary;* g, *the height of the primary mirror surface above the bottom of the cell;* h, *the height of the cell's mounting bolts above the bottom of the cell.*

tube diameter, as well as enable you to build a telescope that delivers the maximum performance possible (Fig. A1).

Three specifications of a focusing mount are important to your design: the total distance the drawtube can travel, the drawtube's minimum height (*b* in the diagram), and the width of the drawtube's lower opening. Ideally, the total travel should be as large as possible, the minimum height should be small, and the opening should be large. Almost every manufacturer offers the same basic focusing mount for standard (1.25″) oculars, with a drawtube travel of about 2 m316″, and a minimum height of 3″. The drawtube aperture can vary; look for about 1.5″. Such a mount should be more than adequate to your purposes.

When you consider the implications of that 2 m316″ of travel, you will see that the dimensions of the focusing mount, size of the diagonal, and diameter of the tube are interrelated. The drawtube travel limits your margin for error in positioning the Newtonian focus. The location of the focus *a* is determined in turn by the position of the secondary *d* (which in turn determines the position of the focusing mount). If you miss that margin, the focal plane will not fall in the drawtube. The margin is even smaller, because different oculars require a half-inch or more travel above and below the focal plane.

The focal plane should fall slightly more than ½″ above the minimum drawtube height. This allows for the use of a barlow lens, which generally focuses about that far inside the focal plane; the longer travel above the focus allows the use of long, low-power eyepieces.

An important consideration in telescope design is light loss. Few telescopes actually pass all of the light from the objective to the observer; often, an undersized diagonal or narrow drawtube chokes off the beam of light. The observational effect of this is a slight dimming at the edges of a low-power field, one you are not likely to notice, but worth avoiding, if possible. A realistic goal is to ensure that a prime-focal plane ¾″ across receives 100% of the light available. Your focal-plane disc is likely to be larger than that, but if the peripheries of the disc are dimmed slightly, the loss is not worth fixing: few eyepieces can see a disc larger than ¾″. To determine the size of the widest disc your system can produce (whether or not you have an eyepiece capable of using it), multiply the true field of view in degrees (as determined by the apparent field of the widest-angle eyepiece you plan to use) by the objective's focal length, and divide the sum by 57.3.

$$df = ft \times F/57.3$$

The most likely causes of light loss at the edges of the field are (1) a narrow tube, (2) an undersized diagonal, or (3) a narrow drawtube. To know what sizes you need for your telescope, the most reliable way is to make a *ray trace.* This is a diagram of the paths taken by light rays within a telescope. You can make a life-sized ray trace for your system by taping together pieces of graph paper and sketching in the objective, its optical axis, and prime focus (Fig. A2).

To locate the prime focus, you need to know the precise focal length of your objective. Focal lengths vary by several percent, so you must measure yours rather than rely on the nominal value. Use a tape measure to find the

A2 Ray trace of a reflector. Line f–f′ *is the size of the focal plane. Lines from* O *to* f *and* O′ *to* f′ *define the edges of the light path. The minor axis of the secondary* (s–s′) *and the drawtube opening* (d–d′), *when set at their proper distances from the objective, must be wide enough to enclose the light cone.*

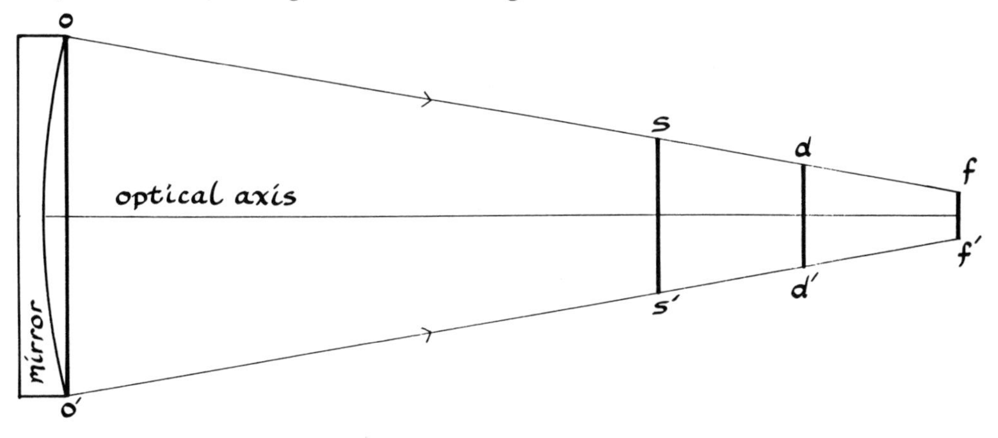

distance from the mirror's surface to the tiny, sharply focused image of the moon. Do not use the sun for this test; it can blind or burn you. Don't use a terrestrial light source either; it will give a misleading result.

When you have located the prime focus, draw it in as a ¾″ line perpendicular to the optical axis. Then draw in rays of light from the edges of the objective to the edges of the prime-focal disc. To determine the minimum width of the tube, sketch in light rays from the edges of the field. These enter the telescope not parallel to the optical axis, but converging slightly, at an angle equal to one-half the true field of the widest-angle eyepiece you plan to use. A 40-mm Kellner in a 6″, *f*/5 gives a true field of about 2.3°, so rays of light enter the telescope at angles as large as 1.15° from the optical axis. The diagram pictured here shows a ray trace for a 6″, *f*/5.

Once you have determined the minimum width of the tube, order the smallest-diameter tube possible. An oversize tube could require you to use an oversize diagonal. To check the size of your diagonal, you must first know where to place it. The distance from primary to secondary is the primary's focal length minus the correct height of the Newtonian focus (*a* plus *b*) in the first diagram minus ½ the outer diameter of the telescope tube. Find that distance along the optical axis of your ray trace, and place a line there to represent the diagonal. The line should be inclined 45° to the axis; its length should equal the major axis of the diagonal. If the cone of light at that point intersects the slanted face of the diagonal, fine. If it doesn't, try moving the diagonal slightly off the optical axis, away from the drawtube. Such a diagonal offset is common in well-designed telescopes, and stems from the lopsided geometry of the intersection between a cone and a plane: the cone is wider on the plane's lower side. If you need to offset the diagonal, make a note of the amount of offset required (Fig. A3). Order a diagonal holder smaller than your diagonal, so that you can build in the offset by attaching the secondary to its holder off-center, without one edge of the holder peeping out from behind the diagonal. Find the minimum size diagonal necessary to intercept the light cone and order it.

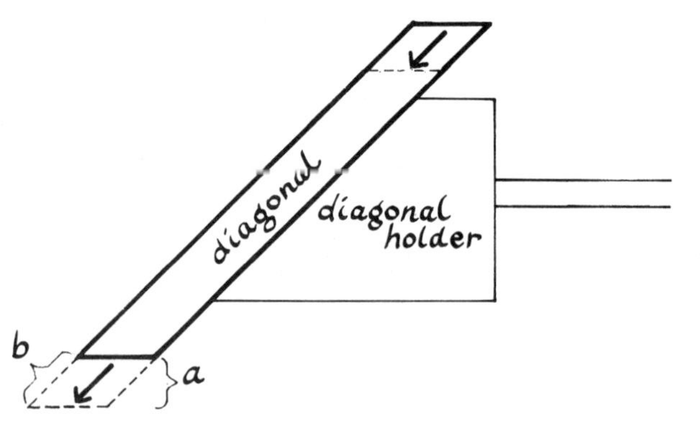

A3 Offsetting the diagonal. To create an offset equal to a, *slide the diagonal along its major axis by distance* b.

The drawtubes of most focusers are roughly the same size; if you order one too small to pass the light cone of your telescope, you can either raise the focuser slightly by shimming it with strips of wood, order another one, or ignore the problem. Each approach has its advantages; choose the one that suits your inclinations.

The ray trace should give you the exact optical dimensions of your telescope. Transfer those dimensions to your working plans, and mark the tube for drilling. Drill mounting holes for the primary mirror cell and determine where the primary's reflecting surface will lie in relation to those holes. Measure up the tube from that point to locate the opening for the drawtube. Locate the mounting bolts for the spider in the same way. A note of caution here: most spiders allow you to adjust the distance between spider and diagonal holder; set this distance to no more than ½″ greater than the minimum. An overextended diagonal holder will vibrate. If you choose to install a finder, drill holes for its mounting rings at this time, taking care to keep the rings on a line parallel to the tube's optical axis. A good location for the finder is about one-sixth of the tube's circumference away from the focusing mount (Fig. A4).

With all the drilling complete, clean up the dust around your work area and clean the inside and outside of the tube carefully. If the interior of your tube is not black, coat it now with flat, black paint (not glossy); if paint won't

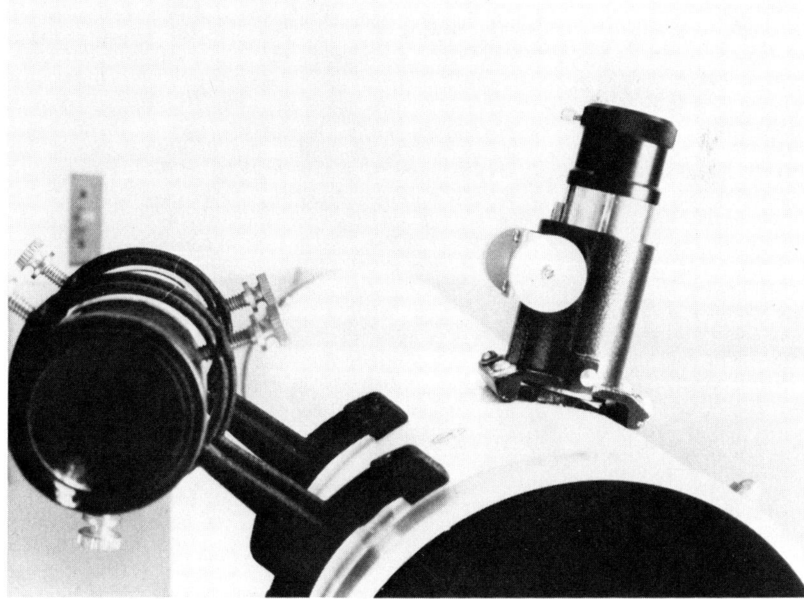

A4 Focusing mount and finder installed.

stick, glue black construction paper. Attach the secondary to its holder, using either the clips provided with the holder or glue. If you use glue, you must use a silicon-based adhesive; ordinary glues are rigid and will warp the diagonal when it expands and contracts. If you need to offset the diagonal, you can glue it onto its holder off-center. If you use clips, you can still build in the offset by shortening one arm of the spider. If neither method seems practical, you can either order a larger diagonal or decide not to worry about it.

Install the spider, focuser, and finder rings. Then insert the primary mirror cell, without its mirror, to check the fit. If it fits, install the diagonal holder with diagonal, and the primary cell with the objective in place. Voilà: a telescope. To use it, see the instructions on collimation in chapter 11. Before you can put it to serious use, however, you may want to put it on a mount. Appendix 2 has instructions for such a project, which is even simpler, and a good deal less expensive.

APPENDIX

A Dobsonian Mount

2

Integrally balanced, uncannily smooth, easy and inexpensive to build, the Dobsonian mount is the best all-around choice for the beginning star-gazer. You can use it while building (or saving for) something more ambitious, but it may turn out to be the most satisfactory mount you ever own—the kind you keep coming back to long after you own one more complicated. See Figure 7.7 for a photo of the Dobsonian.

The idea is simple: altazimuth configuration, with large, smooth bearing surfaces (Fig. A5). The azimuth bearing is a sheet of formica, easily obtainable from lumber or building-supply stores that sell kitchen countertops. The part you want is the scrap left over when they cut out the hole for the kitchen sink (ask for a "kitchen sink cutout"). The size of cutout you want depends on the size of your telescope's tube. For this design (and for all others), let's assume a tube outer diameter of 8″; if your tube varies from this, adjust the relevant figures to match. This design is currently in use with telescopes up to 29″, so feel free to expand these dimensions to suit your needs.

The azimuth bearing surfaces must be wider than the altitude bearing surfaces. For an 8″ tube, the latter will be between 10″ and 11″ apart, so the minimum width of your azimuth formica surface should be about a foot. In the middle of the cutout drill a 1″ perpendicular hole. Glue a 1″ dowel, about 3″ long, into this hole. Rub it with paraffin or other wax until it is smooth. (You may want to touch up this waxing from time to time, to keep the mount turning smoothly, but since this post bears no weight, it's not vital.)

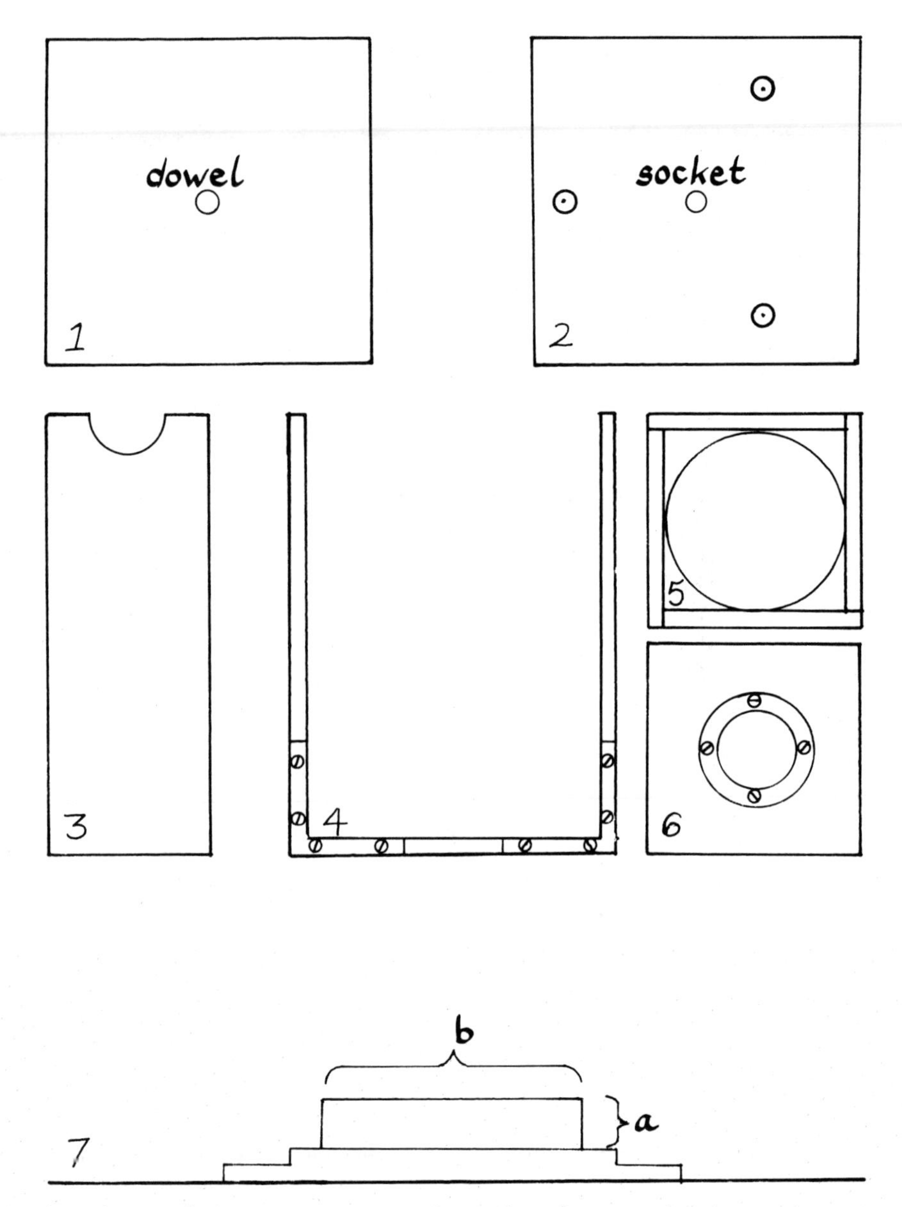

A5 Plans for a Dobsonian mounting: 1, *base plate (formica "kitchen sink cutout");* 2, *upper base plate, showing placement of Teflon or tack glides;* 3, *fork, side view;* 4, *fork, front view. Notice angle-bracket reinforcement;* 5, *cradle, end view, with tube in place;* 6, *cradle, side view, showing placement of PVC flange;* 7, *detail of cradle, showing dimensions of flange.*

To complete the azimuth bearing, cut a disc of sturdy (½″ or thicker) plywood, a foot or so across. It should be wider than your altitude bearings. In the center of this plate, bore another 1″ hole (the center is the point exactly beneath the midpoint of the telescope's optical axis, so you don't need to worry where you put this hole, so long as you place the telescope over it later). The dowel in the base plate should fit snugly but still turn freely in it. If it binds at all, rasp out the hole, then polish with fine sandpaper. The color to paint your telescope mount is a matter of taste. Traditionalists favor glossy white, but bright colors and stained or natural varnished wood are gaining popularity. Black or dark colors are probably not a good idea, simply so you don't trip over your mount in the dark (red, by the way, appears as one of the darkest colors at night). The most important consideration is to seal the wood against moisture so it won't warp. When you've chosen a color, paint the azimuth disc and set it aside.

For the altitude fork, use ½″ or sturdier plywood, or 1″ pine boards, 8″ wide. Keep the tines as short as possible. To know how long they must be, determine the center of gravity of your tube assembly. Rest the tube (with an eyepiece in the drawtube) on a piece of pipe. Shift the tube back and forth until it teeters, and mark the balance point. Measure from the mark to the back end of the tube; the distance between the center of your altitude bearings and the base of your fork needs to be longer than this measurement so that your telescope can swing freely through the fork without bumping its tail.

The other dimensions of the fork depend on the size of the telescope's cradle. The cradle is simply a box, open at two ends, that fits snugly around the tube. The cradle's width is therefore the tube's outer diameter plus two board thicknesses. For ¾″-thick boards (1″ nominal lumber), the cradle will be 8″ (tube outer diameter) plus 2 × ¾″, for a total of 9½″ wide; it should be as long as it is wide. Knock this box together, using a carpenter's square to keep it reasonably true. On two opposite sides of the cradle, center and attach a pair of PVC closet flanges.

A closet flange is a piece of plumbing used to attach a pipe to a flat surface; you're looking for the kind made of polyvinyl chloride, a plastic. Ask your hardware salesperson for "a pair of large PVC closet flanges," preferably 6-inch or larger diameter. Attaching these flanges finishes the cradle.

Before turning your attention to the fork, measure the thickness of the flanges' bases (*a* on the diagram), and double that figure. If that thickness is ¾″, then the total width of the cradle, which is equal to the length of the fork's base, is 9.5″ plus 2 × ¾″, or 11″. Cut two pieces of lumber to that length for the base, and two more for the arms, long enough to allow clear swing-through for the telescope tube. If you're using plywood, all the pieces in the fork should be the same width; that width should be an inch more than the diameter of the flange's bearing surfaces (*b* on the diagram). Glue the base pieces together, face to face, making a double-thickness board; fasten the two arms to either end. Reinforce these joints with metal braces, making certain that any bracing you use does not interfere with the tube's swing-through.

Two more steps complete the fork. At the end of each arm, cut a semicircle the same diameter as the bearing surface of the flanges. The PVC flanges rest in these cutouts. When making these and other cuts, work from a single, clearly drawn line on the wood, cutting slowly, with a fine, narrow blade. The more slowly you work, the closer the cut will approximate a circle. About 45° on either side of the bottom of these cuts, place a Teflon pad. To find Teflon, look in the Yellow Pages under "Plastics," and shop around for the best price. If you have no large plastics supplier nearby, you can order Teflon by mail (see the supplier's index). Or you can line the cut with a strip of felt, which works well enough. You can get felt in 50-cent scraps at most fabric and sewing stores. Wood glue should hold the felt in place.

To attach the fork to the azimuth plate, set its base on the upper azimuth bearing plate, centered directly over the center of the azimuth plate. Make sure the fork is centered over the pivot hole in the upper azimuth plate; otherwise, its motion in azimuth will be eccentric. Use glue and screws to join the two. Paint the fork, and let it dry.

To complete the azimuth bearing, attach three friction bearings to the lower surface of the azimuth plate. These bearings work best if they are Teflon; a kind of furniture glider called "tack glides" will also work well for telescopes of 8″ and under. Attach the Teflon or gliders at 120° intervals on a circle centered on the 1″ hole in the upper plate. This circle should be as wide as you can fit on the plate, to keep the base of this stubby tripod you're now putting under the scope as broad as possible.

When you have everything in place, put the telescope in the cradle. If the fit isn't tight enough, lay down strips of duct tape until the cradle holds the tube firmly. You should be able to rotate the tube in the cradle, but it should not slip under ordinary use. Set the tube horizontal, and move it back and forth in the cradle until it balances. Set the tube vertical, and rotate it until the weight of the focusing mount (with eyepiece) balances the weight of the finder. If you have no finder, simply locate the focuser directly above one tine of the fork. Now give the whole assembly a spin. Does it move freely? Is the fork on-center? Does the tube stay balanced wherever you point it? Continue to adjust the cradle until it does. Once it's balanced, you might want to put a few reference marks on the tube and the cradle, so you can restore it to balance easily whenever it gets knocked out of line.

This mount works best when it's level. You can install a leveling system by setting three large "T-nuts" at the edges of the lower azimuth plate. (T-nuts are threaded sockets that fit into holes drilled in wood, allowing you to bolt wooden pieces together. They tend to pull out of their sockets, unless you epoxy them in.) Set the T-nuts about equal distances apart, and screw large bolts into them from the bottom, forming a tripod. You can glue crutch tips, or small wooden feet, to the heads of the bolts, so they won't sink into soft ground. You needn't carry along a carpenter's level; just raise the side of the mount that the telescope tends to swivel toward (the downhill side) until the swiveling stops.

APPENDIX

Units of Measurement 3

Amateur astronomy in America suffers from our ambivalence over the metric system: a telescope's objective, for instance, is almost always given in inches, while eyepieces are almost always metric (except for their barrels, which are in inches again!). This uneasy coexistence of new and old systems is starting to spread through the rest of our lives. Most of us have a pretty firm sense of a meter as something slightly longer than a yard, and a kilogram as a little more than two pounds; a kilometer is a pretty short mile, and a liter is a hefty quart. More abstract units, like degrees of temperature, are harder to keep in mind: 100 degrees Celsius will boil water, but is 27 degrees comfortably warm or a bit too cool? Most of us raised on the English system still have a better intuitive feel for the size of an English unit than for the metric equivalents. On the other hand, the advantages of the metric system are enormous: its even, decimal increments make calculation much easier (quick—how many teaspoons in a pint?); everybody else in the world—including the English—uses it.

This book, like its audience, uses a mixed system. In the chapters dealing with equipment, the bias is toward English units. There is little need to complicate the construction projects described there with a set of conversion tables. When we come to descriptions of celestial objects, however, the issue is less clear-cut. On the scale at which many astronomical phenomena occur, the difference between kilometers and miles, degrees Celsius and Fahrenheit, all dissolve into the Very Large. And on the scale of astronomical units, light years, and parsecs, there is no disagreement. Angular measurements, of course, exist independently of either system.

Use the following equivalents to help you make any conversions between the English and metric systems:

· ENGLISH TO METRIC ·

If you know:	*divide by:*	*to find:*
inches	0.394	centimeters (cm)
feet	3.28	meters (m)
miles	0.62	kilometers (km)
ounces	0.04	grams (gm)
pounds	2.20	kilograms (kg)
tons	1.10	tonnes (tn)

· METRIC TO ENGLISH ·

If you know:	*divide by:*	*to find:*
centimeters	2.54	inches
meters	0.30	feet
kilometers	1.61	miles
grams	28.35	ounces
kilograms	0.45	pounds
tonnes	0.9	tons

· WITHIN THE METRIC SYSTEM ·

1 Ångstrom	$= 10^{-8}$ cm
1 micron	$= 10^{-4}$ cm
1 m	$= 10^{2}$ cm
1 km	$= 10^{3}$ m
1 astronomical unit (AU)	$= 1.5 \times 10^{8}$ km
1 light year (ly)	$= 9.46 \times 10^{12}$ km $= 6.36 \times 10^{4}$ AU
1 parsec (pc)	$= 3.09 \times 10^{13}$ km $= 2.07 \times 10^{5}$ AU $= 3.26$ l.y.

TEMPERATURE

Most temperatures in this book are given in degrees on the Fahrenheit and Kelvin scales. The Kelvin scale is used commonly to express temperatures beyond the terrestrial range. A single degree on this scale is the same as a Celsius (Centigrade) degree. The zero point on the Kelvin scale, however, is much lower: 0 Kelvin is Absolute Zero. This is the temperature at which all molecular motion stops; there is simply no heat. The present overall temperature of the universe—the lingering heat of the Big Bang—is around 3 K. Water freezes at 273 K, and boils at 373 K. Room temperature is around 294 K, and your body heat is about 310 K. There is no upper limit to this scale; it suffices to measure the body heat of stars (13,000,000 K, at the core of the sun), supernovae (5,000,000,000 K)—even the Big Bang itself (about 1,000,000,000,000 K).

To convert among the Fahrenheit, Celsius, and Kelvin scales, use the following formulae:

If you know:		***to find:***
degrees Fahrenheit	subtract 32, multiply by 5/9	Celsius
degrees Celsius	multiply by 9/5, add 32	Fahrenheit
degrees Celsius	add 273	Kelvin
degrees Kelvin	multiply by 9/5, subtract 460	Fahrenheit
degrees Fahrenheit	add 460, multiply by 5/9	Kelvin

SCIENTIFIC NOTATION

When dealing with very large numbers, it is often convenient to express them as powers of ten, a system called scientific (or exponential) notation. A power of ten is the number of times you must multiply ten by itself to reach the general size of the number. For instance, 100 is 10×10, or 10^2 ("ten squared"); 1,000 is $10 \times 10 \times 10$, or 10^3, and so on. The number of times you multiply ten is also called its exponent.

This system is also useful for numbers that are not exponents of ten. For instance, three hundred can be written as 3×10^2; 326 is 3.26×10^2. For small numbers like 326, scientific notation makes little sense. To express 17 quadrillion, 968 trillion, however, is much easier if you lump all of the zeroes such a figure contains into the exponent; 17,968,000,000,000,000 becomes 17.968×10^{15}.

To express very large numbers, where the precise amount is less important than the general size, we often talk about the number's "order of magnitude," which refers simply to the size of the exponent; 1,234,678, for instance, is on the order of 10^6, or a million.

The same system applies to very small numbers. One-tenth (0.1) can also be written as 10^{-1}; one thousandth is 10^{-3}. In this system, 0.00000674, is 6.74×10^{-6}: you subtract one from the exponent for every digit you move the decimal point to the right. The number one is 10^0.

Calculation in this system is simplified. To add numbers of different exponents, insert the significant digits of the smaller number into the larger:

1.65×10^4 (16,500) plus 3×10^2 (300) equals 1.68×10^4.

Subtraction reverses the process.

To multiply, multiply the significant digits as usual, then *add* the exponents together:

2×10^5 times 3×10^2 ($200{,}000 \times 300$) equals 6×10^7

To divide, divide the significant digits, then *subtract* exponents:

8×10^{12} divided by 2×10^3 (8,000,000,000,000/2,000) equals 4×10^9.

APPENDIX

On Astrophotography 4

As you may have noticed, there is no chapter on astrophotography in this book. There are several reasons for this. First, astrophotography costs more than a beginner should spend. Second, it assumes you already know your way around a telescope and the sky, and are familiar with an unrelated body of knowledge pertaining to photography. Finally, a person engaged in astrophotography is not doing the activity that is the subject of this book: looking at the sky. Having said as much, it's necessary to state the other side of the question. Astrophotography is fun. It's a way of extending the reach of your telescope into space, by increasing the total amount of light you can gather into an image. The photographs produced by thousands of amateurs can be as beautiful as those taken at the great observatories. They can also, through programs such as PROBLICOM, add to the scientific value of amateur astronomy. Astrophotography can add a whole new dimension to your enjoyment of astronomy—not to mention an entire galaxy of new gadgets to covet, build, buy, and use.

For the beginner, however, those gadgets, along with the other, special techniques and equipment, can put the entire process out of reach. Even if you can afford to start out with a complete astrophotographic observatory, my advice is to wait until you have a clearer idea of what your interests and talents in astronomy are. Advice like that is always easier to give than to take. If you find that you really want to get started taking pictures of the stars right away, there is one project that any amateur can take on with minimal investment of equipment and time, one that will give you a good taste of the challenges waiting in this field. You will also quickly build up a valuable collection of photographs.

This project involves nothing more complicated than taking still photographs of the constellations. If this sounds dull, wait until your first photos come back from the lab: you'll be thrilled with the results. And, although it's not too complicated a task to prevent you from getting good results the first night out, neither is it exactly a piece of cake.

The equipment you will need for this project is any 35 millimeter camera with a "B" or "T" shutter setting, a tripod, and a short cable-release (ask at any camera store for the latter—it should cost between five and ten dollars at most). For the tripod, I have often found it more convenient (and less expensive) to substitute a clamp mount. You can use any film, black-and-white or color, slide or print, so long as it is reasonably fast. Look for an ISO rating of at least 400. The faster the film, the better the results.

Get as far away from city lights as you can (I drive out of town to a hill I know, and clamp the camera to the car's window). Shoot the entire roll at different apertures and exposures. With your aperture wide open, take exposures of five seconds, ten, fifteen, etc., up to a minute. Then two minutes, five, ten, twenty, up to an hour or more. Then repeat the process at an aperture of f/5 or so, then again at f/16. Don't try to cover too many regions of the sky. Take some of your shots toward the north celestial pole, and the trailing images of the stars as the earth rotates during the exposure will form neat circles. Exposures of the equatorial regions will form the longest, straightest trails. *Take notes,* recording the aperture and duration of each exposure. It is almost impossible to remember to do this, because you're going to be much more involved with wiping dew off the lens, watching out for airplanes flying into the field of view, shivering, etc., but force yourself to do it between each exposure, or you will find you've wasted your time.

When you've exhausted your film or yourself, get the film to a reliable lab. Include a note warning the developer that the pictures will have very dark backgrounds. If you don't do this, one of two things is likely to happen. They may mistake your hours of effort for the work of a complete fool who forgot to take his lens-cap off, and, instead of printing your negatives, simply insert a helpful little pamphlet on how to take snapshots. Worse, they may cut each of your negatives in half, because they couldn't tell where the borders of the images were.

If all goes well, when you get your prints back, compare the results with your notes. Some shots will be completely black. These will be the shorter exposures at the slowest f/ratios. Others, however, will be completely white: long exposures at fast ratios. Long exposures that allow too much light to reach the film suffer from a variety of odd effects, but the one you will notice the most is the fogging of the film from the excess of natural and artificial skyglow. Consult your notes to find the combination of f/ratio and exposure that shows detail yet avoids fogging.

With your notes in hand, you should be able to deduce the proper f/ratio for a variety of exposures. Pay close attention to the shorter exposures. On those prints, you will notice that it takes between 30 seconds (near the equator) and a minute (near the pole) for the stars to show any trails at all. Keeping the exposures under a minute, it is still possible to record stars fainter than sixth magnitude. With color film, the brighter emission nebulae appear a brilliant red. On the next dark night, head out with your camera and your notes, and take fifteen-second to 120-second exposures of individual constellations. Take notes again. Try to be consistent in the orientation of your photographs: if you keep one side of frame toward the north celestial pole on all your shots, they will be more useful to you in the long run (and much easier to recognize when you first try to sort them out).

As you gain experience, you will find that it is possible to add a short telephoto lens to zoom in on regions like Orion's sword or the starclouds of the summer Milky Way. If you have an equatorially-mounted telescope, you might want to strap a telephoto lens "piggy-back" to the barrel of the telescope, and manually steer the camera through an exposure of several minutes. Keep a bright star centered in a high-power eyepiece while you turn the telescope in right ascension, and you may be able to keep the image on film from smearing. At this point, however, you're one very small step away from full-scale astrophotography, and beyond the purlieus of this book. The Bibliography gives some suggestions on where you can look for more advice. Good luck.

APPENDIX

5

The Messier Catalogue

by Alan Dyer

The columns contain: Messier's number (M); the constellation; the object's New General Catalogue (NGC) number; the type of object (OC = open cluster, GC = globular cluster, PN = planetary nebula, EN = emission nebula, RN = reflection nebula, G = galaxy (with the type of galaxy also listed); the 1980 co-ordinates; the visual magnitude (unless marked with a "p" which indicates a photographic magnitude). The "Remarks" column contains comments on the object's appearance and observability. Most data are from the Skalnate Pleso *Atlas of the Heavens* catalog; occasionally from other sources.

All these objects can be seen in a small telescope (60 mm refractor, for instance), with M74 and M83 generally considered to be the most difficult. The most southerly M-objects are M6 and M7 in Scorpius, with M54, M55, M69, and M70 in Sagittarius almost as far south.

M	*Con*	*NGC*	*Type*	R.A. (1980) h	m	Dec. °	′	*m*	*Remarks*
The Winter Sky									
1	Tau	1952	PN	5	33.3	+22	01	8.4	Crab Neb.; supernova remnant
45	Tau	—	OC	3	46.3	+24	03	1.4	Pleiades; RFT object
36	Aur	1960	OC	5	35.0	+34	05	6.3	best at low magnification
37	Aur	2099	OC	5	51.5	+32	33	6.2	finest of 3 Aur. clusters
38	Aur	1912	OC	5	27.3	+35	48	7.4	large, scattered group
42	Ori	1976	EN	5	34.4	−05	24	—	Orion Nebula
43	Ori	1982	EN	5	34.6	−05	18	—	detached part of Orion Neb.
78	Ori	2068	RN	5	45.8	+00	02	—	featureless reflection neb.
79	Lep	1904	GC	5	23.3	−24	32	8.4	20 cm scope needed to resolve
35	Gem	2168	OC	6	07.6	+24	21	5.3	superb open cluster
41	CMa	2287	OC	6	46.2	−20	43	5.0	4°S. of Sirius; use low mag.
50	Mon	2323	OC	7	02.0	−08	19	6.9	between Sirius and Procyon
46	Pup	2437	OC	7	40.9	−14	46	6.0	rich cl.; contains PN NGC 2438
47	Pup	2422	OC	7	35.6	−14	27	4.5	coarse cl.; 1.5°W. of M46
93	Pup	2447	OC	7	43.6	−23	49	6.0	smaller, brighter than M46
48	Hya	2548	OC	8	12.5	−05	43	5.3	former "lost" Messier object
The Spring Sky									
44	Cnc	2632	OC	8	38.8	+20	04	3.7	Beehive Cl.; RFT object
67	Cnc	2682	OC	8	50.0	+11	54	6.1	"ancient" star cluster
40	UMa	—	—	12	34.4	+58	20	9.0	two stars; sep. 50″
81	UMa	3031	G-Sb	9	54.2	+69	09	7.9	very bright spiral
82	UMa	3034	G-Pec	9	54.4	+69	47	8.8	the "exploding" galaxy
97	UMa	3587	PN	11	13.7	+55	08	12.0	Owl Nebula
101	UMa	5457	G-Sc	14	02.5	+54	27	9.6	large, faint, face-on spiral
108	UMa	3556	G-Sc	11	10.5	+55	47	10.7	nearly edge-on; near M97
109	UMa	3992	G-Sb	11	56.6	+53	29	10.8	barred spiral; near γ UMa
65	Leo	3623	G-Sb	11	17.8	+13	13	9.3	bright elongated spiral
66	Leo	3627	G-Sb	11	19.1	+13	07	8.4	M65 in same field
95	Leo	3351	G-SBb	10	42.8	+11	49	10.4	bright barred spiral
96	Leo	3368	G-Sbp	10	45.6	+11	56	9.1	M95 in same field

105	Leo	3379	G-E1	10	46.8	+12	42	9.2	very near M95 and M96
53	Com	5024	GC	13	12.0	+18	17	7.6	15 cm scope needed to resolve
64	Com	4826	G-Sb	12	55.7	+21	48	8.8	Black Eye Galaxy
85	Com	4382	G-SO	12	24.3	+18	18	9.3	bright elliptical shape
88	Com	4501	G-Sb	12	30.9	+14	32	10.2	bright multiple-arm spiral
91	Com	4548	G-SBb	12	34.4	+14	36	10.8	not the same as M58
98	Com	4192	G-Sb	12	12.7	+15	01	10.7	nearly edge-on spiral
99	Com	4254	G-Sc	12	17.8	+14	32	10.1	nearly face-on spiral
100	Com	4321	G-Sc	12	21.9	+15	56	10.6	face-on spiral; star-like nuc.
49	Vir	4472	G-E4	12	28.8	+08	07	8.6	very bright elliptical
58	Vir	4579	G-SB	12	36.7	+11	56	9.2	bright barred spiral
59	Vir	4621	G-E3	12	41.0	+11	47	9.6	bright elliptical near M58
60	Vir	4649	G-E1	12	42.6	+11	41	8.9	bright elliptical near M59
61	Vir	4303	G-Sc	12	20.8	+04	36	10.1	face-on barred spiral
84	Vir	4374	G-E1	12	24.1	+13	00	9.3	bright elliptical
86	Vir	4406	G-E3	12	25.1	+13	03	9.7	M84 in same field
87	Vir	4486	G-E1	12	29.7	+12	30	9.2	nearly spherical galaxy
89	Vir	4552	G-E0	12	34.6	+12	40	9.5	resembles M87; smaller
90	Vir	4569	G-Sb	12	35.8	+13	16	10.0	bright spiral; near M89
104	Vir	4594	G-Sb	12	38.8	−11	31	8.7	Sombrero Galaxy
3	CVn	5272	GC	13	41.3	+28	29	6.4	contains many variables
51	CVn	5194	G-Sc	13	29.0	+47	18	8.1	Whirlpool Galaxy
63	CVn	5055	G-Sb	13	14.8	+42	08	9.5	Sunflower Galaxy
94	CVn	4736	G-Sbp	12	50.1	+41	14	7.9	very bright and comet-like
106	CVn	4258	G-Sbp	12	18.0	+47	25	8.6	large, bright spiral
68	Hya	4590	GC	12	38.3	−26	38	8.2	15 cm scope needed to resolve
83	Hya	5236	G-Sc	13	35.9	−29	46	10.1	very faint and diffuse
102	Dra	5866	G-E6p	15	05.9	+55	50	10.8	small, edge-on galaxy
5	Ser	5904	GC	15	17.5	+02	11	6.2	one of the finest globulars

The Summer Sky

13	Her	6205	GC	16	41.0	+36	30	5.7	spectacular globular cl.
92	Her	6341	GC	17	16.5	+43	10	6.1	9°NE. of M13; bright
9	Oph	6333	GC	17	18.1	−18	30	7.3	smallest of Oph. globulars
10	Oph	6254	GC	16	56.0	−04	05	6.7	rich cl.; M12 3.4° away
12	Oph	6218	GC	16	46.1	−01	55	6.6	loose globular
14	Oph	6402	GC	17	36.5	−03	14	7.7	20 cm scope needed to resolve
19	Oph	6273	GC	17	01.3	−26	14	6.6	oblate globular
62	Oph	6266	GC	16	59.9	−30	05	6.6	unsymmetrical; in rich field
107	Oph	6171	GC	16	31.3	−13	02	9.2	small, faint globular
4	Sco	6121	GC	16	22.4	−26	27	6.4	bright globular near Antares
6	Sco	6405	OC	17	38.9	−32	11	5.3	best at low magnification
7	Sco	6475	OC	17	52.6	−34	48	3.2	excellent in binoculars
80	Sco	6093	GC	16	15.8	−22	56	7.7	very compressed globular
16	Ser	6611	EN	18	17.8	−13	48	—	Star-Queen Neb. w/ open cl.
8	Sgr	6523	EN	18	02.4	−24	23	—	Lagoon Neb. w/cl. NGC 6530
17	Sgr	6618	EN	18	19.7	−16	12	—	Swan or Omega Nebula
18	Sgr	6613	OC	18	18.8	−17	09	7.5	sparse cluster; 1°S. of M17
20	Sgr	6514	EN	18	01.2	−23	02	—	Trifid Nebula
21	Sgr	6531	OC	18	03.4	−22	30	6.5	0.7°NE. of M20
22	Sgr	6656	GC	18	35.2	−23	55	5.9	low altitude dims beauty
23	Sgr	6494	OC	17	55.7	−19	00	6.9	bright, loose cluster
24	Sgr	—	—	18	17	−18	27	4.6	Milky Way patch; binoc. obj.
25	Sgr	I4725	OC	18	30.5	−19	16	6.5	bright but sparse cluster
28	Sgr	6626	GC	18	23.2	−24	52	7.3	compact globular near M22
54	Sgr	6715	GC	18	53.8	−30	30	8.7p	not easily resolved
55	Sgr	6809	GC	19	38.7	−31	00	7.1p	bright, loose globular
69	Sgr	6637	GC	18	30.1	−32	23	8.9	small, poor globular
70	Sgr	6681	GC	18	42.0	−32	18	9.6	small globular; 2°E. of M69
75	Sgr	6864	GC	20	04.9	−21	59	8.0	small, remote globular
11	Sct	6705	OC	18	50.0	−06	18	6.3	superb open cluster
26	Sct	6694	OC	18	44.1	−09	25	9.3	bright, coarse cluster

56	Lyr	6779	GC	19	15.8	+30	08	8.2	within rich field
57	Lyr	6720	PN	18	52.9	+33	01	9.3	Ring Negula
71	Sge	6838	GC	19	52.8	+18	44	9.0	loose globular cl.
27	Vul	6853	PN	19	58.8	+22	40	7.6	Dumbbell Nebula
29	Cyg	6913	OC	20	23.3	+38	27	7.1	small, poor open cl.
39	Cyg	7092	OC	21	31.5	+48	21	5.2	very sparse cluster

The Autumn Sky

2	Aqr	7089	GC	21	32.4	−00	54	6.3	20 cm scope needed to resolve
72	Aqr	6981	GC	20	52.3	−12	39	9.8	near NGC 7009 (Saturn Neb.)
73	Aqr	6994	OC	20	57.8	−12	44	11.0	group of 4 stars only
15	Peg	7078	GC	21	29.1	+12	05	6.0	rich, compact globular
30	Cap	7099	GC	21	39.2	−23	15	8.4	noticeable elliptical shape
52	Cas	7654	OC	23	23.3	+61	29	7.3	young, rich cluster
103	Cas	581	OC	01	31.9	+60	35	7.4	3 NGC clusters nearby
31	And	224	G-Sb	00	41.6	+41	09	4.8	Andromeda Gal.; large
32	And	221	G-E2	00	41.6	+40	45	8.7	companion gal. to M31
110	And	205	G-E6	00	39.1	+41	35	9.4	companion gal. to M31
33	Tri	598	G-Sc	01	32.8	+30	33	6.7	large, diffuse spiral
74	Psc	628	G-Sc	01	35.6	+15	41	10.2	faint, elusive spiral
77	Cet	1068	G-Sbp	02	41.6	+00	04	8.9	Seyfert gal.; star-like nuc.
34	Per	1039	OC	02	40.7	+42	43	5.5	best at very low mag.
76	Per	650	PN	01	40.9	+51	28	12.2	Little Dumbbell Neb.

APPENDIX

6

The Finest N.G.C. Objects + 20

by Alan Dyer

The first four sections of the following list contain 110 of the finest NGC objects that are visible from mid-northern latitudes. The arrangement is similar to that used in the preceding Messier Catalogue. A telescope of at least 15 cm aperture will likely be required to locate all these objects. The last section is for those wishing to begin to extend their deep-sky observing program beyond the basic catalogue of Charles Messier or the brightest objects of the New General Catalogue. It is a selected list of 20 "challenging" objects, and is arranged in order of right ascension.

Abbreviations used are OC = open cluster, GC = globular cluster, PN = planetary nebula, EN = emission nebula, RN = reflection nebula, E/RN = combination emission and reflection nebula, DN = dark nebula, SNR = supernova remnant, G = galaxy (the Hubble classification is also listed with each galaxy). Magnitudes are visual; exceptions are marked with a "p" indicating a photographic magnitude. Sizes of each object are in minutes of arc, with the exception of planetary nebulae which are given in seconds of arc. The number of stars (*) and, where space permits, the Shapley classification is also given for star clusters in the Remarks column.

No.	NGC	Con	Type	R.A. (1950) Dec.				m_v	Size	Remarks
The Autumn Sky				h	m	°	′			
1	7009	Aqr	PN	21	01.4	−11	34	9.1	44″ × 26″	Saturn Nebula; bright oval planetary
2	7293	Aqr	PN	22	27.0	−21	06	6.5	900″ × 720″	Helix Nebula; very large and diffuse
3	7331	Peg	G-Sb	22	34.8	+34	10	9.7	10.0 × 2.3	large, very bright spiral galaxy
4	7789	Cas	OC	23	54.5	+56	26	9.6	30	200*; faint but very rich cluster
5	185	Cas	G-EO	00	36.1	+48	04	11.7	2.2 × 2.2	companion to M31; quite bright
6	281	Cas	EN	00	50.4	+56	19	—	22 × 27	large, faint nebulosity near γCas.
7	457	Cas	OC	01	15.9	+58	04	7.5	10	100*; Type e—intermediate rich
8	663	Cas	OC	01	42.6	+61	01	7.1	11	80*; NGC 654 and 659 nearby
9	7662	And	PN	23	23.5	+42	14	9.2	32″ × 28″	star-like at low mag.; annular, bluish
10	891	And	G-Sb	02	19.3	+42	07	10.9p	11.8 × 1.1	faint, classic edge-on with dust lane
11	253	Scl	G-Scp	00	45.1	−25	34	8.9	24.6 × 4.5	very large and bright but at low alt.
12	772	Ari	G-Sb	01	56.6	+18	46	10.9	5.0 × 3.0	diffuse spiral galaxy
13	936	Cet	G-SBa	02	25.1	−01	22	10.7	3.3 × 2.5	near M77; NGC 941 in same field
14a	869	Per	OC	02	15.5	+56	55	4.4	36	Double Cluster; superb!

14b	884	Per	OC	02	18.9	+56	53	4.7	36	Double Cluster; superb!
15	1023	Per	G-E7p	02	37.2	+38	52	10.5p	4.0 × 1.2	bright, lens-shaped galaxy; near M34
16	1491	Per	EN	03	59.5	+51	10	—	3 × 3	small, fairly bright emission nebula
17	1501	Cam	PN	04	02.6	+60	47	12.0	56″ × 58″	faint, distinctive oval; darker centre
18	1232	Eri	G-Sc	03	07.5	−20	46	10.7	7.0 × 5.5	fairly bright, large face-on spiral
19	1300	Eri	G-SBb	03	17.5	−19	35	11.3	5.7 × 3.5	large barred spiral near NGC 1232
20	1535	Eri	PN	04	12.1	−12	52	10.4	20″ × 17″	blue-grey disk
The Winter Sky										
21	1907	Aur	OC	05	24.7	+35	17	9.9	5	40*; nice contrast with nearby M38
22	1931	Aur	EN	05	28.1	+34	13	—	3 × 3	haze surrounding 4 stars
23	1788	Ori	E/RN	05	04.5	−03	24	—	8 × 5	fairly bright but diffuse E/R neb.
24	1973+	Ori	E/RN	05	32.9	−04	48	—	40 × 25	near M42 and M43; often neglected
25	2022	Ori	PN	05	39.3	+09	03	12.4	28″ × 27″	small, faint but distinct, annular
26	2194	Ori	OC	06	11.0	+12	50	9.2	8	100*; Type e; faint but rich
27	2158	Gem	OC	06	04.3	+24	06	12.5	4	40*; same field as M35; nice contrast

28	2392	Gem	PN	07	26.2	+21	01	8.3	47″ × 43″	Clown-Face Nebula; very bright
29	2244	Mon	OC	06	29.7	+04	54	6.2	40	16*; in centre of Rosette Nebula
30	2261	Mon	E/RN	06	36.4	+08	46	var.	5 × 3	Hubble's Variable Nebula
31	2359	CMa	EN	07	15.4	−13	07	—	8 × 6	fairly bright; NGC's 2360 & 2362 nearby
32	2438	Pup	PN	07	39.6	−14	36	11.8	68″	within M46 open cluster
33	2440	Pup	PN	07	39.9	−18	05	10.3	54″ × 20″	almost starlike; irregular shape at HP
34	2539	Pup	OC	08	08.4	−12	41	8.2	21	150*; Type f—fairly rich
35	2403	Cam	G-Sc	07	32.0	+65	43	8.9	17 × 10	bright, very large; visible in binocs.
36	2655	Cam	G-S	08	49.4	+78	25	10.7	5.0 × 2.4	bright ellipse w/ star-like nucleus

The Spring Sky

37	2683	Lyn	G-Sb	08	49.6	+33	38	9.6	8.0 × 1.3	nearly edge-on spiral; very bright
38	2841	UMa	G-Sb	09	18.6	+51	12	9.3	6.4 × 2.4	classic elongated spiral; very bright
39	2985	UMa	G-Sb	09	46.0	+72	31	10.6	5.5 × 5.0	near M81 and M82
40	3077	UMa	G-E2p	09	59.4	+68	58	10.9	2.3 × 1.9	small elliptical; companion to M81/82
41	3079	UMa	G-Sb	09	58.6	+55	57	11.2	8.0 × 1.0	edge-on spiral, NGC 2950 nearby

42	3184	UMa	G-Sc	10	15.2	+41	40	9.6	5.6 × 5.6	large, diffuse face-on spiral
43	3675	UMa	G-Sb	11	23.5	+43	52	10.6	4.0 × 1.7	elongated spiral; same field as 56 UMa
44	3877	UMa	G-Sb	11	43.5	+47	46	10.9	4.4 × 0.8	edge-on; same field as Chi UMa
45	3941	UMa	G-Sa	11	50.3	+37	16	9.8	1.8 × 1.2	small, bright, elliptical shape
46	4026	UMa	G-E8	11	56.9	+51	12	10.7	3.6 × 0.7	lens-shaped edge-on; near γ UMa
47	4088	UMa	G-Sc	12	03.0	+50	49	10.9	4.5 × 1.4	nearly edge-on; 4085 in same field
48	4111	UMa	G-SO	12	04.5	+43	21	9.7	3.3 × 0.6	bright, lens-shaped, edge-on spiral
49	4157	UMa	G-Sb	12	08.6	+50	46	11.9	6.5 × 0.8	edge-on, a thin sliver, 4026+4088 nearby
50	4605	UMa	G-Scp	12	37.8	+61	53	9.6	5.0 × 1.2	bright, distinct, edge-on spiral
51	3115	Sex	G-E6	10	02.8	−07	28	9.3	4.0 × 1.2	"Spindle Galaxy"; bright, elongated
52	3242	Hya	PN	10	22.4	−18	23	9.1	40″ × 35″	"Ghost of Jupiter" planetary
53	3344	LMi	G-Sc	10	40.7	+25	11	10.4	7.6 × 6.2	diffuse, face-on spiral
54	3432	LMi	G-Sc	10	49.7	+36	54	11.4	5.8 × 0.8	nearly edge-on; faint flat streak
55	2903	Leo	G-Sb	09	29.3	+21	44	9.1	11.0 × 4.6	very bright, large elongated spiral
56	3384	Leo	G-E7	10	45.7	+12	54	10.2	4.4 × 1.4	same field as M105 and NGC 3389

57	3521	Leo	G-Sc	11	03.2	+00	14	9.5	7.0 × 4.0	very bright, large spiral
58	3607	Leo	G-E1	11	14.3	+18	20	9.6	1.7 × 1.5	NGC 3605 and 3608 in same field
59	3628	Leo	G-Sb	11	17.7	+13	53	10.9	12.0 × 1.5	large, edge-on; same field as M65/M66
60	4214	CVn	G-1rr	12	13.1	+36	36	10.3	6.6 × 5.8	large irregular galaxy
61	4244	CVn	G-S	12	15.0	+38	05	11.9	14.5 × 1.0	large, distinct, edge-on spiral
62	4449	CVn	G-1rr	12	25.8	+44	22	9.2	4.1 × 3.4	bright rectangular shape
63	4490	CVn	G-Sc	12	28.3	+41	55	9.7	5.6 × 2.1	bright spiral; 4485 in same field
64	4631	CVn	G-Sc	12	39.8	+32	49	9.3	12.6 × 1.4	very large, bright, edge-on; no dust lane
65	4656	CVn	G-Sc	12	41.6	+32	26	11.2	19.5 × 2.0	same field as 4631; fainter, smaller
66	5005	CVn	G-Sb	13	08.5	+37	19	9.8	4.4 × 1.7	bright elongated spiral; near αCVn
67	5033	CVn	G-Sb	13	11.2	+36	51	10.3	9.9 × 4.8	large, bright spiral near NGC 5005
68	4274	Com	G-Sb	12	17.4	+29	53	10.8	6.7 × 1.3	NGC 4278 in same field
69	4494	Com	G-E1	12	28.9	+26	03	9.6	1.3 × 1.2	small, bright elliptical
70	4414	Com	G-Sc	12	24.0	+31	30	9.7	3.2 × 1.5	bright spiral; star-like nucleus
71	4559	Com	G-Sc	12	33.5	+28	14	10.6	11.0 × 4.5	large spiral; coarse structure
72	4565	Com	G-Sb	12	33.9	+26	16	10.2	14.4 × 1.2	superb edge-on spiral with dust lane

73	4725	Com	G-Sb	12	48.1	+25	46	8.9	10.0 × 5.5	very bright, large spiral
74	4361	Crv	PN	12	21.9	−18	29	11.4	18″	12^{m}8 central star
75	4216	Vir	G-Sb	12	13.4	+13	25	10.4	7.4 × 0.9	nearly edge-on; two others in field
76	4388	Vir	G-Sb	12	23.3	+12	56	11.7p	5.0 × 0.9	edge-on; near M84 and M86
77	4438	Vir	G-S	12	25.3	+13	17	10.8	8.0 × 3.0	paired with NGC 4435
78	4473	Vir	G-E4	12	27.3	+13	42	10.1	1.6 × 0.9	NGC 4477 in same field
79	4517	Vir	G-Sc	12	29.0	+00	21	12.0	8.9 × 0.8	faint edge-on spiral
80	4526	Vir	G-E7	12	31.6	+07	58	10.9	3.3 × 1.0	between two 7^{m}0 stars
81	4535	Vir	G-Sc	12	31.8	+08	28	10.4p	6.0 × 4.0	near M49
82	4697	Vir	G-E4	12	46.0	−05	32	9.6	2.2 × 1.4	small, bright elliptical
83	4699	Vir	G-Sa	12	46.5	−08	24	9.3	3.0 × 2.0	small, bright elliptical shape
84	4762	Vir	G-Sa	12	50.4	+11	31	11.0	3.7 × 0.4	flattest galaxy; 4754 in same field
85	5746	Vir	G-Sb	14	42.3	+02	10	10.1	6.3 × 0.8	fine, edge-on spiral near 109 Virginis
86	5907	Dra	G-Sb	15	14.6	+56	31	11.3	11.1 × 0.7	fine, edge-on spiral with dust lane
87	6503	Dra	G-Sb	16	49.9	+70	10	9.6	4.5 × 1.0	bright spiral
88	6543	Dra	PN	17	58.8	+66	38	8.7	22″	luminous blue-green disk
The Summer Sky										
89	6207	Her	G-Sc	16	41.3	+36	56	11.3	2.0 × 1.1	same field as M13 cluster
90	6210	Her	PN	16	42.5	+23	53	9.2	20″ × 13″	very star-like blue planetary

91	6369	Oph	PN	17	26.3	−23	44	9.9	28″	greenish, annular, and circular
92	6572	Oph	PN	18	09.7	+06	50	8.9	16″ × 13″	tiny oval; bright blue
93	6633	Oph	OC	18	25.1	+06	32	4.9	20	wide-field cluster; IC4756 nearby
94	6712	Sct	GC	18	50.3	−08	47	8.9	2.1	small globular near M26
95	6819	Cyg	OC	19	39.6	+40	06	10.1	6	150*; faint but rich cluster
96	6826	Cyg	PN	19	43.4	+50	24	9.4	27″ × 24″	Blinking Planetary Nebula
97	6960	Cyg	SNR	20	43.6	+30	32	—	70 × 6	Veil Nebula (west component)
98	6992–5	Cyg	SNR	20	54.3	+31	30	—	78 × 8	Veil Nebula (east component)
99	7000	Cyg	EN	20	57.0	+44	08	—	120 × 100	North America Neb.; binoc. obj.
100	7027	Cyg	EN	21	05.1	+42	02	10.4	18″ × 11″	very star-like H II region
101	6445	Sgr	PN	17	47.8	−20	00	11.8	38″ × 29″	small, bright and annular; near M23
102	6818	Sgr	PN	19	41.1	−14	17	9. 9	22″ × 15″	"Little Gem"; annular; 6822 nearby
103	6802	Vul	OC	19	28.4	+20	10	11.0	3.5	60*; small, faint but rich
104	6940	Vul	OC	20	32.5	+28	08	8.2	20	100*; Type e; rich cluster
105	6939	Cep	OC	20	30.4	+60	28	10.0	5	80*; very rich; 6946 in same field
106	6946	Cep	G-Sc	20	33.9	+59	58	9.7p	9.0 × 7.5	faint, diffuse, face-on spiral

107	7129	Cep	RN	21	42.0	+65	52	—	7 × 7	small faint RN; several stars inv.
108	40	Cep	PN	00	10.2	+72	15	10.5	60″ × 38″	small circular glow; 11^{m}.5 central star
109	7209	Lac	OC	22	03.2	+46	15	7.6	20	50*; Type d; within Milky Way
110	7243	Lac	OC	22	13.2	+49	38	7.4	20	40*; Type d; within Milky Way

Challenge Objects

1	246	Cet	PN	00	44.6	−12	09	8.5	240″ × 210″	large and diffuse; deceptively difficult
2	1275	Per	G	03	16.4	+41	20	12.7	0.7 × 0.6	small and faint; exploding gal.; Perseus A
3	1432/35	Tau	RN	03	43.3	+23	42	—	30 × 30	Pleiades nebl'y; brightest around Merope
4	1499	Per	EN	04	00.1	+36	17	—	145 × 40	California Neb.; very large and faint
5	IC434/35/ B33/2023	Ori	E/R/DN	05	38.6	−02	26	—	60/3/10	complex of nebl'y S. of zeta Ori., B33 is famous dark Horsehead Neb.; difficult
6	IC431/32/ NGC2024	Ori	E/RN	05	39.4	−01	52	—	4/6/30	complex of nebl'y N. of zeta Ori., NGC2024 is easy but masked by glow from zeta.
7	IC 443	Gem	SNR	06	13.9	+22	48	—	27 × 5	v. faint supernova remnant NE. of η Gem.

8	J 900	Gem	PN	06	23.0	+17	49	12.2	12″ × 10″	bright but star-like; oval at high mag.
9	2237/46	Mon	EN	06	29.6	+04	40	—	60	Rosette Neb.; very large; incl. NGC2244
10	2419	Lyn	GC	07	34.8	+39	00	11.5	1.7	most distant known Milky Way GC (2 × 10^5 l.y.)
11	5897	Lib	GC	15	14.5	−20	50	10.9	7.3	large, but faint and loose globular cl.
12	B 72	Oph	DN	17	21.0	−23	35	—	30	Barnard's dark S-Nebula; RFT needed
13	6781	Aql	PN	19	16.0	+06	26	11.8	106″	pale version of M97; large, fairly bright
14	6791	Lyr	OC	19	19.0	+37	40	11	13	large, faint but very rich cl.; 100+*
15	M1-92	Cyg	RN	19	34.3	+29	27	11	0.2 × 0.1	Footprint Neb.; bright but star-like; double
16	6822	Sgr	G-Irr	19	42.1	−14	53	11.0	16.2 × 11.2	Barnard's Gal.; member Local Grp.; faint
17	6888	Cyg	SNR?	20	10.7	+38	16	—	18 × 12	Crescent Neb.; small faint arc near γ Cyg.
18	IC 5146	Cyg	RN	21	51.3	+47	02	—	12 × 12	Cocoon Neb.; faint; at end of long dark neb.
19	7317–20	Peg	G's	22	34	+33	42	14–15	—	Stephan's Quintet; ½°SSW. of NGC 7331
20	7635	Cas	EN	23	18.5	+60	54	—	4 × 3	Bubble Neb.; v. faint; ½°SW. of M52

Seasonal Star Charts

APPENDIX 7

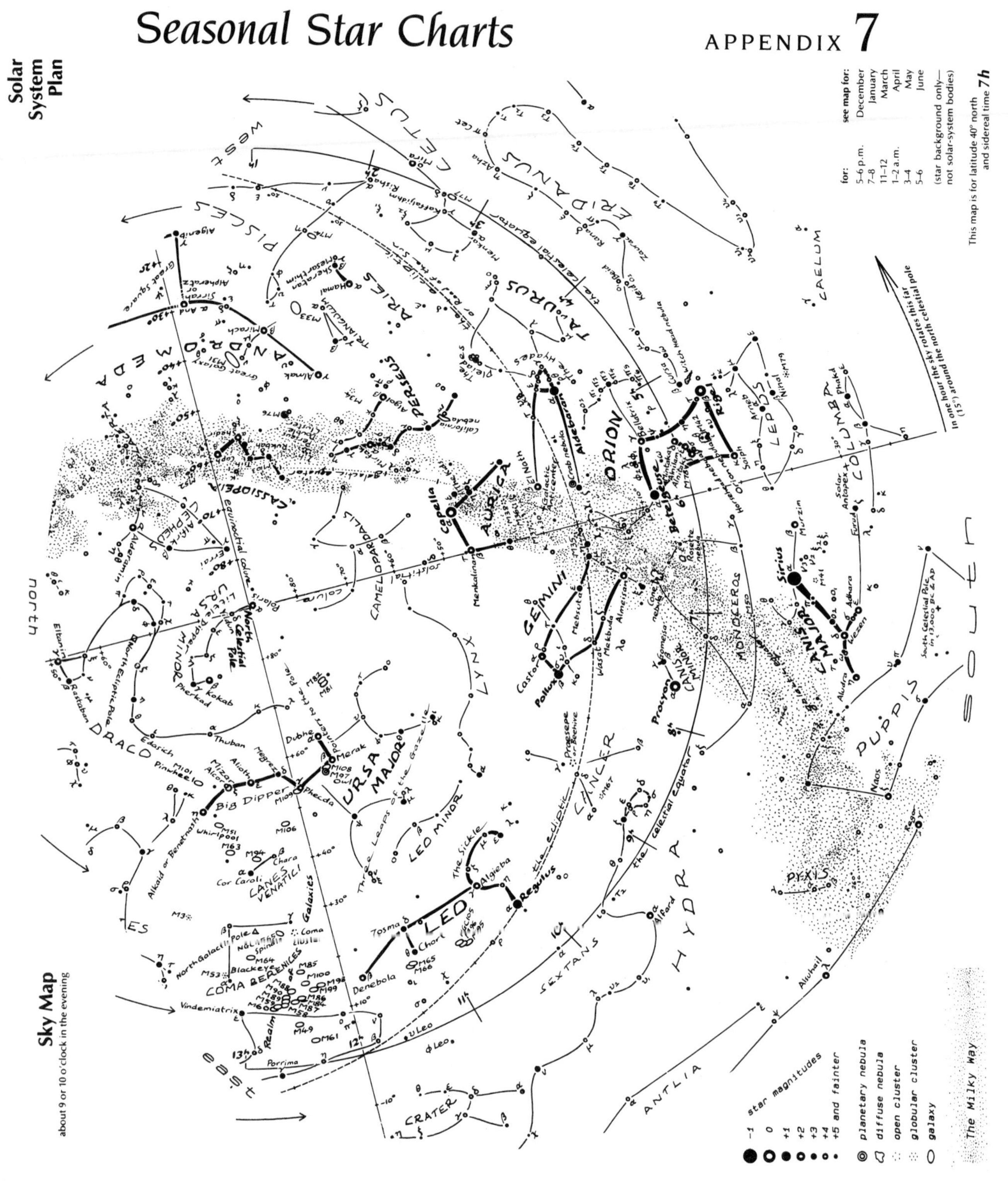

February

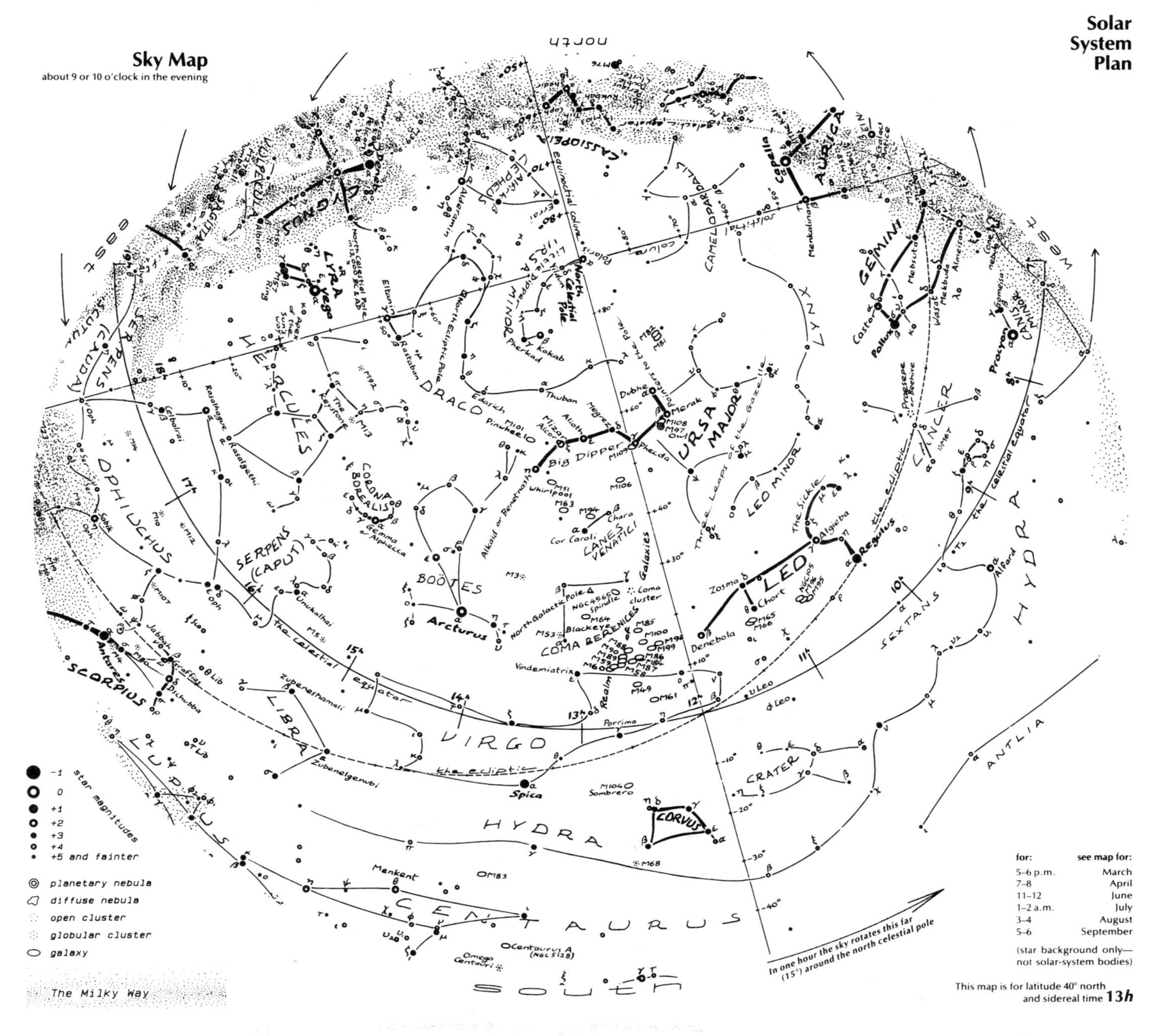
Sky Map
about 9 or 10 o'clock in the evening
Solar System Plan
north
west
east
south
HYDRA
ANTLIA
SEXTANS
CRATER
CORVUS
CENTAURUS
VIRGO
LIBRA
LUPUS
SCORPIUS
OPHIUCHUS
SERPENS (CAUDA)
SERPENS (CAPUT)
SCUTUM
HERCULES
LYRA
CYGNUS
VULPECULA
SAGITTA
DRACO
BOÖTES
CORONA BOREALIS
COMA BERENICES
CANES VENATICI
URSA MAJOR
URSA MINOR
LEO MINOR
LEO
CANCER
CANIS MINOR
GEMINI
AURIGA
LYNX
CAMELOPARDALIS
CASSIOPEIA
CEPHEUS
Procyon
Pollux
Castor
Capella
Menkalinan
Regulus
Algieba
Zosma
Chort
Denebola
Alphard
Arcturus
Spica
Porrima
Vindemiatrix
Antares
Vega
Alderamin
Albireo
Mizar
Alcor
Alioth
Megrez
Dubhe
Merak
Phecda
Thuban
Kochab
Pherkad
Edasich
Eltanin
Rastaban
Rasalhague
Rasalgethi
Cebalrai
Sabik
Unukalhai
Menkent
Zubenelgenubi
Zubeneschamali
Cor Caroli
Chara
Gemma or Alphecca
Alkaid or Benetnasch
Polaris
North Celestial Pole
North Ecliptic Pole
North Galactic Pole
equinoctial colure
solstitial colure
the celestial equator
the ecliptic
Pointers to the Pole
Big Dipper
Little Dipper
The Sickle
The Keystone
Three Leaps of the Gazelle
Apex of the Sun's way
Praesepe or Beehive
Coma cluster
Galaxies
Whirlpool
Pinwheel
Blackeye
Owl
Sombrero
M13
M57 Ring
Centaurus A (NGC 5128)
Omega Centauri
In one hour the sky rotates this far (15°) around the north celestial pole
star magnitudes
−1
0
+1
+2
+3
+4
+5 and fainter
planetary nebula
diffuse nebula
open cluster
globular cluster
galaxy
The Milky Way
for: see map for:
5–6 p.m. March
7–8 April
11–12 June
1–2 a.m. July
3–4 August
5–6 September
(star background only—not solar-system bodies)
This map is for latitude 40° north and sidereal time 13h

May

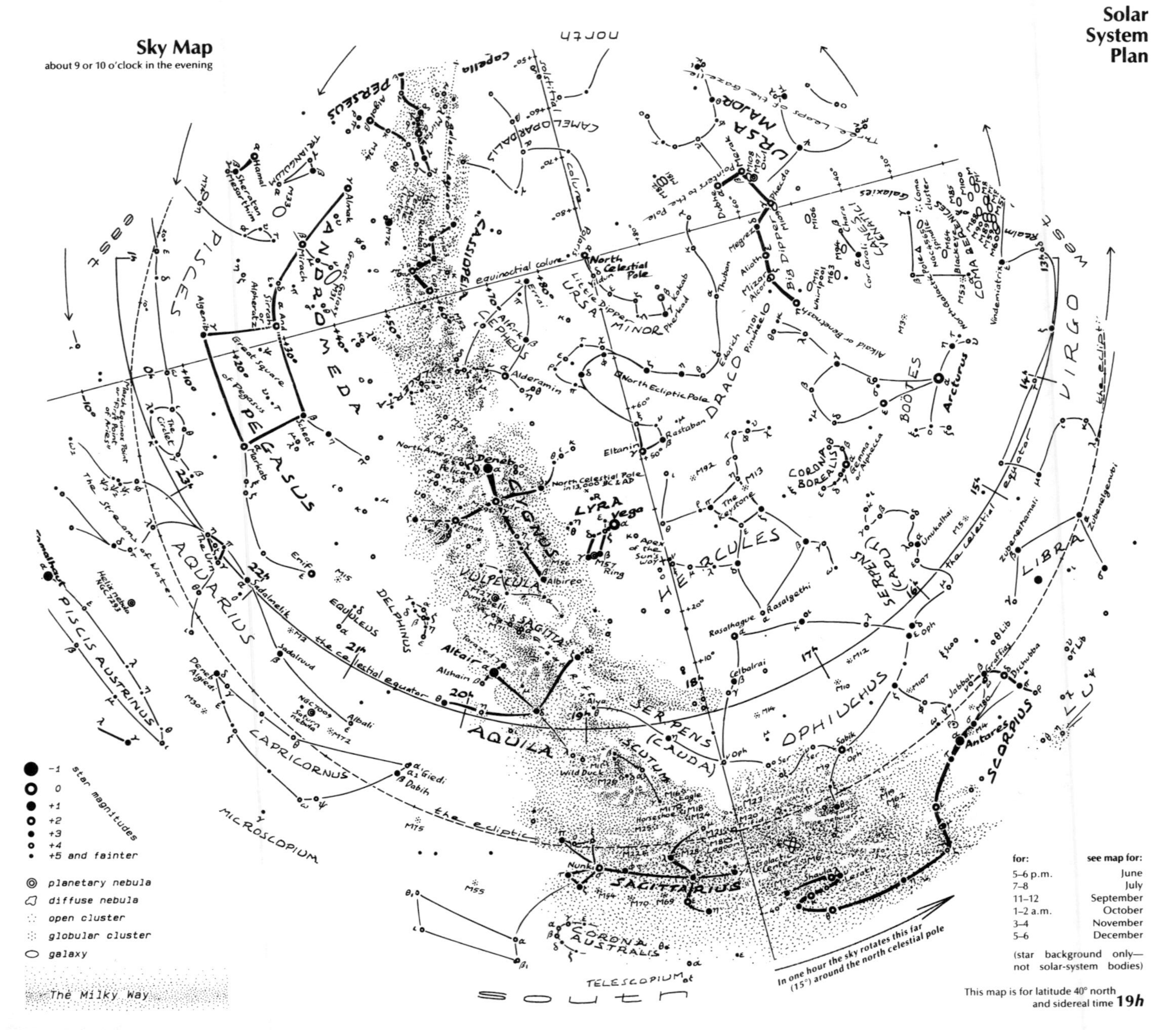
Sky Map
about 9 or 10 o'clock in the evening
north
south
east
west
PERSEUS
CAMELOPARDALIS
URSA MAJOR
Three Leaps of the Gazelle
CASSIOPEIA
TRIANGULUM
ANDROMEDA
PISCES
PEGASUS
Great Square of Pegasus
CEPHEUS
URSA MINOR
North Celestial Pole
Polaris
Little Dipper
Big Dipper
Pointers to the Pole
DRACO
North Ecliptic Pole
CANES VENATICI
COMA BERENICES
BOÖTES
Arcturus
VIRGO
CORONA BOREALIS
HERCULES
The Keystone
LYRA
Vega
CYGNUS
Deneb
North America
Pelican
VULPECULA
Dumbbell
SAGITTA
DELPHINUS
EQUULEUS
AQUILA
Altair
Altair
Apex of the Sun's way
SERPENS (CAPUT)
SERPENS (CAUDA)
SCUTUM
OPHIUCHUS
LIBRA
SCORPIUS
Antares
SAGITTARIUS
CORONA AUSTRALIS
TELESCOPIUM
MICROSCOPIUM
CAPRICORNUS
AQUARIUS
PISCIS AUSTRINUS
Fomalhaut
Helix nebula NGC7293
The Circlet
March Equinox Point or "First Point of Aries"
The Streams of Water
The Urn
the celestial equator
the ecliptic
equinoctial colure
solstitial colure
galactic equator
Great Galaxy M31
Capella
Algol
Almak
Mirach
Hamal
Sheratan
Mesarthim
Algenib
Markab
Scheat
Enif
Sadalmelik
Sadalsuud
Albireo
Alderamin
Errai
Thuban
Kochab
Pherkad
Dubhe
Merak
Phecda
Megrez
Alioth
Mizar
Alcor
Eltanin
Rastaban
Rasalhague
Rasalgethi
Unukalhai
Zubenelgenubi
Zubeneshamali
Vindemiatrix
Cor Caroli
Nunki
star magnitudes
−1
0
+1
+2
+3
+4
+5 and fainter
planetary nebula
diffuse nebula
open cluster
globular cluster
galaxy
The Milky Way
for: 5–6 p.m. see map for: June
7–8 July
11–12 September
1–2 a.m. October
3–4 November
5–6 December
(star background only—not solar-system bodies)
In one hour the sky rotates this far (15°) around the north celestial pole
This map is for latitude 40° north and sidereal time 19h

August

Bottom of Square is along +15° line.
Top is along +30° line.

Sky Map

about 9 or 10 o'clock in the evening

Solar System Plan

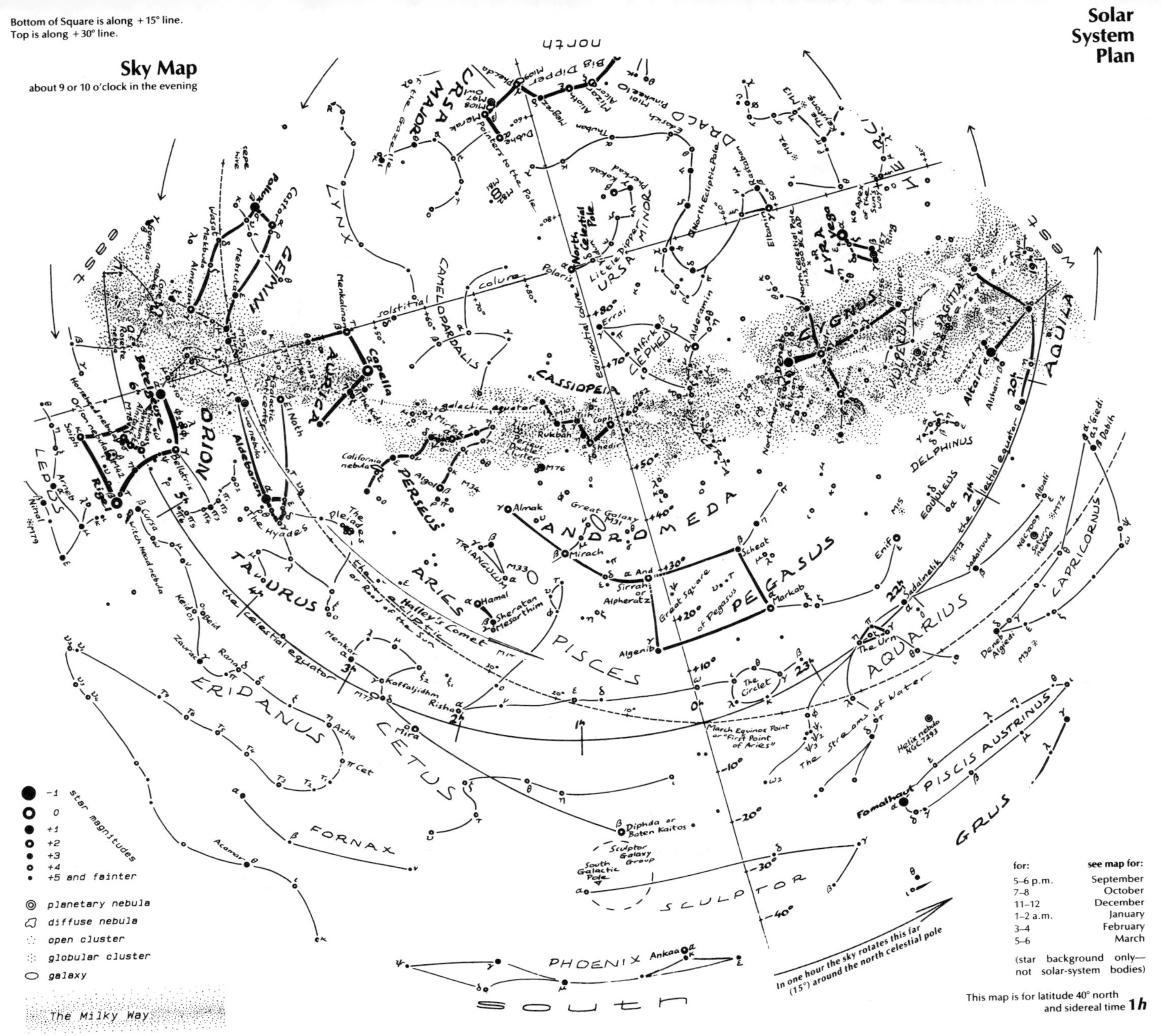

for:	see map for:
5–6 p.m.	September
7–8	October
11–12	December
1–2 a.m.	January
3–4	February
5–6	March

(star background only—not solar-system bodies)

This map is for latitude 40° north and sidereal time **1 *h***

November

APPENDIX

8

The Constellations

Visible from latitude 40° north:

Name		*Season*	*Genitive*	*Abbreviation*
Andromeda	"maiden in chains"	fall	Andromedae	And
Antlia	"the air pump"	spring	Antliae	Ant
Aquarius*	"the water-carrier"	spring	Aquarii	Aqr
Aquila	"the eagle"	summer	Aquilae	Aql
Aries*	"the ram"	fall	Arietis	Ari
Auriga	"the charioteer"	winter	Aurigae	Aur
Boötes	"the herdsman"	summer	Bootis	Boo
Caelum	"the chisel"	winter	Caeli	Cae
Camelopardalis	"the giraffe"	winter	Camelopardalis	Cam
Cancer*	"the crab"	spring	Cancri	Cnc
Canes Venatici	"the hunting dogs"	spring	Canum Venaticorum	CVn
Canis Major	"the big dog"	winter	Canis Majoris	CMa
Canis Minor	"the little dog"	winter	Canis Minoris	CMi
Capricornus*	"the sea-goat"	fall	Capricorni	Cap
Cassiopeia	"the queen in her chair"	fall	Cassiopeiae	Cas
Centaurus	"the centaur"	summer	Centauri	Cen
Cepheus	"the king"	fall	Cephei	Cep
Cetus	"the whale"	fall	Ceti	Cet
Columba	"the dove"	winter	Columbae	Col
Coma Berenices	"Berenice's hair"	spring	Comae Berenices	Com

Corona Australis	"the southern crown"	summer	Coronae Australis	CrA
Corona Borealis	"the northern crown"	summer	Coronae Borealis	CrB
Corvus	"the crow"	spring	Corvi	Crv
Crater	"the cup"	spring	Crateris	Crt
Cygnus	"the swan"	summer	Cygni	Cyg
Delphinus	"the dolphin"	summer	Delphini	Del
Draco	"the dragon"	summer	Draconis	Dra
Equuleus	"the colt"	fall	Equulei	Equ
Eridanus	"the river"	winter	Eridani	Eri
Fornax	"the forge"	fall	Fornacis	For
Gemini*	"the twins"	winter	Geminorum	Gem
Grus	"the crane"	fall	Gruis	Gru
Hercules		summer	Herculis	Her
Hydra	"the sea monster"	spring	Hydrae	Hya
Lacerta	"the lizard"	fall	Lacertae	Lac
Leo*	"the lion"	spring	Leonis	Leo
Leo Minor	"the small lion"	spring	Leonis Minoris	LMi
Lepus	"the hare"	winter	Leporis	Lep
Libra*	"the scales"	summer	Librae	Lib
Lupus	"the wolf"	summer	Lupi	Lup
Lynx	"the lynx"	winter	Lyncis	Lyn
Lyra	"the lyre"	summer	Lyrae	Lyr
Microscopium	"the microscope"	summer	Microscopii	Mic
Monoceros	"the unicorn"	winter	Monocerotis	Mon
Ophiuchus	"the snake handler"	summer	Ophiuchi	Oph
Orion	"the hunter"	winter	Orionis	Ori
Pegasus	"the flying horse"	fall	Pegasi	Peg
Perseus		fall	Persei	Per
Pisces*	"the fish"	fall	Piscium	Psc
Piscis Austrinus	"the southern fish"	fall	Piscis Austrini	PsA
Puppis	"the poop-deck"	winter	Puppis	Pup
Pyxis	"the compass"	spring	Pyxidis	Pyx
Sagitta	"the arrow"	summer	Sagittae	Sge
Sagittarius*	"the archer"	summer	Sagittarii	Sgr
Scorpius*	"the scorpion"	summer	Scorpii	Sco
Sculptor	"the sculptor's studio"	fall	Sculptoris	Scl
Scutum	"the shield"	summer	Scuti	Sct
Serpens	"the snake"	summer	Serpentis	Ser
Sextans	"the sextant"	spring	Sextantis	Sex
Taurus*	"the bull"	winter	Tauri	Tau
Triangulum	"the triangle"	fall	Trianguli	Tri
Ursa Major	"the great bear"	spring	Ursae Majoris	UMa
Ursa Minor	"the little bear"	summer	Ursae Minoris	UMi
Virgo*	"the virgin"	spring	Virginis	Vir
Vulpecula	"the little fox"	summer	Vulpeculae	Vul

*zodiacal constellations

The following constellations are south circumpolar from mid-northern latitudes:

Apus	Norma
Ara	Octans
Carina	Pavo
Chameleon	Phoenix
Circinus	Pictor
Crux	Reticulum
Horologium	Telescopium
Hydrus	Triangulum Australe
Indus	Tucana
Mensa	Vela
Musca	Volans

APPENDIX

9

Suppliers

The following companies make or distribute equipment for amateur astronomers. Since I have not been able to sample or test everything offered by the companies listed below, this listing is purely for reference; it does not constitute an endorsement of any of the companies listed, or their products.

Ad Libs Astronomics. Distributors of name brand telescopes and accessories. Catalog $1. 2140 Melrose Court, Norman, Okla. 73069.

Astro Cards. Handy little index cards printed with finder charts of Messier, NGC objects, and double stars. "Free catalog" $0.37. P.O. Box 35, Natrona Heights, Pa. 15065.

Bushnell. A division of Bausch & Lomb, a quality optics firm; recently offered a 4″ catadioptric, in addition to their regular line of binoculars. Free brochure. "Criterion 4000," 620 Oakwood Ave., West Hartford, Conn. 06110.

Celestron International. For two decades the country's leading manufacturer of Schmidt-Cassegrains and their accessories. In response to recent challenges to their dominant position they have adopted a very aggressive approach to marketing. As part of that approach, you can now call them, toll-free, at 1-800-421-1526, from 8 A.M. to 4 P.M. Pacific Time, and request more information. Their address is P.O. Box 8578-TH, 2835 Columbia St., Torrance, Calif. 90503.

Coulter Optical Co. Manufacturers of large reflecting optics, especially for Dobsonians. They enjoy an excellent reputation for quality at good prices, and I only wish they would revive their discontinued line of smaller optics. Write Dept. TH, P.O. Box K, Idyllwild, Calif. 92349 for more information on their Dobsonian line.

Daystar Filter Corp. Manufacturers of narrow-band filters for solar observation, and arguably the best LPR filter. Free information. P.O. Box 1290, Pomona, Calif. 91769.

E & W Optical. Fused quartz diagonals, for those who want their secondaries to be better than their primaries. Also Pyrex, for those who don't. Free price list. 2420 East Hennepin Ave., Minneapolis, Minn. 55413.

Edmund Scientific Corp. The country's largest purveyors of scientific gewgaws, all profusely illustrated and poorly described in an astonishing catalog. Their parts for telescope-makers are hideously overpriced, but they do offer an extremely portable 4.25″ RFT of a clever and durable design, and a line of good eyepieces (the RKE brand). Free catalog. 101 E. Gloucester Pike, Barrington, N.J. 08007.

Edwin Hirsch. Dealer in name brands of telescopes and accessories. 168 Lakeview Dr., Tomkins Cove, N.Y. 10986.

The Image Point. Manufacturers of heavy-duty equatorial mounts, and dealers in name brands. Brochure $1. 831 N. Swan, Tucson, Ariz. 85711.

A. Jaegers. An institution. They have been running the same ad for as long as I can remember, and will send you a delightfully old-fashioned (ca. 1953), cluttered catalog of miscellaneous optics. A good source for lenses for refractors and kits for mirror grinding. Catalog, $0.25. 691S Merrick Rd., Lynbrook, N.Y. 11563.

Meade Instruments. A major manufacturer of telescopes and accessories for the amateur. I have used a wide range of their products and have always been pleased with their quality and prices. Gorgeous catalog $3. 1675 Toronto Way, Costa Mesa, Calif. 92626.

New Frontier Enterprises. A variety of gadgets. Send self-addressed, stamped envelope for product list. P.O. Box 176, Gilberts, Ill. 60136.

North Star Telescope Co. A good source of equipment, such as cells, spiders, cardboard tubing, for the amateur telescope maker, at bargain prices. I have used some of their equipment, and liked it. Catalog $1. 3542 Elm St., Toledo, Ohio 43608.

Kenneth F. Novak & Co. Manufacturers of high-quality, innovative parts for amateur telescope makers, including perhaps the best spider or diagonal holders on the market. They now carry parts that fit the large Dobsonian optics sold by Coulter. Free catalog. Box 69V, Ladysmith, Wis. 54848.

Omega Systems. *Teflon!* Package of precut pieces for one Dobsonian, $7.95, including shipping; 8″ × 10″ sheet, $17.95. If they had only been in business four years ago, my life would have been easier. . . . N.W. Plaza Station, P.O. Box 1194, St. Ann, Mo. 63074.

Optica b/c Co. A cornucopia of stuff for the amateur telescope buyer, user, and maker, listed in four catalogs: (1) telescopes and accessories, (2) astrophotography, (3) publications, and (4) *ATM,* goods for the telescope maker. $1 each, $3 the set. Sales/Service Division, 4100 MacArthur Blvd., Oakland, Calif. 94619.

Optron Systems. Dealers in name brands of telescopes and accessories. Free price list. Eugene L. Cisneros, 704–706 Charcot Ave., San Jose, Calif. 95131.

Orion Telescope Center. Major dealer in name brands and accessories, including electric socks. Catalog $1. P.O. Box 1158-T, Santa Cruz, Calif. 95061.

Parkes/Scope City. Discounts on Celestron, Meade, and Edmund telescopes, and a range of accessories for amateurs. Catalog package, which duplicates those of above manufacturers, also contains "over $200 of valuable coupons"; $6. 679 Easy Street (!), Dept. TH, Simi Valley, Calif. 93065.

Questar Corp. Makers of exceptionally fine Maksutov telescopes, that first impressed me with the fact that the moon is not attached to anything. Catalog, $2. Box C, Dept. 105, New Hope, Pa. 18938.

R.V.R. Optical. Dealers in a wide range of name brands, with an emphasis on refractors. Catalog $3. P.O. Box 62, Eastchester, N.Y. 10709.

Santini Opticians. Dealers in name brands, emphasizing binoculars. 2878 Merrick Rd., Bellmore, N.Y. 11710.

Scott Optical. Large-aperture mirrors. 4628 E. Cornell, Fresno, Calif. 93703.

Star Instruments. A wide range of optics for Newtonian and Cassegrain telescopes, including Dobsonians. Catalog $1. 3641 E. Fox Lair Dr., Flagstaff, Ariz. 86001.

Star-Liner Co. An established manufacturer of complete Newtonian and Cassegrain telescopes, with very high quality mountings. Catalog $2. 1106 S. Columbus, Tucson, Ariz. 85711.

Star Tracker. Dealers in name brands and accessories. A very informative catalog, $2 and worth it. 3093 Walnut, Boulder, Colo. 80302.

Swift Instruments. Manufacturers of refractors. Catalog $2. Takahashi Systems Division, 952 Dorchester Ave., Boston, Mass. 02125.

Tele-Optics. Dealers in name brands. Free price list, $3 catalog package (which will probably duplicate the Meade, Celestron, Bushnell, and Orion catalogs). 2026 8th Ave., SE, Calgary, Alberta T2G 0N8, Canada.

Telescope Systems. Parts for the amateur telescope maker. Free catalog. P.O. Box 340738, Dallas, Tex. 75234.

Telescopics. Optics, components, and accessories. Catalog $2. P.O. Box 98, La Canada, Calif. 91011.

Tele-Vue Optics. Manufacturers of extraordinary eyepieces and other optical accessories for the amateur. Their Plössl eyepieces are superb. 20 Dexter Plaza, Pearl River, N.Y. 10965

Texas Nautical Repair Co., is, believe it or not, a dealer in name brands of telescopes. 3209 Milam, Houston, Tex. 77006.

Roger W. Tuthill, Inc. Manufacturer of the "Solar Skreen" Mylar filter, an inexpensive solar filter that I prefer, and designer of a number of imaginative accessories for the Celestron line, in which he also deals. Send 8.5″ × 11″ envelope with 3 oz. worth of postage for catalog. Worth having even if you don't want a Celestron. 11 Tanglewood Lane, Box 1086-TH, Mountainside, N.J. 07092.

University Optics. Major dealer in equipment for the amateur, including eyepieces, components for ATMs, filters, and so on. Free catalog. P.O. Box 1205, 2122 E. Delhi Rd., Ann Arbor, Mich. 48106.

Willmann-Bell, Inc. Primarily distributors of books, atlases, and catalogs on astronomical subjects. Their catalog is the most useful astronomical bibliography you can own; $1. P.O. Box 3125, Richmond, Va. 23235.

Zephyr Services. Computer programs (BASIC, 32 K, listings or on disc for Apple or IBM PC) for a variety of general astronomical problems—the first of what will be by the time you read this a booming subset of the amateur astronomy industry. 306 TH. Homewood, Pittsburgh, Pa. 15208.

Bibliography

The following is only a partial list of the vast literature on a vast subject. This selection favors recent books—although some classics in the field cannot be overlooked—and books for the layman—although some more advanced texts are also included. I have tried to give prices, current as this book goes to press, for those volumes you may wish to own, especially observer's guides and charts. Finally, a bibliography of the steady stream of articles on astronomy in popular scientific periodicals would be a book in itself—so instead, I'll recommend the back issues of *Scientific American,* which should keep you occupied for a decade or so.

MAGAZINES

Astronomy. A monthly magazine, geared mainly toward the beginner, profusely illustrated. $18/yr. P.O. Box 92788, Milwaukee, Wis. 53202.

Comet News Service. A quarterly newsletter, with special bulletins to announce recent discoveries. $5/yr. P.O. Box TDR No. 92, Truckee, Calif. 95734.

Deep Sky. A quarterly magazine devoted to observing galaxies and nebulae. $12/yr. P.O. Box 92788, Milwaukee, Wis. 53202.

The Griffith Observer. A monthly newsletter. $5/yr. Griffith Observatory, 2800 E. Observatory Rd., Los Angeles, Calif. 90027.

McDonald Planetarium News. A monthly newsletter. $4.50/yr. University of Texas Astronomy Department, RLM 15.220, Austin, Tex. 78712.

Mercury. A bimonthly magazine, especially useful for West Coast amateurs interested in amateur activities. $18/yr. Astronomical Society of the Pacific, 1290 24th Ave., San Francisco, Calif. 94122.

Scientific American. While not specifically an astronomy magazine, practically every issue has something for the astronomer; back issues are a mine of historic and authoritative articles on almost anything of importance to astronomy in this century. Available on newsstands. Subscription Manager, *Scientific American*, P.O. Box 5969, New York, N.Y. 10017.

Sky & Telescope. The oldest monthly astronomy magazine, with something for everybody. Less geared toward beginners than *Astronomy*, and perhaps slightly more substantial. $18/yr. Sky Publishing Corp., 49 Bay State Rd., Cambridge, Mass. 02138.

Sky Calendar. Monthly newsletter, featuring attractive, simple charts of the sky and planets. $5/yr. Abrams Planetarium, Michigan State University, East Lansing, Mich. 48824.

Telescope Making. A quarterly magazine devoted to building telescopes. $12/yr. P.O. Box 92788, Milwaukee, Wis. 53202.

EPHEMERISES

The Astronomical Almanac. Formerly titled *The American Ephemeris*, this annual is now jointly produced by the U.S. Naval Observatory and its British counterpart. It contains highly technical numerical data on the behavior of anything that moves in the sky. You probably won't want to buy this hardbound volume, but your library should subscribe. It's available from the U.S. Government Printing Office, Washington, D.C.

The Astronomical Calendar. Published annually, it is a beautifully illustrated description of what's due in the coming year. Complete with monthly sky charts. A bargain at $8 c/o Department of Physics, Furman University, Greenville, S.C. 29613.

The Observer's Handbook. (Annual) A compact, informative, authoritative source for almost everything in the sky. Complete information on times of sunrise and sunset, the moon and planets, and variable stars. One recent issue published a list of the known meteorite impact scars in North America, with a day-to-day listing of events. No amateur should try to get through January without it. The price goes up annually—it should still be under $10 by the time you read this. Royal Astronomical Society of Canada, 124 Merton St., Toronto, Ontario M4S 2Z2, Canada.

CHARTS

Chandler, D. *The Night Sky.* The best planisphere on the market, available for three northern latitudes: 30–40, 20–32, and 38–50 degrees. $5 from Sky Publishing (see address for *Sky & Telescope*, above).

Levitt, I. M., and Marshall, R. *Star Maps for Beginners.* New York: Simon and Schuster. $2.95.

Map of the Heavens. An attractive wall chart, crammed with information. The National Geographic Society, Washington, D.C. 20036. $4.

Norton's Star Atlas and Reference Handbook. In its seventeenth or eighteenth edition, this atlas has been in circulation since the turn of the century. In some respects still maddeningly antique, despite constant revision, it is still perhaps the most informative and convenient chart for beginning amateurs to use at the telescope. $30.00 from Sky Publishing (see above).

Tirion, W. *Sky Atlas 2000.0.* The new standard in charts for the amateur, showing stars to magnitude 8.0, and more deep-sky objects than you will log for many years to come. Available in three editions, bound or as separate sheets: desk (large format, black stars on white), field (large format, white stars on black), and deluxe (too large to hold open without a *large* chart stand; in six colors). $34.95 for the deluxe, $15.95 for the desk and field.

OBSERVER'S GUIDES

Burnham, R. *Burnham's Celestial Handbook.* A terrific amount of information, from representations of constellations in ancient coins, to the relationship between quasars and a Wagner motif, as well as all things celestial, engagingly and authoritatively told. If you bought no other guide but this one, you wouldn't do too badly. Arranged alphabetically by constellation, in three volumes. New York: Dover Books, 1979. $8.95 each vol.

Howard, N. E. *The Telescope Handbook and Star Atlas.* Useful information on telescope function and design, with a set of moderately detailed charts. New York: T. Y. Crowell, 1975. $21.10.

Muirden, J. *The Amateur Astronomer's Handbook.* "A guide to exploring the heavens," from one of the leading British amateurs. Useful information, emphasizing the amateur's role as a scientist. New York: Crowell, 1974. $9.95.

Ottewell, G. *The Astronomical Companion.* Companion to the same author's *Astronomical Calendar,* and with all of the same virtues, giving clear pictorial and verbal explanations of the entire range of astronomical phenomena. Same publisher and address as the *Calendar,* $12.

Pasachoff, J. *A Field Guide to the Stars and Planets.* With a great deal of information, including whole-sky charts, a complete photographic atlas of the moon and sky, and numerous tables of double and variable stars, clusters, and so on. One of the Peterson Field Guides. Boston: Houghton Mifflin. $12.95.

Sherrod, P. C. *The Complete Manual of Amateur Astronomy.* A thorough explanation of the methods available to amateurs interested in systematic gathering of data, primarily on solar system objects. A good source book for group activities. Englewood Cliffs, N.J.: Prentice-Hall. $10.95.

Sidgwick, J. B. *The Amateur Astronomer's Handbook.* The authoritative source, and a monument to the old, British school of *very* serious amateur astronomy: an exhaustive treatise on the telescope, in rigorous, scientific detail. Heavy reading, but an invaluable source book. The bibliography, especially, gives an exhaustive review of the technical literature on optics over the past several decades. The fourth edition was revised by Muirden. New York: Dover, 1981. $7.95.

———. *Observational Astronomy for Amateurs.* The companion volume to the above, and equally chunky with data, especially for planetary observers, with another exhaustive technical bibliography. Also in a fourth edition, revised by Muirden. New York: Enslow Publishers, 1982. $5.95.

The Webb Society Deep-Sky Observer's Handbooks. Detailed information on every class of deep-sky object, in five volumes: 1. *Double Stars,* $7.95; 2. *Planetary and Gaseous Nebulae,* $9.95; 3. *Open and Globular Clusters,* $13.95; 4. *Galaxies,* $15.95; 5. *Clusters of Galaxies,* $16.95. Available from Sky Publishing, Willmann-Bell, and other distributors.

THE CONSTELLATIONS

Allen, R. H. *Star Names, Their Lore and Meaning.* The encyclopedic text on the subject, giving all the known names, etymologies, and backgrounds on the names of the constellations and the stars. New York: Dover Books.

Graves, R. *The Greek Myths,* 2 vols. Not a book on astronomy, but it's the best version I know of the stories behind many of the constellations. Spare, clean retellings of the great myths, complete with a structural, anthropological exegesis. Penguin Books.

Rey, H. A. *The stars, a new way to see them.* Rey's redrawings of the constellations are still considered a scandal in some quarters, and I must admit that I don't see Ursa Major the way he does, but this is still perhaps the best guide to the constellations for the naked-eye observer. Boston: Houghton Mifflin, 1976. $6.95.

AMATEUR TELESCOPE MAKING

Berry, Richard. *Build Your Own Telescope.* Five fully-illustrated plans that any amateur can follow. New York: Charles Scribner's Sons, 1985. $24.95.

Brown, Sam. *All About Telescopes.* A thoroughly illustrated, step-by-step guide to making and using a telescope, with very simple materials; more advanced designs require machine work. $9.95 from Edmund Scientific.

Howard, N. E. *Standard Handbook for Telescope Making.* rev. ed. A good introduction to the fine art of mirror grinding; not a complete source, but neither is it so forbidding as some books on the subject. New York: Harper & Row, 1984. $15.95.

Ingalls, A., ed. *Amateur Telescope Making,* 3 vols. Not an integrated textbook, but a massive collection of articles by amateur telescope makers, assembled over the past half-century. This set is the most valued text in the field. New York: *Scientific American.*

Texereau, J. *How to Make a Telescope.* 2nd ed. Look for this volume at used bookstores or your local library—I believe it's now out of print. This is the only book I know that covers the entire process of grinding, testing, and coating a mirror in complete detail. Forbiddingly technical in places, and somewhat perfectionist in approach, but still well worth finding. New York: Willmann–Bell, 1984. $19.95.

LIGHT AND TELESCOPES

Adler, I. *The Story of Light.* Harvey House, 1971.

Conn, G. K. *Atoms and Their Structure.* New York: Oxford University Press, 1974.

Page, T., and Page, L. W. *Telescopes.* New York: Macmillan, 1974.

Paul, Henry E. *Binoculars and All-Purpose Telescopes.* New York: American Photographic Book Publishing Co.

ASTROPHOTOGRAPHY

Astrophotography Exposure Guide. Astronomics, 227 Linwood St., Lynn, Mass. 01905.

Paul, Henry E. *Outer Space Photography for the Amateur,* New York: American Photographic Book Publishing Co., 1976.

THE MOON

Alter, D. *Pictorial Guide to the Moon.* New York: Crowell, 1973.

Atlas and Gazetteer of the Near Side of the Moon. Washington, D.C.: U.S. Government Printing Office, 1971.

Cadogan, P. *The Moon, Our Sister Planet.* New York: Cambridge University Press, 1981.

French, B. *The Moon Book.* Penguin Books, 1977.

Gamow, G., and Stubbs, H. G. *Moon.* New York: Harper and Row, 1971.

Kopal, Z. *New Photographic Atlas of the Moon.* New York: Taplinger, 1970.
Marsden, B. G., and Cameron, A. G. *The Earth-Moon System.* New York: Plenum, 1966.
Wilkins, H. P., and Moore, P. A. *The Moon.* New York: Macmillan, 1955.

THE SOLAR SYSTEM

Alexander, A. F. O'D. *The Planet Saturn.* New York: Dover Books, 1980.
———. *The Planet Uranus.* Faber and Faber, 1965.
Beatty, J. K.; O'Leary, B.; and Chaikin, A.; eds. *The New Solar System,* 2nd ed. New York: Cambridge University Press, 1982.
Brandt, J., and Chapman, C. R. *Introduction to Comets.* New York: Cambridge University Press, 1981.
Briggs, G. A., and Taylor, F. W. *The Cambridge Photographic Atlas of the Planets.* New York: Cambridge University Press, 1982.
Burns, J. A. *Planetary Satellites.* Tucson: University of Arizona Press, 1977.
Chapman, C. R. *The Inner Planets.* New York: Charles Scribner's Sons, 1977.
Francis, P. *Planets.* Penguin Books, 1981.
Gehrels, T., ed. *Asteroids.* Tucson: University of Arizona Press, 1979.
———. *Jupiter.* Tucson: University of Arizona Press, 1976.
Grosser, M. *The Discovery of Neptune.* Cambridge, Mass: Harvard University Press, 1962.
Hartmann, W. K. *Moons and Planets: an introduction to planetary science.* 2nd ed. Belmont, Calif: Wadsworth, 1983.
———, and Raper, O., eds. *The New Mars.* NASA SP-337. Washington, D.C.: U.S. Government Printing Office, 1974.
Kopal, Z. *The Solar System.* New York: Oxford University Press, 1972.
Lowell, P. *Mars and Its Canals.* New York: Macmillan, 1960.
Lyttleton, R. *Mysteries of the Solar System.* New York: Oxford University Press, 1968.
McCall, G. *Meteorites and Their Origins.* Halsted, 1973.
Moore, P. *A Guide to the Planets.* New York: W. W. Norton, 1954.
Morrison, D., ed. *Satellites of Jupiter.* Tucson: University of Arizona Press, 1981.
———. *Voyager to Jupiter.* NASA SP-439. Washington, D.C.: U.S. Government Printing Office, 1980.
NASA. *The Planets Uranus, Neptune, and Pluto.* NASA SP-8103. Washington, D.C.: U.S. Government Printing Office, 1971.
———. *Viking Orbiter Views of Mars.* NASA SP-441. Washington, D.C.: U.S. Government Printing Office, 1980.
Nourse, A. E. *Asteroids.* Watts, 1975.
Ozima, M. *The Earth, Its Birth and Growth.* New York: Cambridge University Press, 1981.
Page, T., and Page, L. W. *Origin of the Solar System.* New York: Macmillan, 1966.
———. *Wanderers in the Sky.* New York: Macmillan, 1965.
Peek, B. *The Planet Jupiter.* New York: Macmillan, 1958.
Sears, D. *The Nature and Origin of Meteorites.* New York: Oxford University Press, 1978.
Tombaugh, C., and Moore, P. *Out of the Darkness: The Planet Pluto.* Stackpole Books, 1980.
Wasson, J. *Meteorites.* New York: Springer-Verlag, 1974.
Whipple, F. L. *Orbiting the Sun.* Cambridge, Mass.: Harvard University Press, 1981.
Wood, J. A. *The Solar System.* Englewood Cliffs, N.J.: Prentice-Hall, 1979.

THE SUN

Eddy, J. *A New Sun: The Solar Results from Skylab.* NASA SP-402. Washington, D.C.: U.S. Government Printing Office, 1979.

Gamow, G. *A Star Called the Sun.* New York: Viking, 1964.

Gibson, E. *The Quiet Sun.* NASA SP-303. Washington, D.C.: U.S. Government Printing Office, 1973.

Menzel, D. *Our Sun.* Cambridge, Mass.: Harvard University Press, 1959.

Moore, P. *Sun.* New York: W. W. Norton, 1969.

THE STARS, GALAXIES, AND COSMOLOGY

Aller, L. *Atoms, Stars, and Nebulae.* Cambridge, Mass.: Harvard University Press, 1971.

Blaauw, A., and Schmidt, M., eds. *Galactic Structure.* Chicago, Ill.: University of Chicago Press, 1965.

Bok, B., and Bok, P. *The Milky Way,* 5th. ed. Cambridge, Mass: Harvard University Press, 1981.

Brandt, J. *The Sun and Stars.* New York: McGraw-Hill, 1966.

Calder, N. *Violent Universe.* New York: Viking, 1970.

Chandrasekhar, S. *An Introduction to the Study of Stellar Structure.* New York: Dover Books, 1957.

Coleman, J. *Relativity for the Layman.* New York: Macmillan, 1957.

Davies, P. *The Runaway Universe.* New York: Penguin Books, 1980.

Dickson, F. *The Bowl of Night.* Cambridge, Mass.: Massachusetts Institute of Technology Press, 1969.

Dufay, J. *Galactic Nebulae and Interstellar Matter.* New York: Dover Books, 1968.

Ferris, T. *Galaxies.* San Francisco: Stewart, Tabori, and Chang, 1981.

———. *The Red Limit.* New York: William Morrow, 1977.

Gamow, G. *One, Two, Three . . . Infinity.* New York: Bantam Books, 1971.

Gardner, M. *Relativity for the Millions.* New York: Macmillan, 1966.

Glasby, J. *Variable Stars.* Cambridge, Mass.: Harvard University Press, 1968.

Golden, F. *Quasars, Pulsars, and Black Holes: A Scientific Detective Story.* New York: Charles Scribner's Sons, 1976.

Harrison, E. *Cosmology: The Science of the Universe.* New York: Cambridge University Press, 1981.

Hodge, P. *Galaxies and Cosmology.* New York: McGraw-Hill, 1966.

Hoyle, F. *Galaxies, Nuclei, and Quasars.* New York: Harper & Row, 1965.

Hubble, E. *The Realm of the Nebulae.* New York: Dover Books, 1958.

Iaki, S. *The Milky Way, an Elusive Road for Science.* New York: Science History Publications, 1972.

Jastrow, R. *Red Giants and White Dwarfs.* New York: Harper & Row, 1971.

Kaufmann, W. *The Cosmic Frontiers of General Relativity.* Boston: Little, Brown, 1977.

———. *Relativity and Cosmology.* New York: Harper & Row, 1973.

Lynds, B., ed. *Dark Nebulae, Globules, and Protostars.* Tucson: University of Arizona Press, 1970.

Motz, L. *The Universe: Its Beginning and End.* New York: Charles Scribner's Sons, 1975.

Nicolson, I, and Moore, P. *Black Holes in Space.* Orbach and Chambers, 1974.

O'Connell, D., ed. *Stellar Populations.* New York: Wiley-Interscience, 1958.

Page, T., and Page, L. *Evolution of Stars.* New York: Macmillan, 1967.
———. *Stars and Clouds of the Milky Way.* New York: Macmillan, 1968.
Sandage, A. *The Hubble Atlas of Galaxies.* Washington, D.C.: Carnegie Foundation, 1961.
Shapley, H. *The Inner Metagalaxy.* New Haven, Conn.: Yale University Press, 1957.
———. *Galaxies,* 3rd ed. Cambridge, Mass.: Harvard University Press, 1972.
Schatzman, E. *The Origin and Evolution of the Universe.* New York: Basic Books, 1965.
Shipman, H. *Black Holes, Quasars, and the Universe,* 2nd ed. New York: Houghton Mifflin, 1980.
Silk, J. *The Big Bang,* San Francisco: W. H. Freeman, 1980.
Strohmeier, W. *Variable Stars.* New York: Pergamon Press, 1973.
Sullivan, W. *Black Holes.* New York: New York Times Books, 1979.
Weinberg, S. *The First Three Minutes.* Basic Books, New York: 1976.
Whitney, C. *The Discovery of Our Galaxy.* New York: Knopf, 1971.

GENERAL ASTRONOMY

Abell, G. *Exploration of the Universe,* 4th ed. Philadelphia: Saunders College Publications, 1982.
Berman, Louis, and Evans. *Exploring the Cosmos,* 3rd ed. Boston: Little, Brown, 1980.
Jastrow, R., and Thompson, M. *Astronomy: Fundamentals and Frontiers,* 3rd ed. New York: J. Wiley, 1977.
Mitton, S., ed. *The Cambridge Encyclopedia of Astronomy,* New York: Crown, 1977.
Moore, P. *The A–Z of Astronomy.* New York: Charles Scribner's Sons, 1977.
Pasachoff, J. *Contemporary Astronomy,* 2nd ed. Philadelphia: Saunders College Publications, 1981.
Sagan, C. *Cosmos.* New York: Random House, 1980.
Shapley, H., ed. *A Source Book in Astronomy, 1900–1950.* Cambridge, Mass.: Harvard University Press, 1960.
———. *Star Clusters.* Cambridge, Mass.: Harvard University Press, 1930.
———, and Howarth, H., eds. *A Source Book in Astronomy.* New York: McGraw-Hill, 1929.

HISTORY OF ASTRONOMY

Abetti, G. *The History of Astronomy.* New York: Henry Schuman, 1952.
Doig, P. *A Concise History of Astronomy.* London: Chapman and Hall, 1950.
Hoyle, F. *From Stonehenge to Modern Cosmology.* San Francisco: W. H. Freeman, 1972.
Pannekoek, A. *A History of Astronomy.* London: Allen and Unwin, 1961.

WEATHER

Baird, M. *Weather Forecasting for Astronomy.* Winmark Press, 1982.
Spilhaus, A. F. *Weathercraft.* New York: Viking Press, 1951.
Sutton, O. *Understanding Weather.* Penguin Books, 1960.
Thompson, P., and O'Brien, R. *Weather.* Life Science Library. New York: Time-Life Books, 1965.

Glossary

Aberration: an imperfection in an optical component, causing distortion of an image.

Absolute Magnitude: the magnitude of an object at a standard distance of 10 parsecs.

Airy Disc: irreducible disc of light at the image of a star; caused by diffraction at the objective, the size of this disc determines a telescope's resolution.

Airy Rings: concentric rings of light around the image of a bright star; caused by diffraction and interference.

Albedo: brightness of sunlight reflected from a planetary surface; expressed as a percentage of the total incident sunlight.

Altazimuth Coordinates: a system to express the position of an object in terms of its compass bearing and altitude above the horizon; the coordinates vary with time and the location of observation.

Altazimuth Mounting: a telescope mounting in which the telescope pivots on horizontal and vertical axes.

Altitude: the distance of an object above the horizon, in angular measure.

Aluminizing: coating an optical surface with a microscopic film of evaporated aluminum, to make it reflect light.

Amplification Factor: amount of magnification caused by the curve of a Cassegrain secondary.

Amplitude: the height of a wave; the range of a variable star's light curve.

Angular Diameter: the diameter of a celestial object, expressed in units of arc.

Angular Measure: a system of expressing the apparent size of objects on the celestial sphere, in units of arc (degrees, minutes, and seconds).

Angstrom: 10^{-10} meter.

Apastrion: in the orbit of a body around another star, the point farthest from that star; the point of widest actual separation in the orbit of a binary star.

APERTURE: the width of the objective of a telescope; the light-gathering opening of a telescope.

APERTURE MASK: any device that restricts or otherwise alters the size or shape of a telescope's aperture.

APHELION: the point farthest from the sun in a planet's orbit.

APOCHROMATIC TRIPLET: a refracting lens comprising three elements, offering almost complete freedom from chromatic aberration.

APOGEE: in the orbit of the moon or an artificial satellite around the earth, the point in its orbit farthest from the earth.

APPARITION: the intervals between successive conjunctions of a planet, during which it is visible in our sky.

APPULSE: close apparent approach of two or more planets, two or more planets and a star; a near-miss occultation.

ARC-MINUTE: unit of arc or angular measure, $\frac{1}{60}$ of a degree.

ARC-SECOND: unit of arc or angular measure, $\frac{1}{60}$ of an arc-minute.

ASHEN LIGHT: (1) sunlight, reflected from earth, often visible as a faint glow illuminating the dark area of the crescent moon; (2) dim glow of sunlight refracted by Venus's atmosphere, rarely visible illuminating dark area of crescent disc.

ASTERISM: conspicuous pattern of stars within a constellation, e.g., the Big Dipper in Ursa Major, the teapot in Sagittarius.

ASTEROID: a small, rocky, or metallic body, orbiting the sun within the solar system; a minor planet.

ASTEROID BELT: a region of the solar system, between the orbits of Mars and Jupiter, where most asteroids orbit.

ASTIGMATISM: optical aberration, preventing precise focus.

ASTRONOMICAL UNIT: a unit of length, used in measuring distances within the solar system or in the region of other stars. Equal to mean radius of earth's orbit: 149,600,000 kilometers, or 93,000,000 miles. Abbreviated A.U.

AURORA: glow, caused by solar wind accelerating charged particles into the upper atmosphere, visible in a halo around the magnetic poles of the earth, Jupiter, and probably the other Jovian planets.

AVERTED VISION: an observational technique used to see dim objects by looking to one side of them, so that the image falls on more sensitive periphery of retina.

AZIMUTH: compass bearing, expressed in degrees of arc, for example, 90° = east.

BAFFLE: in a Cassegrain telescope, collars placed around the secondary and the central perforation in the primary, to prevent stray light from impinging on the image, washing out contrast.

BARLOW: a negative lens, placed in the light path of a telescope ahead of the eyepiece, to increase the effective focal length; a means of increasing the magnification of any eyepiece.

BARRED GALAXY: a type of spiral galaxy, in which the spiral arms trail from a bar extending through the central bulge.

BARYCENTER: in any orbital system, the center of gravity of that system; the point about which a double star, or planet and satellite, revolve.

BIG BANG: an explosion which in current theory started the universe, 12 to 20 billion years ago.

BINARY: a multiple star.

BLACKBODY: a theoretical, "perfect" radiator of energy; a star behaves like a blackbody.

BLACK HOLE: the most extreme of several possible stellar endstates, in which a massive star collapses to a sphere of zero radius, with an escape velocity greater than the speed of light.

Blue Shift: Doppler shift of lines in the spectrograph of an object toward the blue end of the spectrum, indicating that the object is moving toward the observer.

Cassegrain Reflector: a reflecting telescope in which light reflects from a (usually) paraboloidal primary, then from a (usually) hyperboloidal secondary, which redirects the light back through a hole ("perforation") bored in the center of the primary.

Cassini Division: an apparent gap, actually a region of lower particle density, in the rings of Saturn.

Catadioptric: a telescope using a combination of lenses and mirrors to form an image, for example, a Schmidt-Cassegrain.

Celestial Equator: great circle on the celestial sphere directly above all points on earth's equator; the circle of zero declination.

Celestial Meridian: great circle on the celestial sphere connecting celestial poles and zenith.

Celestial Poles: the two points on the celestial sphere directly over the earth's poles; the north and south celestial poles.

Celestial Sphere: the sky, imagined as a sphere enclosing the earth, with the center of the earth for its center; a convention used in mapping the apparent positions of objects in the sky, without regard to their actual distance from the earth.

Cell: part of a telescope that supports the objective in the tube.

Cepheid Variable: a class of variable star, having a regular period, and subject to a period-luminosity relationship.

Chromatic Aberration: a flaw in a refracting optical system, in which light of different colors focuses at different distances from a lens.

Chromosphere: the layer of the solar atmosphere between the photosphere and corona, approximately 2,000 kilometers thick, with a temperature ranging from 6,000° to 100,000° Kelvin.

Circumpolar: stars or the region of the sky between the celestial pole and the horizon; north circumpolar stars never set; south circumpolar stars never rise above the southern horizon. The size of the region depends on the observer's latitude.

Colatitude: the difference between your latitude and 90°; also equal to the angular distance between the zenith and the north celestial pole, or the distance along a meridian from the south horizon to the celestial equator.

Collimation: alignment of telescope optics.

Color Index: measurement of a star's color and therefore spectral class, derived by subtracting the star's magnitude at one wavelength from another.

Coma: (1) optical aberration common to short-focus paraboloids, in which star-images at the edges of the field elongate into teardrop or fan-shapes; (2) haze of gases and plasma surrounding the head of a comet.

Comes: any dim, stellar object visible in the field around a bright star.

Comet: a member of the solar system, consisting of a mountain-sized chunk of frozen water and gases, and some dust. Most travel on highly elongated orbits between the inner solar system and Oort Cloud.

Comparison Star: star of known brightness used to gauge the brightness of a variable star.

Conduction: method of heat transfer among stationary molecules, most efficient within solids.

Conjunction: (1) alignment of objects on the celestial sphere, in which two or more bodies (planets, sun, star) share the same right ascension; (2) inferior conjunction: the conjunction of an inferior planet with the sun, between the earth and the sun; (3) superior conjunction: conjunction of an inferior planet with the sun, on the far side of the sun.

CONSTELLATION: (1) a region of the celestial sphere, usually containing a conspicuous grouping of stars. The celestial sphere is divided into 88 constellations; (2) a conspicuous grouping of stars.

CONVECTION: method of heat transfer by motion of heated molecules, most efficient within liquids and gases. Does not work in vacuum.

CONVECTION ZONE: outermost of sun's internal layers, comprising the outer 15% of its radius, within which energy migrates from the core primarily by convection. Temperatures in this zone range up to 1.5 million degrees Kelvin.

CORE: central region of sun, where thermonuclear fusion occurs; temperature at core is 10–15 million degrees Kelvin.

CORONA: outermost region of the sun's atmosphere, lying above the chromosphere and extending as far as 21,000,000 kilometers out into space. Temperatures in the corona range up to 2 million degrees Kelvin.

CORRECTING PLATE: in a Schmidt or Schmidt-Cassegrain telescope, the thin lens at the front of the tube that compensates for the spherical aberration of the primary mirror.

CRADLE: in a telescope, the part that supports the tube, connecting it to the mount.

CRATER: geographical feature common to terrestrial planets and asteroids, a usually circular indentation with raised walls, caused by meteor impact.

CULMINATION: (1) passage of a circumpolar star across the celestial meridian; (2) upper culmination: culmination above the pole; (3) lower culmination: culmination below the pole.

CURVATURE OF FIELD: optical aberration, in which an image focuses clearly, but not on a plane surface, making clear focus of the entire image impossible.

CYCLE: time required for any repetitive process to repeat; a wave or a variable star will pass through successive cycles of its process; equivalent to period.

DARK NEBULA: a nebula not illuminated by nearby stars, appearing as a black, usually starless region of the sky.

DAWES'S LIMIT OF RESOLUTION: value of the smallest visible angular separation between two objects, such as stars, for telescopes of given aperture; important in observing double stars.

DAY: (1) time required for the earth (or any other planet) to complete one rotation; (2) sidereal day: a day measured with respect to the celestial sphere, on earth approximately 4 minutes shorter than a solar day; (3) synodic day: a day measured with respect to the sun, a solar day.

DECLINATION: a coordinate of the celestial sphere, measuring the location north or south of the celestial equator in units of angular measure; equivalent to latitude. Constant over time for all observer locations.

DEFINITION: the quality of sharpness and clarity of a telescopic image.

DEGREE: unit of angular measure, $1/360$ of a circle.

DEWCAP: large tube attached to the front of a refracting telescope, to protect the objective from dew and stray light.

DIAGONAL: secondary mirror of a Newtonian telescope, used to redirect the beam of light from the objective out the side of the tube to the observer.

DIAGONAL HOLDER: the part of a telescope that supports the diagonal at the proper location and angle within the tube.

DIFFRACTION: a phenomenon of light, in which light waves propagate around obstructions, bending slightly toward the obstruction as they pass; it causes interference effects in telescopes and sets absolute limit on telescopic resolution.

Diffraction Grating: species of filter used in a telescope or spectroscope to disperse light into a spectrum; usually a thin sheet of glass or plastic, engraved with many fine parallel lines.

Diffraction Limited: said of optics, indicating that the quality of their image is limited by diffraction, rather than by flaws in manufacture; an expression of high quality in optics.

Diffraction Spikes: thin rays of light appearing around the images of bright stars in reflecting telescopes, caused by the diffraction of light around the arms of the spider.

Diffuse Nebula: a cloud of gas and dust in interstellar space, frequently illuminated by nearby stars.

Distance Modulus: difference between absolute and apparent magnitude of a star or other distant object; useful in determining distances to objects too far to gauge by the parallax method.

Distortion: optical aberration in which the image focuses sharply but with its parts displaced, causing the image to appear to bulge in or out from a plane.

Dobsonian: a Newtonian telescope, usually of large aperture, with a thin objective supported around its lower circumference as well as at its back; and with an alt-azimuth mounting, in which Teflon sheets slide on large, smooth plastic surfaces. Becoming increasingly popular with amateurs seeking inexpensive, large-aperture instruments.

Doppler Effect: a change of the frequency of a wave depending on the motion of the source; frequency increases for approaching objects, decreases for receding objects, by an amount depending on the velocity. A very important tool in understanding large-scale form, motion, and the history of the universe.

Drawtube: part of a telescope; a sliding metal tube holding the eyepiece; it's usually controlled by the rack-and-pinion gear.

Dwarf: a luminosity class of star, comprising essentially those stars, such as the sun, that lie on the main sequence.

Eccentricity: a measure of the departure of an ellipse, especially a planetary orbit, from a perfect circle (which has eccentricity equal to 0).

Eclipse: generally, the passage of one body in front of another. Specific kinds of eclipse include: (1) lunar, in which the earth passes between the moon and sun, casting the moon into shadow; (2) solar, in which the moon passes between the earth and sun, blocking sunlight from a small region of the earth's surface; (3) eclipses of planetary satellites, in which a satellite passes into the shadow of a planet. Lunar and solar eclipses can be either partial, in which surface of moon or sun is not entirely obscured, or total, in which obscuration is complete. Solar eclipses can also be annular, in which a small rim of the sun's circumference remains visible around the circumference of the moon.

Eclipsing Binary: a class of variable star, in which one member of a multiple star system eclipses another, causing the total light of system as seen from earth to decrease.

Ecliptic: a great circle on the celestial sphere that is the apparent path of the sun over the course of the year; also marks the plane of the solar system, on or near which most planetary orbits lie.

Ecliptic Poles: points on the celestial sphere 90° from the ecliptic.

Effective Focal Length: focal length of telescope as modified by amplification factor of secondary (in Cassegrain), barlow lens, or certain photographic techniques.

Ejecta: debris catapulted over a planetary surface by meteor impact.

Electromagnetic Radiation: energy, carried by photons in waves, in the form of interacting electrical and magnetic fields. Visible light, as well as radio waves, microwaves, infrared and ultraviolet waves, X rays, and gamma rays are forms of electromagnetic radiation.

Elongation: (1) the angular distance of a planet from the sun, or of a satellite from its primary, usually used in reference to the inferior planets; may be either east, in which case the planet appears in the evening sky, or west, in which case the planet appears in the morning sky; (2) greatest elongation: the moment when elongation of an inferior planet is greatest, placing it at highest altitude above an observer's horizon.

Emission Nebula: a nebula that glows by the fluorescence of its own atoms, which have been energized by the light of nearby stars.

Ephemeris: a publication giving times of celestial events such as eclipses, conjunctions and sunrise that involve solar system objects; usually published annually.

Epoch: the year for which the coordinate system of a chart or set of coordinates is accurate.

Equation of Time: the amount by which actual solar time varies from mean solar time, owing to variations in the earth's orbital speed; the difference between clock-noon and the time of the sun's transit.

Equatorial Mounting: a method of supporting a telescope so that the tube pivots on two perpendicular axes, one of which is parallel to the earth's rotational axis; a necessity for long-exposure astrophotography; a convenience for visual observers.

Equatorial Platform: platform pivoted so that a telescope resting upon it will follow the course of a star across the sky; for example, a Poncet mounting.

Equinox: (1) the moment at which the sun's declination equals 0 degrees; (2) vernal equinox: first day of spring, approximately 20 March; (3) autumnal equinox: first day of fall, approximately 20 September.

Erecting Prism: a prism, or more commonly a diagonal mirror, inserted in the light path of a refractor or Cassegrain telescope, to turn the image rightside up and bring the eyepiece to a more comfortable angle for observing.

Eruptive Variable: a star whose magnitude changes suddenly, such as a nova.

Escape Velocity: the speed needed to escape the gravitational attraction of a planet or star.

Exit Pupil: a disc-shaped region above the eyepiece of a telescope where light rays transmitted from an eyepiece converge; the size of that region; size dependent upon diameter of objective divided by magnification of system.

Extended Object: any celestial object that is not a point source of light; nonstellar celestial objects.

Eye Lens: the lens in an eyepiece closest to the observer's eye.

Eyepiece: system of lenses used to magnify the image formed by a telescope's objective.

Eye Relief: the distance between the exit pupil and the eye lens of telescope.

Faculae: plages seen at the limb of the sun.

Field: (1) the image visible in an eyepiece, or the size of that image; (2) true field: the angular diameter of the region of sky visible in an eyepiece, dependent on apparent field divided by magnification; (3) apparent field: angle through which an observer's eye moves when scanning from one edge of field to other; the illusion of spaciousness of an eyepiece's image, dependent on the eyepiece design.

Field of Full Illumination: the portion of a telescope field, usually the inner half, receiving the full amount of light gathered by the objective; restricted by several factors—usually by eyepiece design, an undersized secondary, or the drawtube.

Filament: a solar prominence seen from above.

Filter: an optical accessory used to block certain wavelengths of light, or to reduce the total amount of light reaching the objective or eyepiece. Useful in planetary observation, to increase image contrast. Essential in direct telescopic solar observation to prevent the blinding of the observer.

Finderscope: a small, usually refracting telescope, with a wide field and low magnification, that is attached to a larger telescope; used to locate objects.

First Point of Aries: in the celestial coordinate system, the point of right ascension equal to zero hours, and declination zero degrees; currently in Pisces, south of the eastern edge of the Great Square of Pegasus; location of the sun at vernal equinox.

Flare Star: a variable star, usually a dim, red dwarf, capable of sudden, brief increases in brightness, apparently because of the equivalent of a solar flare.

Focal Length: the distance between the objective and the prime focus of a telescope; determines magnification produced by the eyepieces used with a telescope; also determines the size of the image produced at the focal plane.

Focal Plane: the region in a telescope where light rays focused by the objective converge to their smallest area; the place where the image forms.

Focal Ratio: ratio between the focal length and the aperture of a telescope, expressed as f/n, where n = ratio; for example, $f/5$, for a 6-inch telescope of 30-inches focal length.

Focus: (1) the convergence of light rays gathered by a telescope objective to a small disc or point; the place in a telescope where this occurs; (2) adjustment of the eyepiece within a telescope to give a sharp, clear image.

Focuser: any device used to hold and adjust the eyepiece and bring it to the focal plane of a telescope.

Focusing Mount: focuser.

Forbidden Lines: lines in the spectrograph of an object indicative of high temperatures and low gas pressures, difficult to achieve on earth, relatively common in some kinds of emission nebulae.

Frequency: of a wave, the number of crests (cycles) passing an observer each second; measured in cycles per second, or hertz (hz).

Galactic Bulge: the central region of a spiral galaxy, roughly spherical in shape, comprising primarily population II stars, and containing the galactic nucleus.

Galactic Cluster: open cluster.

Galactic Corona: region of hot plasma surrounding a galaxy, apparently containing a significant proportion of a galaxy's mass.

Galactic Halo: thinly populated outer regions of a galaxy, the domain of globular clusters and dead stars.

Galactic Nucleus: the center of a spiral galaxy, frequently the location of violent, explosive activity, and possibly of a supermassive black hole; appears almost starlike in some cases.

Galaxy: extremely large collections of anywhere from a few million to a trillion stars, all bound by gravity to elliptical orbits around a galactic center; occurring in three distinct forms (elliptical, irregular, and spiral); galaxies contain most of the detectable matter in the universe; billions of them populate the known universe.

Gamma Ray: most energetic form of electromagnetic radiation, usually produced by the annihilation of an electron and a positron in the process of thermonuclear fusion; also produced by other extremely high-temperature phenomena, such as the acceleration of matter into a black hole.

Ghost: in a telescope, false images of a bright object such as a planet, caused by internal reflection in the lenses of an eyepiece.

Giant: a luminosity class of star, comprising stars that have evolved off the main sequence, swollen and brightened to many times the size and brightness of the sun.

Globular Cluster: a compact gravitational association of as many as one million stars of population II, usually found in a halo around a galaxy.

Great Circle: a circle on the celestial sphere having the center of the earth as its center, dividing the celestial sphere into two equal parts; the celestial equator, ecliptic, and celestial meridian are all great circles.

HI Region: a cloud of thin, cool, neutral hydrogen gas, common in the interstellar medium.

HII Region: a cloud of thin, hot, ionized hydrogen gas, glowing by fluorescence of its atoms.

Heliocentric: sun-centered, as in a coordinate system used to locate sunspots, or in the Copernican model of the solar system.

Hertz: unit of frequency; one cycle per second (hz).

Hertzsprung-Russell Diagram: a chart, plotting brightness on the vertical axis, and spectral class or color index on the horizontal axis; stars plotted on the H-R diagram fall into distinct regions of the chart, the most populous of which is called the main sequence. The diagram offers valuable information on stellar evolution and behavior.

Horizon: great circle 90 degrees from the zenith, defining the apparent boundary between the sky and earth.

Hour Angle: angle, measured to the west, between celestial meridian and an hour circle passing through a given object.

Hour Circle: any great circle on the celestial sphere passing through the celestial poles. Also called a meridian.

Hubble Recession Constant: the amount by which the speed of recession of cosmically distant objects increases with their increasing distance. Now thought to be equal to something between 50 and 100 kilometers per second per megaparsec.

Hubble Sequence: a system of classifying spiral galaxies by their compactness, and elliptical galaxies by their eccentricity; originally proposed as a scheme of galactic evolution, since discarded.

Huygens's Principle: process by which waves diffract around obstacles; if each point on a wave front is considered as a point emitter, the combined effect of which is the next wave front, then the point next to an obstacle, by radiating over an arc 180 degrees around itself, will cause its end of the next wave front to veer slightly off-course.

Image: (1) light rays focused by an objective, by reflection, or refraction; (2) real image: an image that can be projected on a screen (telescopic images are real); (3) virtual image: an image that cannot be projected or touched (the image in a flat mirror is virtual).

Inferior Planet: a planet with an orbit around the sun smaller than Earth's; Mercury or Venus.

Integrated Magnitude: the magnitude an extended, especially a diffuse, object would have if its light radiated from a single point.

INTERFERENCE: (1) in waves, the various ways in which two or more waves encountering each other interact; (2) constructive interference: production of a single wave of greater intensity by two or more waves; (3) destructive interference: production of a weak or flat wave by two or more waves.

INVERSE-SQUARE LAW: law applying to phenomena, such as gravity and electromagnetic radiation, which act at a distance, describing the relationship between a phenomenon's strength and the distance at which it acts. For phenomena subject to this law, the strength decreases exponentially as the distance increases, being one-fourth as strong at double the distance, one-ninth at triple the distance, and so on.

ION: an atom with an electrical charge caused by the loss of one or more of its electrons; common in gases exposed to high levels of electromagnetic radiation. Ionized: said of a substance in which the atoms have become ions; a plasma is ionized.

JOVIAN PLANET: a planet, like Jupiter, composed primarily of light, gaseous elements, such as hydrogen, helium, methane, and ammonia; Jupiter, Saturn, Uranus, and Neptune are Jovian planets.

JULIAN DATE: a system of timekeeping that does not observe years, months, or weeks, but keeps consecutive track of the days since noon, 4,713 B.C., Greenwich Mean Time; used primarily in observation of variable stars.

KIRKWOOD GAP: any of several regions of the asteroid belt where few asteroids are found; caused by the gravitational influence of Jupiter.

LENS: a transparent optical element with curved surfaces, used to collect and focus light.

LIBRATION: process by which the moon appears to wobble within its spin-orbital lock, displaying approximately 18% of its hidden side, owing to variation in its orbital speed between perigee and apogee.

LIGHT CURVE: a graph showing the changing brightness of an object (such as a variable star) over time.

LIGHT-GRASP: the ability of a telescope to collect light, dependent on aperture; of prime importance in observing extended objects.

LIGHT POLLUTION: artificial brightening of the sky background, usually caused by artificial light backscattered from haze, smog, or other particles suspended in the air; a serious hindrance to observers in populated areas.

LIGHT YEAR: a measure of *distance,* used on the interstellar scale; the distance traversed by light in a vacuum in one year; 9.46×10^{12} kilometers, or 6×10^{12} miles.

LIMB: the edge of the disc of a moon, planet, or sun.

LIMB DARKENING: the darkening of a limb of a planet or sun, owing to absorption of light by its atmosphere, which is thickest at the limb.

LIMITING MAGNITUDE: magnitude of the dimmest object visible in a telescope or displayed on a chart; a measure of the light-grasp of telescope.

LOCAL GROUP: a cluster of approximately two dozen galaxies, spanning several million light years, of which our own is a member, along with M.31 in Andromeda, and M.33 in Triangulum.

LPR FILTER: light-pollution-rejection filter; a filter that blocks wavelengths of artificial light, darkening light-polluted sky background, increasing the contrast between sky and dim objects, especially emission nebulae.

LUMINOSITY: measure of the amount of light produced by a star or other radiating object.

LUMINOSITY CLASS: classification of a star by size and light output; the classes are: Ia, bright supergiants; Ib, supergiants; II, bright giants; III, giants; IV, subgiants; V, dwarfs (main-sequence stars); there is also a subdwarf classification, for underluminous stars.

LUNATION: the period from new moon to new moon; a synodic month.

MAGELLANIC CLOUDS: two irregular galaxies, visible to the naked eye from the southern hemisphere; nearby satellites of the Milky Way.

MAGNETOPAUSE: the border of a magnetosphere.

MAGNETOSHEATH: a thin, planar region in a magnetotail, where auroral disturbances may originate.

MAGNETOSPHERE: a region around a planet or star within which its magnetic field is the dominant magnetic force.

MAGNETOTAIL: extended region of a planet's magnetosphere, downwind (in the solar wind) from the sun. It gives the magnetosphere a characteristic teardrop shape.

MAGNIFICATION: enlargement of the image produced at the prime focus of a telescope; the factor by which a telescope reduces the apparent distance of an object.

MAGNITUDE: measure of the apparent brightness of a star or other celestial object, in which dimmer objects have higher values; brighter objects have lower absolute values, extending in several cases below zero into negative figures. A difference of 1 magnitude equals a difference in light intensity of 2.512 times; a difference of 5 magnitudes (approximately the span from the brightest to the dimmest stars visible to the naked eye) equals a difference in light intensity of 100 times.

MAIN SEQUENCE: region of the Hertzsprung-Russell diagram containing the majority of stars; stars on the main sequence are of luminosity class V (dwarfs), and produce energy primarily by the fusion of hydrogen at their cores; most stars spend most of their lives on the main sequence.

MAKSUTOV: a type of catadioptric telescope, in which a spherical primary is corrected by a thick correcting plate, called a meniscus plate; as a Cassegrain, the design offers a long effective focal length in an extremely compact tube.

MARE: see maria.

MARIA: "seas" on the surface of the moon, actually regions of geologically recent upwellings of lava from the lunar interior, caused by impact of large asteroids; visible to the naked eye as dark regions of moon, "the man in the moon."

MASS-LOSS: the process by which giant and supergiant stars evaporate a significant percentage of their outer atmospheres; the initial stage in the creation of a planetary nebula, and an important check on the occurrence of supernovae.

MERIDIAN: any great circle passing through the poles, either of the celestial sphere or of a planet or sun; in common usage, the celestial meridian; also, often the central meridian of the visible disc of a planet.

MESSIER CATALOG: a listing of nebulae, galaxies, and star clusters visible in amateur telescopes, including most of the brightest such objects in the heavens; assembled by the eighteenth-century French astronomer Charles Messier, the catalog lists from 103 to 110 objects, depending on the inclusion of a number of objects cataloged by Messier's contemporaries.

METEOR: a "shooting star"; the flash produced by the passage of a meteoroid through the atmosphere; occurring either in concentrated showers or as random, sporadic events.

METEORITE: a meteoroid that has struck the earth's surface.

METEOROID: a small piece of solid matter, rocky, ashen, or metallic in composition, in orbit around the sun; remnants of the original nebula from which the solar system formed, or of decayed comets, or fragmented asteroids.

METEOR SHOWER: a stream of meteoroids in orbit around the sun, usually marking the path of a decayed comet; when the earth passes through such a stream, many of its members strike the atmosphere and burn, producing occasionally spectacular displays of meteors.

MINOR PLANET: an asteroid.

MOUNTING: any device that supports a telescope's tube assembly; ideally, such a device should allow smooth motion of the tube to view all points of the sky and support the telescope without vibration.

MOVING GROUP: a group of stars, more or less widely scattered across the sky, which share the same motion through space; not necessarily a constellation or asterism.

NADIR: point on the celestial sphere directly below the observer (that is, in the sky above the far side of the earth); opposite of zenith.

NEBULA: (1) an extended region of thin gases, primarily hydrogen, among the stars; (2) any of several objects, such as galaxies or clusters, which appears cloudy to the naked eye or in the telescope (archaic).

NEUTRON STAR: one of several possible stellar endstates, common to stars somewhat more massive than the sun; a fantastically compressed object, in which the subatomic particles of a stellar core are gravitationally compressed into a sphere of neutrons only a few miles across; often, a pulsar.

NGC: the New General Catalog, a listing of over 7,000 (with its supplemental Index Catalogs [IC], over 10,000) nebulous objects, many of which are visible in amateur telescopes.

NODE: (1) intersection between two great circles on the celestial sphere, especially between the orbit of a planet and the ecliptic, or between the ecliptic and the celestial equator; the ascending node of a planet's orbit is the node at which it passes north of the ecliptic; the descending node is the node where it passes south of it. (2) of a wave, the point where its energy is momentarily zero.

NOVA: a star experiencing an explosive increase in luminosity, probably owing to the collapse of matter accumulated in a shell above its surface. A nova can be rapid or slow to decline, and also recurrent, repeating its outbursts at intervals ranging from months to centuries.

NUCLEUS: (1) of an atom: the extremely compact central region, occupied by protons and neutrons, the scene of thermonuclear reactions; (2) of a comet: the central region of a comet's head, the solid body of ice, rock, and dust from which the rest of the comet (coma and tail) evaporates; (3) of a spiral galaxy: the extremely condensed region at the center of the galactic bulge, frequently the scene of poorly understood high-energy phenomena.

NULL-TESTED: shown to exhibit no detectable aberrations under one of several common tests of optical quality; a "perfect" optical surface.

NUTATION: "nodding," the very slight (9 arc-seconds), 19-year wobble of the earth's poles under the gravitational influence of the moon.

OBJECTIVE: the main light-gathering and light-focusing optical element of a telescope; a telescope's primary mirror or main lens.

OCCULTATION: the eclipse of one distant celestial body by a nearer, such as of a star or planet by the moon. In a grazing occultation, the eclipse is partial, and the limb of the nearer object skims the occulted one.

OCCULTATION PATH: the region of earth's surface from which an occultation is visible.

OCCULTING BAR: any device inserted into an eyepiece to block light from part of the field, such as from a bright star; useful in observing dim, close companions of bright objects, such as planetary satellites or multiple stars.

OCULAR: an eyepiece.

OORT CLOUD: the region in the outer solar system, beyond the orbit of Pluto, where millions of comets orbit the sun.

Open Cluster: a close aggregation of several dozen to several hundred stars, usually of population I, found in the spiral arms of the galaxy; may be extremely condensed or very thinly scattered, and occasionally may show nebulosity; the Pleiades is an open cluster; a galactic cluster.

Opposition: the moment when a planet passes through a line drawn from the sun through the earth; the position opposite the sun on the celestial sphere; also said occasionally of stars or constellations having right ascension plus or minus 12 hours of the sun's.

Optical Axis: in a telescope, a line drawn from the center of the objective, perpendicular to the plane defined by its circumference; defines the center of the telescope's light path.

Optical Double: a pair of stars appearing close together on the celestial sphere, although actually at great distances from each other in space; a "false" double star.

Orbital Resonance: a property of the orbits of two or more bodies in which the orbital period of one body is a proper fraction or whole multiple of the orbital period of another, for example, having an orbital period that is one-half, one-third, one-fifth that of another; also called orbital harmonics, it is the force responsible for the Kirkwood gaps, and probably for the gaps in planetary rings.

Parabolizing: altering the shape of an optical surface to a paraboloid, in order to correct spherical aberration; necessary in optics shorter than *f*/10.

Parallax: apparent shift in the position of a nearby object (relative to distant ones) when seen from two different positions; observation of parallax in stars (the stellar parallax) is a vital technique in determining the distance of stars, useful to a distance of approximately 300 light years.

Parsec: a unit of length used over interstellar distances, equal to the distance to a star showing a parallax of one arc-second; 3.26 light years.

Penumbra: (1) the outer region of shadow cast by a planet; (2) the outer region of a sunspot, not as dark as the central region.

Periastrion: in the orbit of a body around a star, the point closest to that star.

Perigee: in the orbit of a body around the earth, the point closest to the earth.

Perihelion: in the orbit of a body around the sun, the point closest to the sun.

Period: the time required for a repetitive process to repeat once, such as the period of a planet's orbit, or the light variation of a variable star; a cycle.

Period-Luminosity Relationship: in a Cepheid variable star, a predictable relationship between its luminosity and the period of its light variation; an important tool in measuring distances to nearby galaxies.

Phase: (1) of a planet, the amount of its disc illuminated by sunlight, given as a percentage of total area of disc; the moon, Mercury, Venus, and Mars show discernible phases; (2) of a wave, in relation to another wave, the difference between the time of their maximum (or minimum) amplitudes, given in degrees, 1/360th of the wave's cycle.

Photon: the quantum of electromagnetic energy; not a particle exactly, best thought of as an isolated bit of a wave.

Photosphere: the visible surface of the sun, the layer at which the process of heat transport gives over almost entirely from convection to radiation, and the scene of sunspots; temperature approximately 6,000 degrees Kelvin.

Plage: a bright area of the photosphere, occasionally visible (with proper filtration) in the vicinity of sunspots.

PLANETARY NEBULA: an emission nebula formed as a result of mass-loss and other processes around giant stars at the end of their lives; visible, usually as a ring of faintly glowing gas, around extremely hot, freshly formed white dwarfs; for example, the Ring Nebula (M.57) in Lyra.

PLANETISMAL: aggregation of dust, collected out of the proto-solar nebula, which evolved into a planet or asteroid.

PLASMA: any matter heated to the degree that its atoms ionize into a collection of positively and negatively charged subatomic particles, subject to the influence of electromagnetic fields.

POLARIZATION: the alignment of all of the waves in a beam of light (or other wave phenomenon) into the same plane; a filtering technique useful in reducing light intensity or detecting magnetic fields in emission nebulae.

POPULATION I: that portion of the stellar population of a galaxy formed in cosmically recent times from gas and dust enriched by elements produced by older stars; found in the arms of spiral galaxies; the sun is a population I star, as are most stars visible to the naked eye.

POPULATION II: stars formed about 10 billion years ago and exhibiting little or no metallic elements in their spectra; found in the bulge of spiral galaxies, in globular clusters, and in elliptical galaxies.

POSITION ANGLE: in a telescopic field, the angle, measured in degrees east of due north, of any line extended across the field; of double stars, the angle of the line from primary to secondary; of any extended object, the angle of the line of its greatest extension; abbreviated p.a.

POWER: magnification.

PRECESSION: (1) the lunisolar precession: a long-term (about 26,000 years), sizable (23.5 degree radius) rotation of the earth's poles around the ecliptic poles, caused by the gravitational influence of the sun and moon, causing the nodes of the ecliptic and celestial equator to shift west around the ecliptic by approximately one hour of right ascension per 1,000 years; (2) planetary precession: a much smaller precession caused by the gravitational attraction of the other planets. The combined effect of these two precessions causes an annual westward shift of the first point of Aries of 50.26 seconds of right ascension.

PRIMARY: (1) the most massive or luminous member of a multiple star system; (2) in any orbital system, the most massive member.

PRIMARY MIRROR: of a reflecting telescope, the objective mirror.

PRIME FOCUS: of a telescope, the location of the focal plane; in a reflecting telescope, the prime focus falls inside the tube, necessitating a secondary mirror to bring it outside the tube. Also called first focus.

PROGRADE: (1) rotation or revolution in a direction counterclockwise when viewed from above the object's north pole; the direction of rotation or revolution of the majority of bodies in the solar system; (2) of the apparent motion of a planet along the ecliptic, motion from west to east.

PROMINENCE: an arc of plasma extending from the photosphere into the sun's outer atmosphere. These may be eruptive—highly energetic ejections of matter, associated with sunspot regions—or quiescent—long-lived phenomena caused by cooling plasma falling inward from the corona. Prominences are apparently caused by disorders in the sun's magnetic field.

PROPER MOTION: (1) of a star: its motion independent of the celestial sphere, being one component (namely, that across our line of sight) of its actual motion through space; (2) of a sunspot or marking on the atmosphere of a planet: the motion of that mark relative to the rest of the disc, independent of the rotation of the body as a whole.

Pulsar: a neutron star radiating electromagnetic energy in rapid pulses.

Pulsating Variable: a variable star whose variations are caused by changes in its surface temperature and diameter; the kinds of pulsating variable include the Cepheid, long-period, semiregular, and irregular variable stars.

Quadrature: aspect of planetary orbit when the angle between a line drawn from the sun to the planet and from the sun to the earth equals 90 degrees; western quadrature finds the planet at quadrature west of the sun, on the meridian at sunrise; eastern quadrature finds the planet at quadrature east of the sun, on the meridian at sunset.

Quasar: poorly understood objects at the borders of the observable universe, radiating as much energy as an entire galaxy from a region less than a light year across.

Radiant: of a meteor shower, the point on the celestial sphere from which meteors appear, through an effect of perspective, to radiate; the location of a stream of meteoroids producing a meteor shower.

Radiation: (1) emission of electromagnetic energy; (2) a method of heat-transfer through emission of infrared electromagnetic energy, the only method that works across a vacuum.

Radiation Zone: the region of the sun's interior, extending from core to convection zone, where heat migrates outward mostly via radiation.

Ramsden Disc: exit pupil.

Ray: (1) a beam of light; (2) on the moon, a light-colored streak of ejecta radiating from a crater.

Rayleigh's Limit: size of the smallest aberration of an optical surface that will produce a noticeable defect in the image; measured in units of a wavelength of light (usually green light), this limit is one-eighth wave for reflecting surfaces, one-half wave for refractors.

Ray Trace: diagram of the path of light rays through a telescope or eyepiece.

Red Shift: Doppler shift of lines in the spectrograph of an object toward the red end of the spectrum, indicating motion away from the observer ("in recession"); most objects at cosmic distances display a red shift, indicating that the universe is expanding.

Reflection: the bouncing of light away from a surface.

Reflection Nebula: a nebula that shines by light from nearby stars reflected off dust grains rather than by fluorescence.

Reflector: a telescope that collects and focuses light with mirrors, which reflect.

Refraction: the bending of light rays as they pass from one transparent medium to another.

Refractive Index: the measure of optical density of a transparent medium, determining the amount by which it will refract a beam of light.

Refractor: a telescope that collects and focuses light with lenses, which refract.

Resolution: the ability of a telescope to distinguish between closely separated objects, such as two nearby stars; measured in units of arc, it depends directly on aperture.

Retrograde: opposite of prograde.

Retrograde Motion: of a planet's apparent motion across the celestial sphere, an optical illusion of reverse motion, displayed around the time of opposition, when the planet stops, moves from east to west, and stops again before resuming prograde (easterly) motion.

Revolution: motion of one body around another; orbital motion.

RFT: rich-field telescope; any telescope of *f*-ratio less than *f*/5; RFTs are useful primarily for their wide field, bright image, and photographic speed; may also be more portable and convenient to use than ordinary telescopes.

RICH-FIELD TELESCOPE: see RFT.

RIGHT ASCENSION: the celestial coordinate equivalent to longitude, measuring location east of the first point of Aries; measured in hours, minutes, and seconds, along the celestial equator. Constant over time for all observer locations.

RILLE: a canyon, crevasse, or gorge in the lunar surface, frequently meandering, like a river.

ROCHE'S LIMIT: the distance from a planet, star, or other massive body at which a satellite crumbles under tidal stress; depends on the density and composition of the orbiting body as well as on the distance from and mass of the primary.

ROTATION: spin; motion of a body, such as a planet, around an axis passing through its center.

SATELLITE: any body that orbits a planet; a moon is a natural satellite; in orbit, the space shuttle is an artificial satellite.

SCHMIDT-CASSEGRAIN: the most popular kind of catadioptric telescope, in which a thin correcting plate corrects the aberration of a spherical mirror, and a convex secondary reflects the light path through a perforation in the center of the primary; mass-produced in apertures from 4 inches to 14 inches, operating usually at a focal ratio of *f*/11, they give good quality, highly magnified images with a short, portable tube.

SCHRÖTER'S EFFECT: an apparent distortion of the terminator of Venus, owing to refraction of sunlight within the Venusian atmosphere.

SECONDARY: the lesser member of a double star system; secondary mirror.

SECONDARY MIRROR: in a reflecting telescope, the small mirror that deflects the light rays gathered by the objective to the outside of the tube.

SEEING: stability of earth's atmosphere, as it affects the quality of the telescopic image; unstable air causes poor seeing, degrading resolution of image; usually apparent as blurriness in otherwise satisfactory optics.

SETTING CIRCLES: devices attached to axes of equatorial mounting, indicating the celestial coordinates of a telescope's direction; the hour circle attaches to the polar axis and is marked in hours and usually every sixth minute of right ascension; the declination circle attaches to the declination axis and is marked usually in half-degrees of declination; useful for locating objects too small or dim to locate with a finder, inconveniently placed for star hopping.

SIDEREAL DAY: the time required for one rotation of the earth, measured with respect to the celestial sphere, not the sun; time between successive transits of any star; 23 hours, 56 minutes, 4.091 seconds.

SIDEREAL MONTH: the period of the lunar orbit; a month measured with respect to the celestial sphere; shorter than the synodic month commonly used.

SIDEREAL PERIOD: period of any planetary orbit, measured with respect to the celestial sphere.

SIDEREAL TIME: hour angle of first point of Aries; right ascension of observer's celestial meridian; a system of timekeeping based on the sidereal day; useful because of its link to right ascension, the sidereal time will also tell the observer what portion of the celestial sphere is above the horizon.

SLEWING: rapid motion of a telescope on the axes of its mounting across a wide region of sky.

SOLAR FLARE: an explosive release of energy from the photosphere, rarely visible in amateur telescopes as an almost stellar brightening of the photosphere; solar flares intensify the solar wind, causing disturbances, such as increased auroral activity, in the earth's magnetosphere.

SOLAR TIME: time measured with relation to the sun; mean solar time is a compromise figure, taken by averaging the length of days throughout the year (which varies owing to variation in the earth's orbital speed).

SOLAR WIND: a stream of ions constantly emitted from the sun, flowing outward through the solar system; responsible for formation of aurorae and cometary tails; variations in the strength of the solar wind can distort the plasma tails of comets and affect auroral activity on earth and on the outer planets.

SPECTRAL CLASS: the division of the range of surface temperatures (and hence colors) of the stars, from O (hottest and bluest), through B, A, F, G, K, M (coolest and reddest), and the C; each class is further divided into nine subclasses, expressed by a number after the class letter, for example, G3, the class of our sun.

SPECTRAL LINES: light or dark lines crossing the spectrum of an object imaged through a spectroscope; the precise location and appearance of these lines reveal valuable information on an object's chemical composition, temperature, luminosity class, motion, and distance.

SPECTROGRAPH: photographic recording of the spectrum of an object as displayed by a spectroscope.

SPECTROSCOPE: instrument, vitally important in astronomy, for displaying the light of an object as a spectrum.

SPECTRUM: (1) the visible spectrum: light dispersed by a prism or diffraction grating so that all of its component wavelengths appear as a continuous band of light, ranging from violet at one end, through blue, green, yellow, orange, and red. (2) the electromagnetic spectrum: the entire range of electromagnetic energy, from the extremely low-frequency, long-wavelength radio waves to high-frequency, short-wave gamma rays. Visible light is part of this spectrum.

SPEED: in photography, the length of time required to take a photograph of a given object; often used as an expression of focal ratio, short ratios being relatively "fast"; frequently confused with visual efficiency, which depends on the aperture divided by magnification.

SPHERICAL ABERRATION: an optical aberration, noticeable in spherical objectives shorter than *f*/10, in which the focal length varies over the surface of the objective. Corrected by parabolizing the objective.

SPICULES: short, hedgelike rays of light seen at the limb of the sun; hot gases rising from the photosphere into the chromosphere.

SPIDER: an arrangement of (usually) four thin vanes that supports the secondary of a reflecting telescope at the optical axis.

SPIRAL GALAXY: a galaxy consisting of a thin (approximately 1,000 light years) disc of population I stars, gas, and dust, arranged in arms of greater density that spiral outward from the central bulge that is composed of population II stars.

STANDARD CANDLE: any class of luminous object, the average luminosity of which is used to estimate distance, using the distance modulus method; for example, HII regions, O-class supergiant stars.

STAR HOPPING: a method of locating dim objects in the telescope by moving the telescope one field at a time from a bright guide star to a series of dimmer stars and eventually to the target.

STELLAR ASSOCIATION: a group of stars of similar age and physical properties, all of which had their origin in the same nebula, usually widely scattered but occupying a distinct region in space.

Stellar Endstate: the last form reached in the life of a star; endstates reached by individual stars depend almost entirely on the star's original mass; least massive stars end as black dwarfs, spheres of cold degenerate matter; slightly more massive stars end as white dwarfs, which eventually go black; more massive stars end as neutron stars; the most massive stars (larger, after a supernova explosion, than three solar masses) end as black holes.

Subtend: to span an angle. An extended object subtends an angle of so many degrees.

Sunspot: regions of low temperature in the photosphere, apparently caused by distortions in the sun's magnetic field; the most obvious feature of sun's visible disc, easily visible in properly filtered amateur telescopes.

Sunspot Cycle: a 22-year cycle, as yet poorly understood, during which the number of sunspots rises to a maximum, then dwindles to minimum, accompanied by variations in both the latitude at which most sunspots occur and in the polarity of the sun's magnetic field.

Supergiant: brightest of the stellar luminosity classes, consisting of those stars that have left the main sequence and swollen to sizes that approach that of the orbit of Saturn; supergiants occasionally produce supernovae and collapse into black holes.

Superior Planet: those planets with orbits larger than Earth's: Mars, Jupiter, Saturn, Uranus, Neptune, Pluto, and any more distant planets that may exist.

Supernova: catastrophic explosion of a supergiant star, initiated by the collapse of fusion processes at its core, followed rapidly by implosion of the distended outer layers; supernovae occur in two types, designated I and II, distinguished by the maximum luminosity and the speed at which their brightness declines after maximum; at maximum, type I supernovae (the brighter of the two) can reach absolute magnitudes of −20, equaling the brightness of an entire galaxy of normal stars.

Supernova Remnant: cloud of ionized gases expanding outward from the scene of a supernova explosion, for example, the Crab Nebula (M.1) in Taurus; a pulsar or other remnant of a stellar core remaining after supernova.

Surface Brightness: the brightness per unit area of an extended object.

Synodic Month: the period from new moon to new moon.

Synodic Period: the period between successive oppositions or inferior conjunctions of a planet.

Tail: of a comet, the elongated stream of dust or plasma ejected from the cometary nucleus by the solar wind.

Telecompressor: a telescopic accessory, inserted in the light path ahead of the eyepiece, usually in Schmidt-Cassegrains, to shorten effective focal length.

Terminator: the line dividing the sunlit and dark portions of the disc of the moon or a planet.

Terrestrial Planet: a planet, such as Mercury, Venus, Earth, or Mars, composed primarily of lightweight minerals, usually with a core of a denser metal, such as iron.

Thermonuclear Fusion: process, occurring in the cores of stars, whereby the nuclei of lightweight atoms are combined to form nuclei of heavier atoms, releasing energy in the form of gamma rays and neutrinos; the power source of the stars.

Tide: a lifting of or strain on the material of a planet or star, caused by the gravitational attraction of a nearby satellite, planet, or star.

Totality: the period, during an eclipse, when an observer or an eclipsed body is entirely within the umbra of the eclipsing body.

Transit: the passage of a celestial object through the meridian.

TRANSPARENCY: clarity of sky; absence of dust, haze, smog, or other particles.

TUBE: in a telescope, the part that connects the objective to the focuser, usually but not necessarily a closed tube made of fiber glass, cardboard, or metal.

TUBE ASSEMBLY: the optical and mechanical parts of a telescope, excluding the cradle and mounting, including the objective, cell, spider, diagonal holder, secondary mirror, focusing mount, finderscope, and tube.

TUBE CURRENTS: currents of heated air spiraling up the interior of a telescope tube, usually caused by setting up over a heated surface on a rapidly cooling night, and often degrading the image.

TURN-OFF POINT: position on Hertzsprung-Russell diagram of the oldest member of a cluster of stars still occupying the main sequence; indicates the age of the cluster.

TWILIGHT: the time when the sun reaches a certain angle below the horizon, indicating darkness of sky; civil twilight: sun 6 degrees below the horizon; nautical twilight: sun 12 degrees below the horizon; astronomical twilight: sun 18 degrees below the horizon, the sky is as dark as it can be until morning astronomical twilight.

UMBRA: (1) in an eclipse, the dark, central region of the shadow of the eclipsing body; (2) in a sunspot, the dark, central region of the spot.

UNIVERSAL TIME: the mean solar time at the Greenwich meridian, expressed from 0 to 24 hours, beginning at midnight. Used commonly in astronomy to time events visible from large regions of earth's surface.

VARIABLE STAR: any star varying in magnitude, for a variety of reasons.

WALLED PLAIN: geographical feature of the moon; a large crater.

WAVE: (1) any of a number of phenomena transferring energy from one place to another in pulses of regularly increasing and decreasing intensity, for example, light, sound; the pulses that transfer that energy; (2) measure of optical quality; see Rayleigh's limit.

WAVELENGTH: the distance between pulses of maximum intensity of a wave.

WILSON EFFECT: the visible concavity or convexity of a sunspot, seen most often near the solar limb.

YEAR: (1) the orbital period of the earth, and by extension that of any planet; (2) tropical year: the period between two successive vernal equinoxes, or 365 days, 5 hours, 58 minutes, 46 seconds; (3) sidereal year: the sidereal period of the earth, which differs from the tropical year owing to the lunisolar precession. It is 365 days, 6 hours, 9 minutes, 9.5 seconds long; (4) the calendar year is a convenient approximation of the tropical year, 365.2425 days long; the Gregorian year.

ZENITH: the point on the celestial sphere directly overhead; not fixed relative to the stars, it varies with location of the observer and the time of observation.

ZODIAC: the traditional 12 constellations lying along the ecliptic, within which the sun, moon, and planets usually (but not always) appear.

ZODIACAL LIGHT: a dim glow, best seen around the equinoxes, visible as a cone of light tapering upward from the horizon, and brightening again at the point in the sky opposite the sun. Following the path of the ecliptic, it is caused by sunlight scattered back by dust orbiting the sun on the ecliptic.

ZONE OF AVOIDANCE: the region of the celestial sphere, roughly coincident with the Milky Way, in which few external galaxies appear. It is caused by the obscuring stars, dust, and gas of the Milky Way, not an actual absence of galaxies.

Index